MW00626437

CSR, Sustainability, Ethics & Governance

Series Editors

Samuel O. Idowu, London Metropolitan University, London, United Kingdom

René Schmidpeter, Cologne Business School, Germany

More information about this series at
http://www.springer.com/series/11565

Myria Allen

Strategic Communication for Sustainable Organizations

Theory and Practice

 Springer

Myria Allen
Department of Communication
University of Arkansas
Fayetteville, AR
USA

ISSN 2196-7075   ISSN 2196-7083   (electronic)
CSR, Sustainability, Ethics & Governance
ISBN 978-3-319-18004-5   ISBN 978-3-319-18005-2   (eBook)
DOI 10.1007/978-3-319-18005-2

Library of Congress Control Number: 2015938585

Springer Cham Heidelberg New York Dordrecht London

© Springer International Publishing Switzerland 2016
This work is subject to copyright. All rights are reserved by the Publisher, whether the whole or part of the material is concerned, specifically the rights of translation, reprinting, reuse of illustrations, recitation, broadcasting, reproduction on microfilms or in any other physical way, and transmission or information storage and retrieval, electronic adaptation, computer software, or by similar or dissimilar methodology now known or hereafter developed.
The use of general descriptive names, registered names, trademarks, service marks, etc. in this publication does not imply, even in the absence of a specific statement, that such names are exempt from the relevant protective laws and regulations and therefore free for general use.
The publisher, the authors and the editors are safe to assume that the advice and information in this book are believed to be true and accurate at the date of publication. Neither the publisher nor the authors or the editors give a warranty, express or implied, with respect to the material contained herein or for any errors or omissions that may have been made.

Printed on acid-free paper

Springer International Publishing AG Switzerland is part of Springer Science+Business Media (www.springer.com)

To my grandson, Adjani, and all children everywhere. May the adults in your life embark quickly and boldly to ensure a more resilient and sustainable planet.

Foreword

The Brundtland Commission (1987) of the World Commission on Environment and Development—an arm of the United Nations—was asked to formulate a "global agenda for change." In our own view, we believe that this change has since the report came out continued to show its face globally in all areas of human existence. The report has successfully brought to our consciousness a number of issues on sustainable development and the consequences of persistent irresponsibility on the part of global corporate and individual citizens. Needless to say, nearly 30 years on, a lot has happened in our world with regard to—Our Common Future. For instance, a series of UN Conferences on the Environment have seen the light of the day. The Eight Millennium Development Goals had been set; with a few months to its end period, a number of versions of the Global Reporting Initiative (GRI) have been issued; the G4 being its latest version, the ISO 26000 is now in place; similarly research studies, scholarly conferences, debates, workshops, and a massive increase in the number additions to the literature on sustainability are all now part of our life; these are a few examples of what this change has brought to us. Myria Allen's *Strategic Communication for Sustainable Organizations: Theory and Practice* is the latest addition to the literature on sustainability and sustainable development.

Allen's quest to wanting to understand how sustainability initiatives were being enacted across the USA was the motivation for this book. Armed with the knowledge of the theories of different aspects of the field having taught *Organizational Communication* since the emergence of the Brundtland Report in 1987 and *Environmental Communication* more recently and being involved in different facets of sustainability in academia, she was determined to add to readers' practical knowledge of how sustainability has continued to evolve in the USA. Allen embarked on a few months' Sustainable journey across the United States of America—this book is an account of what she saw!

Having spent some time reading some sections of the book and bearing in mind that our world is facing a series of environmental-related challenges including Global Warming, scarcity of many of our resources, etc., every citizen of the world both corporate and individual would have to learn to innovate in order to use sustainably all our depletable and even nondepletable resources and also learn to process sustainably all our ever increasing wastes. Our failure to do this would make life unbearable for both this generation and future generations of residents of planet Earth, after all this is all about our common future.

We wish to take this opportunity to congratulate Myria Allen for this invaluable addition to books on sustainability and sustainable development. We are delighted to recommend the book unreservedly to all readers worldwide both in academia and in practice. In addition, the book is a valuable resource which all undergraduate and graduate students of Corporate Social Responsibility CSR must have; remember, our common future is involved!

London, UK   Samuel O. Idowu
Cologne, Germany   Rene Schmidpeter

Preface

Why Write This Book?

When he was almost 60, John Steinbeck spent 3 months traveling through America with his standard poodle, Charlie. In *Travels with Charley: In Search of America* (1962) Steinbeck writes about setting out to rediscover his native land. He described himself as an American writer who had been writing from memory since he had not traveled extensively in his home country for 25 years. He wrote:

> I had not heard the speech of America, smelled the grass and trees and sewage, seen its hills and water, its color and quality of light. I knew the changes only from books and newspapers. In short, I was writing of something I did not know about, and it seems to me that in a so-called writer this is criminal (p. 5).

As I approached 60, in the summer of 2013, I journeyed West with my husband, A.J., and our standard poodle, Hector, in search of America. I wanted to see how sustainability initiatives were being played out in a variety of organizations. For almost 2.5 decades, I've taught about organizational communication and, more recently, environmental communication. I know the theories. I've taught future scholars and practitioners, conducted research, published scholarly articles, served on University committees addressing sustainability-focused research and curricular design, and built a small eco-efficient Energy Star certified home. I live down the road from the Sustainability Consortium, a group *Scientific American* magazine identified as among the top 10 World Changing Ideas of 2012. This consortium includes businesses, universities, and nonprofit organizations that together seek to develop science-based tools to advance the measurement and reporting of consumer product sustainability. You will learn more about them later. So, I'm aware of multiple sustainability-focused initiatives occurring in my own back yard.

But I didn't know what was going on in Lincoln, NE, Pierre, SD, Missoula, MT, Boardman, WA, Portland, OR, Boulder, CO, and other parts far distant from my home in Fayetteville, AR. I hadn't recently driven past wind farms, or toured hydroelectric plants, seen the building of pipelines to transport oil from the

Canadian tar sands into the USA, witnessed the community and environment changing impact of fracking around Williston, ND, been awestruck driving by 25,000 acres of surreal hybrid poplar trees at the Forest Stewardship Certified tree farm near Boardman, WA, felt saddened by seeing the magnitude of the bark beetle destruction while camping in the Rocky Mountain lodge pole pine forests, driven on flood damaged roads around Colorado Springs, CO, stayed in Pierre, SD, the town which thought it would always be protected from Missouri River floods until the day it wasn't, or toured Greensburg, KS, a Model Green Community, which rebuilt itself with a sustainability focus after an EF5 tornado destroyed its infrastructure and 90 % of its homes and buildings in 2007. So much of what I saw along our route reminded me of the energy and climate challenges scientists, policy makers, and the general public are increasingly recognizing will shape our species' future.

In deciding what parts of America to visit, we chose to address my long-term desire to follow the Lewis and Clark Corps of Discovery trail. We started our 3-month road trip West by joining up with the trail in Nebraska City, NE. It felt fitting to explore America following the footsteps of these brave and perhaps foolhardy men, and Sacagawea their female, Shoshone guide. Like them, we followed the Missouri, the Clearwater, the Columbia, and the Snake rivers, and camped near Astoria, OR, where the Columbia River enters the Pacific. Passing through the West, the Expedition forever changed the lives of the native peoples representing nearly 50 different tribes living between the Mississippi River and the Pacific. In the 2 years (1804–1806) they traveled, the Lewis and Clark Expedition sent news back to President Thomas Jefferson describing the Louisiana Territory which America had purchased from the French. The Expedition introduced more than 300 flora, fauna, and scientific discoveries previously unknown to people of European descent. Like them, I am writing to share what I saw and heard along my journey.

Along their route, members of the Lewis and Clark Expedition broke away from familiar trails, faced severe challenges, and learned how to make do and to innovate. Now, over 210 years later, humanity is facing unfamiliar territory as challenges brought on by global warming increasingly force us to leave our familiar trails, face challenges, and learn to make do and to innovate. As a species, we are at a crossroads where we must choose to act. As the Lewis and Clark Expedition journeyed toward the Pacific, Meriwether Lewis repeatedly wrote in his journal, "We proceeded on." Working collaboratively, today organizations of all types must join in the journey as we proceed on in response to global warming.

Who Should Read This Book?

Increasingly colleges and universities are developing graduate level programs focusing on sustainability. In 2013, The Association for the Advancement of Sustainability in Higher Education (AASHE) program database showed 1378 sustainability-focused academic programs on 456 campuses in 63 states and provinces. Graduate students in sustainability-focused management, public administration, environmental health and safety, public relations, environmental communication, and organizational communication courses will find this book useful, as will faculty researching in these areas.

Interest in sustainability is growing across all types of organizations, changing existing employees' job duties, and leading to the creation of new job categories. For example, in 2009 municipal government sustainability professionals from across North America formed the Urban Sustainability Director's Network, an organization with over 100 members representing cities and counties in the USA and Canada with a combined resident population of about 50 million. In 2010, the Green Sports Alliance was formed and now represents 180 sports teams and venues from 16 different US sports leagues. The International Society of Sustainability Professionals is a professional association for sustainability professionals formed in 2007 with more than 700 members representing every continent. Often these practitioners are the first in their organization to be tasked with managing their organization's sustainability initiatives. This book is geared toward practitioners in these diverse areas believing they can benefit from theory- and research-based suggestions written in an accessible manner.

Plan for This Book

What makes this book unique is it focuses on the role of communication and theory in enacting sustainability initiatives, supplemented by the scholarly literature and by practitioner insights. Like Kurt Lewin, the German-American psychologist, who is often recognized as the founder of modern social psychology, I believe there is nothing as practical as a good theory. In this book I introduce readers to a range of theories drawn from communication, psychology, social psychology, and management. "Theories, ideas, and concepts do matter, and they matter a great deal. They are not only the foundations on which we build our perceptions and 'constructions' of the world, they also are the basis for policy articulation and formulation" (Page and Proops 2003, p. 3). It is my hope that as you read more theories and hear more stories, your own insight and ability to communicate so as to innovate around sustainability expands. But at the end of the day, a theory can only take you so far when enacting sustainability initiatives.

Each chapter begins with information about the Lewis and Clark Expedition which helps frame that chapter's content. Each chapter ends with thought provoking concluding content. Throughout the chapters I will highlight some *Best Practices* and *Actions Plans* reader might make. Chapter 1 defines sustainability, discusses why it is an important concept, and talks about the role of communication. Chapter 2 identifies two paradigms generally used to describe the belief systems regarding the human–natural environment relationship: the dominant social paradigm and the new ecological paradigm. Communication's role in reinforcing and challenging paradigms is discussed. Paradigms are differentiated from *Discourses* and nine environment-related *Discourses* are identified. The societal shift toward the growing importance of the sustainable development *Discourse* within business (e.g., the business case for sustainability arguments), cities, and universities is described. Criticisms of the ecological modernism and sustainability *Discourses* are reviewed. The chapter ends by discussing forces which influence how environmentally related issues are framed, contested, and reframed. In Chap. 3 legitimacy is defined, various types of legitimacy are described, the benefits of being seen as legitimate and having a positive reputation are discussed, and actions organizations take to manage legitimacy are reviewed. Organizations face changing institutional norms and other challenges when trying to craft a reputation for being sustainable. Issues related to message credibility, green-washing, and corporate sustainability communication are discussed. Almost ½ of the chapter addresses communication directed simultaneously toward multiple internal and external stakeholders which reflects standardization efforts using signs and symbols (i.e., annual public meetings; websites and sustainability reports; reporting frameworks and certifications; LEED certified architecture). The chapter concludes by reviewing stakeholder theory, discussing the need to adapt messages to different stakeholders, and describing the role communication plays in stakeholder engagement. Chapter 4 identifies factors influencing an individual's pro-environmental values and behaviors, explores the tentative link between values, attitudes, and behaviors, and addresses the way communication can be used strategically to influence individual level behavioral change. Literature is reviewed which identifies and discusses pro-environmental values and beliefs, and defines pro-environmental behaviors. Various theories are summarized to help practitioners better understand how to stimulate pro-environmental behaviors. The chapter ends by focusing on concrete message strategies (e.g., gains vs. losses, intrinsic vs. extrinsic appeals). Finally, the importance of interpersonal communication in stimulating and reinforcing pro-environmental behaviors is discussed. Chapter 5 focuses on how transformational organizational changes can occur when external information about sustainability-related initiatives is identified, imported, spread, and integrated throughout an organization. Transformational change begins when key individuals become aware of new processes, technologies, opportunities, constraints, and expectations. Once awareness occurs, the challenge becomes to transform information into useable knowledge. Factors influencing an innovation's adoption are reviewed. The characteristics of change adopters and stages of change are identified. Important communication roles during times of change (e.g., leaders, change agents, sustainability champions), the process of communicating about change, guidance for change communicators, and formal structural (e.g., new roles, new inter- and intraorganizational coordinating structures) and communication efforts to facilitate change efforts are discussed. Pathways (e.g., mission and vision statements, goals and plans, formal communication channels) are identified. Chapter 6 addresses issues related to organizational culture and employee empowerment. Sustainable organizational cultures, learning organizations, and the managerial subculture are discussed. Organizational climate, employee emotions, and goal congruence can influence the extent to which employees embrace and enact an organization's sustainability-related goals, plans, and programs. Hiring decisions, socialization, training, and rewards also influence employees' pro-environmental actions. The chapter ends by highlighting the role informal communication plays in creating, reinforcing, and stabilizing sustainable values and goals. Chapter 7 is about groups and group communication at the organizational and interorganizational levels. Ways to build strong teams representing multiple departments are discussed. Communities of practice and the bona fide group theory are highlighted. Relevant literature on interorganizational groups and cross-sector alliances is reviewed including how to create learning networks at this broader cross-organization level. Ultimately, the chapter focuses on one specific type of interorganizational group—supply chains. Using communication to promote more effective relationships within the supply chain is discussed. Diffusion mechanisms for sustainable supply chain decisions and actions are offered along with potential governing mechanisms. Chapter 8 concludes by presenting a model of the factors influencing organizational sustainability-focused actions and the theories which can help us better understand each factor. It concludes by reviewing literature on wisdom and spirituality in the hope wisdom and spirit can illuminate the paths we take on our joint journey toward sustainability.

Participating Organizations

Although I anthropomorphize organizations (i.e., ascribe human actions to them), this is simply a writer's shortcut. Organizations do not plan, adapt, or persuade. People do. Much of the published scholarship focuses on the sustainability initiatives of for-profit organizations, but sustainability initiatives go far beyond large for-profit businesses. In the following pages, you will meet people who work for large and small businesses, cities, universities, governments, and nongovernmental organizations. Some work as educators, others as organizers, innovators, implementers, and activists. All are or have been instrumental in crafting and/or enacting sustainability focused initiatives throughout their organization's membership and/or supply chain. They talk about how they personally define sustainability and how their organization defines sustainability. They describe their organization's sustainability initiatives. What they shared with me provides us with a glimpse into some of the initiatives being undertaken west of the Mississippi and the associated communication-related challenges and successes being encountered. See Table 1 for a list of the organizations represented.

Table 1 Organizations represented in this book

Type	Name	State
Large company	State Farm (branch office)	Nebraska
	Sam's Club	Arkansas
	Tyson Foods	Arkansas
Small company	Bayern Brewing	Montana
	Neil Kelly: Design/Build	Oregon
	ClearSky Climate Solutions	Montana
	Assurity Life Insurance	Nebraska
Business group	WasteCap Nebraska	Nebraska
	Missoula Sustainability Council	Montana
Nongovernmental group	Ecotrust	Oregon
	Heifer International™	Arkansas
	Arbor Day Foundation	Nebraska
	Natural Resources Defense Council	New York City
	Clinton Climate Initiative Home Energy Affordability Loan (HEAL)	Arkansas
University	University of Arkansas, Fayetteville	Arkansas
	University of Idaho, Moscow	Idaho
	University of Colorado, Boulder	Colorado
	University of Colorado, Denver	Colorado
Sports organization	Portland Trail Blazers	Oregon
	Aspen/Snowmass	Colorado
Government	City of Fayetteville	Arkansas
	City and County of Denver	Colorado
	City of Boulder	Colorado
	City of Portland	Oregon
	State of South Dakota	South Dakota

References

Page, E., & Proops, J. (2003). An introduction to environmental thought. In E. A. Page & J. Proops (Eds.), *Environmental thought* (pp. 1–12). Northhampton, MA: Edward Elgar.

Acknowledgments

This book would not be possible without all my interviewees and the many people who linked me with them. I want to thank my inclusive colleagues at the University of Arkansas for inviting me to join research teams and serve on sustainability-related committees, and my outstanding graduate students for sharing my enthusiasm. Special thanks go to the Rockport Daily Grind brew crew for providing me with a welcoming Texas location where I wrote the first half of the book. Most especially, thanks to my husband, A.J. Allen, who traveled alongside me on the Lewis and Clark trail, ran our household during the 2-year book project, and has been my solid foundation and source of encouragement for over 3.5 decades.

Contents

Chapter 1
Sustainability and Communication

Abstract In response to growing global environmental problems and threats associated with global warming, organizations are recognizing that business-as-usual is no longer sufficient and that it is time to go to scale in responding to impending challenges. Effective communication is absolutely essential. Strategic communication is needed to alert, persuade, and help people enact sustainability initiatives within and between organizations. Strategic communication also orients our consciousness by inviting us to take a particular perspective, by evoking certain values and not others, and by creating referents for our attention and understanding. Sometimes effective communication is characterized by strategic ambiguity. Organizations face multiple challenges in terms of their sustainability-related communication. Some are silenced because of their fear of speaking out. For others, the messages they create are not processed because message recipients are unmotivated. Knowledge regarding the elaboration likelihood model of persuasion and the use of repetition are important tools when designing strategic communication. Communicators need to think critically about where their audience is in their understanding of sustainability and how to present information in an accessible way. Attention to message design and message framing can help. Interview data drawn from Aspen Skiing Company; Heifer International®; the City and County of Denver, CO; the Arbor Day Foundation; and ClearSky Climate Solutions illustrates the concepts being discussed.

Lewis and Clark's Story Expands The public's first glimpse into the Lewis and Clark Expedition's adventures emerged through newspapers and government documents. President Jefferson's *Report to Congress* was published in 1806. One year later, the journal of Corps of Discovery member Patrick Gass was published. The first official publication of the journals occurred in 1814 when 1,417 copies of the two-volume work became available. Almost 30 years later, they reappeared as a part of the Harper Family Library. This edition was reprinted 17 times during the settlement of the territory the Corps of Discovery had traversed. In 1904, during the centennial observation of the expedition, all the known journals were printed together for the first time. In 1999, incorporating newly discovered journals, papers, and maps, a 12-volume work was completed. This example illustrates how, through communication, awareness and understanding expanded over time. The Lewis and

© Springer International Publishing Switzerland 2016   1
M. Allen, *Strategic Communication for Sustainable Organizations*, CSR,
Sustainability, Ethics & Governance, DOI 10.1007/978-3-319-18005-2_1

def. of sustain.

Clark story grew from one report and some newspaper articles to become part of the US national story. Today, books, reports, and articles dealing with sustainability are helping to shape the next chapter in the story of the USA and the world.

1.1 Why Sustainability and Why Now?

— our mission is to fill needs of the poor know limitations young
— ↑ temp ↓ food/safety

In 1969, Paul Ehrlich published his influential book, *The Population Bomb*, which brought attention to growing global environmental problems. Prior to that time, sustainable development was discussed in both the international development and the renewable resource management literature in terms of the concept of maximum sustainable yield. However, the term really captured the international communities' attention after the 1987 publication of the World Commission on Environment and Development's report, *Our Common Future*. Gro Harlem Brundtland, Prime Minister of Norway, chaired the commission's inquiry into pressing interrelated global problems (e.g., energy supply, climate change). Today, the most frequently used definition of sustainability appearing in the scholarly literature focusing on organizations comes from *Our Common Future* and reads, "Sustainable development is development that meets the needs of the present without compromising the ability of future generations to meet their own needs" (WCED 1987, p. 43). The report goes on to make two points that rarely are acknowledged in the business literature:

definition

> It [sustainable development] contains within it two key concepts: the concept of 'needs', in particular the essential needs of the world's poor, to which overriding priority should be given; and the idea of limitations imposed by the state of technology and social organization on the environment's ability to meet present and future needs. ... Perceived needs are socially and culturally determined, and sustainable development requires the promotion of values that encourage consumption standards that are within the bounds of the ecological possible and to which all can reasonably aspire.

Although the report was very clear that there are finite resource limits, at the same time, it offered a vision of the possibility of the simultaneous and mutually reinforcing pursuit of economic growth, environmental improvement, population stabilization, peace, social justice, and intergenerational and global equity, maintainable over the long term. The implicit message was that sustainability and development can coexist and limits can be stretched if resources are managed adequately and wise policies put in place. Sustainable development is an attractive concept because it offers hope. Today, sustainable development "is arguably the dominant global discourse of ecological concern" (Dryzek 2005, p. 145). Those interested in sustainable development often see the natural environment as the responsibility of multiple actors including businesses, national governments, global governing bodies, and nongovernmental organizations (NGOs) whose toolkits include appropriate technologies, voluntary business initiatives, and interorganizational partnerships. *Key Point:* Coordinated collective efforts both internationally and at the grassroots level are necessary, all of which require strategic and effective communication.

In *The Sustainability Handbook*, Blackburn (2007) summarized 36 global trends resulting in the need for organizations to strive toward sustainability. I'm most interested in one, climate change, because it is a game changer. The Intergovernmental Panel on Climate Change released a report entitled *Climate Change 2013: The Physical Science Basis*. Coauthored by 259 authors representing 39 counties, the report described the very high probability by the late twenty-first century of increased temperatures and more heat waves over most land areas; increased frequency, intensity, and/or amount of heavy precipitation; increased intensity and duration of drought; increased intense tropical cyclone activity; and increased extreme high sea level. Very high probability statements are cause for alarm because the scientific method is not set up to talk in absolutes.

"Global climate change has become one of the most pressing issues for industry, government, and civil society in the twenty-first century" (Okereke et al. 2012, p. 10). The United Nations Secretary General also identified climate change as the major, overriding environmental issue of our time. Climate change involves shifting weather patterns which threaten food production, rising sea levels which contaminate coastal freshwater reserves and increase flooding risks, and a warming atmosphere which aids in the spread of tropical pests and diseases. The authors of the United Nations Environment Programme report (2013) write:

> Increasing evidence indicates important tipping points, leading to irreversible changes in major ecosystems and the planetary climate system, may already have been reached or passed. Climate feedback systems and environmental cumulative effects are building across Earth systems demonstrating behaviors we cannot anticipate.

In May 2011, according to the International Energy Agency (2011), the global carbon dioxide (CO_2) emissions from energy use in 2010 were the highest in history. By 2015, they continue to ascend. Given that our global economy is likely to grow over the next 10 years, the window has probably closed on our ability to keep the average global temperature rise below 2 °C. We are overloading our atmosphere with carbon dioxide, which traps heat and steadily drives up the planet's temperature. This carbon frequently comes from the fossil fuels we burn for energy—coal, natural gas, and oil—plus from deforestation. The authors of the United Nations Environment Programme report (2013) write:

> The most dangerous climate changes may still be avoided if we transform our hydrocarbon based energy systems and if we initiate rational and adequately financed adaptation programmes to forestall disasters and migrations at unprecedented scales. The tools are available, but they must be applied immediately and aggressively.

Five features of climate change and sustainability-related problems present a challenge for how we are to proceed:

> (1) indeterminacy—it is impossible to foresee the best course of action; (2) value-ladenness/normativity—values effect behaviors, lifestyles and systems; (3) controversy—full agreement or consensus among and even within all stakeholders is rare, if not impossible; (4) uncertainty—it is impossible to identify the exact impact of a chosen strategy or action; and (5) complexity—a whole range of variables messily interact (Wals and Schwarzin 2012, p. 13).

1. How do we solve it best?
2. values + behaviors
3. we need a concensus
4. uncertainty will follow
5. too many variables

Competing claims exist about what is occurring and what should be done. But we must move forward, even in the face of challenges and uncertainty.

In preparation for the release of the Intergovernmental Panel on Climate Change's (2013) report, the children's charity UNICEF was among the NGOs present who urged governments to heed the report's warning. UNICEF warned that the young will bear the brunt of temperature increases. Children born in 2013 will be 17 in 2030 and 37 in 2050 when the worst impacts of climate change will be in full swing including extreme heat waves, expanded diseases, malnutrition, and economic losses. Another NGO, Oxfam, warns that world hunger will worsen as climate change hits crop production and disrupts incomes while food prices spike. The number of people at risk of hunger is anticipated to double by 2050. Other NGOs anticipate facing similar pressures.

1.1.1 Sustainability and Strategic Ambiguity

The beauty and the curse of the term *sustainability is in its ambiguity*. The term resonates with individuals and groups. However, the authors of *Our Common Future* did not demonstrate its feasibility or provide practical steps for its implementation. In the early 1990s, the Transportation Research Board of the US National Academy of Sciences spent a million dollars just trying to come up with a definition of the concept, with no real success. Even today, a clear definition of the term and a road map for achieving it are lacking. As a result, it can mean different things to different people.

Communication scholar Eric Eisenberg (2007) crafted a seminal essay almost 30 years ago where he discussed a concept he called strategic ambiguity. Speakers intentionally design ambiguous messages. Eisenberg argued strategic ambiguity serves four functions related to a speaker's goals—to promote unified diversity, to facilitate change, to foster deniability related to both task and interpersonal communication, and to preserve privileged positions by protecting a speaker's credibility. Later, he acknowledged that strategic ambiguity can also mask and sustain power abuses. In terms of unified diversity, organizational values are often expressed through the creative use of symbols that allow for multiple interpretations while promoting a sense of unity. We see this use of strategic ambiguity in terms of sustainability occurring in organizations' mission statements, goals, and plans. Leaders often speak at the abstract level about sustainability initiatives so that agreement can occur. Strategic ambiguity regarding sustainability efforts can be used to encourage creativity, minimize conflict, and facilitate change. Such ambiguity creates conditions where organizations can change their operations over time in response to changing environmental conditions. It can be used to build cohesiveness within groups around sustainability initiatives and allow employees to protect their private opinions, beliefs, and feelings, while maintaining their relationships.

In the sustainability-related research, the strategic ambiguity concept was used as a framework for analyzing five documents the New Zealand government designed as guides for the development of genetically modified (GM) organisms (Leitch and Davenport 2007). Conflict exists in New Zealand, and elsewhere, around the use of GM organisms. Groups which feel economic growth should be the primary goal see GM as a way to achieve this goal, while others feel economic growth should be secondary to concerns about the society and the environment. Some see the environment as a set of natural resources to manage and use, while others see it as a complex ecosystem to be protected. Finally, some see genes as proteins to be manipulated to make new organisms, while others see them as a natural or God-given order with which people should not interfere. Leitch and Davenport investigated how the strategic ambiguity discourse strategy explained the use of the keyword *sustainability* in these government documents. A discourse strategy is how people use talk and text to achieve their goals. They concluded that the term *sustainability* lent a coherence to the texts, allowed multiple perspectives and objectives to coexist, and facilitated participation by discourse partners who held conflicting economic, environmental, and cultural/spiritual beliefs. The authors contend it is equally useful in discourse contexts where multiple organizations and individuals are interacting. Strategic ambiguity serves seven functions which can facilitate coalition formation between discourse communities which differ in their focus (i.e., on people, on the planet, or on profit) (Wexler 2009). The lack of a clear-cut definition for sustainability allows stakeholders to challenge existing understandings, adapt quickly, and explore new innovative ideas and practices. Christensen et al. (2015) term this as having a license to critique. *Key Point:* Strategic ambiguity is an important concept for understanding the current state of how people talk about sustainability (and climate change) both within and between organizations. Strategic ambiguity can be useful as groups undertake sustainability journeys along trails covered in fog.

1.2   What Are Organizational Actors Doing?

Repeatedly, my interviewees told me there are no road maps for what they are doing. One of them, Auden Schendler, Vice President of Sustainability at Aspen Skiing Company, writes in his book, *Getting Green Done* (Schendler 2009), about how visionaries say humans can achieve true sustainability where waste and pollution no longer exist, energy comes from wind and light, and ecological catastrophes like climate change and fisheries destruction no longer occur. "The only thing is that nobody knows how to get there. Or rather, some very smart people have drawn maps, but we don't know the quality of the roads. Or if there even are roads." However, even without road maps, people are taking action.

In 2013, more than 500 businesses including giants like General Motors, Nike, Starbucks, Levi Strauss, and Unilever signed a Climate Declaration launched by the business network BICEP (Business for Innovative Climate and Energy Policy), which is coordinated by Ceres, a group focused on mobilizing investors, companies, and public interest groups to accelerate and expand the adoption of sustainable business practices and solutions. The 500 businesses which signed the Climate Declaration generally supported the National Climate Action Plan President Barack Obama articulated on June 25, 2013. This climate action plan involved ways to cut carbon pollution in the USA, better prepare the nation for the impacts associated with climate change, and lead international efforts to address global climate change (President Obama's Plan to Fight Climate Change 2013). Acknowledging that climate change poses a serious threat to business, those who signed the Climate Declaration urged US policy makers to capture economic opportunities associated with addressing climate change (BICEP 2013). Embedded in the preamble to the Climate Declaration (www.ceres.org), we see multiple persuasive appeals:

> Tackling climate change is one of America's greatest economic opportunities of the twenty-first century (and it's simply the right thing to do). What made America great was taking a stand. Doing the things that are hard. And seizing opportunities. The very foundation of our country is based on fighting for our freedoms and ensuring the health and prosperity of our state, our community, and our families. Today those things are threatened by a changing climate that most scientists agree is being caused by air pollution. We cannot risk our kids' futures on the false hope that the vast majority of scientists are wrong. But just as America rose to the great challenges of the past and came out stronger than ever, we have to confront this challenge, and we have to win. And in doing this right, by saving money when we use less electricity, by driving a more efficient car, by choosing clean energy, by inventing in new technologies that other countries buy, and by creating jobs here at home, we will maintain our way of life and remain a true superpower in a competitive world. In order to make this happen, however, there must be a coordinated effort to combat climate change—with America taking the lead here at home. Leading is what we've always done. And by working together, regardless of politics, we'll do it again.

Public sector organizations also are responding. Cities emit a substantial portion of global greenhouse gases related to their energy consumption, building design, transportation infrastructure, and land use (The National League of Cities 2013). Frustrated by partisan gridlock on environmental policy at the US federal level, but concerned about the implications of environmental issues such as global warming on their communities, many local governments are stepping forward. As of 2015, over 1,060 mayors representing a total population of almost 89 million citizens had signed the US Mayors Climate Protection agreement to advance the goals of the Kyoto Protocol, an international agreement which sought to establish binding carbon emission reduction targets. On the global level, 12 megacities, 100 supercities and urban regions, 450 large cities, and 450 small- and medium-sized cities and towns in 86 countries joined the International Council of Local Environmental Initiatives (ICLEI) network. ICLEI helps cities develop climate action plans and implement sustainable development. In the USA, over 528 cities are members, accounting for nearly half of ICLEI's global membership.

Clearly, the time has passed for incremental actions based on short-term pragmatic considerations and for listening to climate change skeptics. Larger actions need to be taken by individuals as citizens and by the organizations they work within. Organizations are realizing that business-as-usual is no longer sufficient.

Preparing for climate change presents both opportunities and challenges. In this section, I share what five of my interviewees said about the need to go to scale: Aspen Skiing Company; Heifer International®; the City and County of Denver, CO; the Arbor Day Foundation; and ClearSky Climate Solutions.

Aspen Skiing Company The Aspen Skiing Company owns Aspen/Snowmass, an expensive winter resort complex, in western Colorado. It includes four ski areas (Aspen Mountain, Aspen Highlands, Buttermilk, and Snowmass). The complex includes two hotels, 13 sit-down restaurants and 11 cafeterias. The company employs 1,000 people in the summer and 3,400 in the winter. Their leaders signed the Climate Declaration.

The Aspen Skiing Company's guiding principles stress humanity (treating people the way they'd like to be treated by modeling authenticity, transparency, courtesy, respect, and humility), excellence (in business, quality, craftsmanship, guest services, and athletic achievement), sustainability (of people, profits, the environment, and the community), and passion (living our core values, embracing life-long learning and meaningful work). Their main goals are to stay in business forever by remaining profitable; solving climate change; treating their community well; and operating in a manner that doesn't harm their local environment. They embrace activism because:

> Climate change is the greatest threat facing humanity, not to mention the ski industry. Because the problem is so big, the fix won't come from changing light bulbs. That's why our #1 priority is using the snow sports community as a lever to drive policy change (http://www.aspensnowmass.com/).

Auden Schendler, their Vice President of Sustainability, told me:

> Our mission initially was to reduce the environmental impact of the company but it's changed to helping protect the Earth and enabling us to stay in business forever. So we are very focused now on climate change, solving climate change at a policy level, big scale. Corporate greening (e.g., eco-efficiencies, recycling) isn't going to do it. If you face a challenge like climate change you have to think differently, on a different scale.

In *Getting Green Done*, Auden talks about how, even if global greenhouse gas emissions are reduced, Aspen will warm by 6 °F by 2100, resulting in a climate similar to Los Alamos, NM, and thereby ending skiing in Aspen. The ski industry and its suppliers, communities, and customers face enormous changes, and the long-term outlook for skiers everywhere is bleak (Seelye 2012). As temperatures rise, scores of ski centers, especially those at lower elevations, will vanish. In 2012, snow-based recreation contributed approximately $67 billion annually to the US economy and supported over 600,000 jobs. Those economic contributions will vanish.

Aspen Ski Company is a Platinum member of the Protect Our Winters (POW) advocacy organization which was started in 2007. Mobilizing like-minded partners is a *Best Practice*. The POW website (http://protectourwinters.org) reads:

> We represent the global snow sports community—there are 23 million of us in the U.S., alone. Clearly, it's time for us all to step up and take responsibility to save a season that fuels our passions but is also the foundation for our livelihoods, our jobs and the economic

vitality of our mountain regions. Protect Our Winters is the environmental center point of the global winter sports community, united towards a common goal of reducing climate change's effects on our sports and local economies. POW was founded on the idea that the collective power of the winter sports community is massive, and if we can all work together, the end result can be revolutionary.

Heifer International® Since 1944, the mission of this nonprofit organization in Little Rock, AR, has been to end hunger and poverty while caring for the Earth. Their target audience is the poorest of the poor. Rather than provide one-time aid, they have provided livestock and environmentally sound agricultural training to people in more than 30 countries who struggle daily for reliable sources of food and income. They have placed 22.6 million families or 114.9 million men, women, and children on a path toward prosperity by giving them the tools and training they need to sustain their lives.

Key to their program is something called "Passing on the Gift®." Families share the training they receive with their neighbors and pass on the first female offspring of the livestock they receive from Heifer International® to another family. This extends the impact of the original gift, allowing a once impoverished family to become donors and full participants in improving their communities. Their 12 Cornerstones for Just and Sustainable Development include accountability, sharing and caring, sustainability and self-reliance, improved animal management, nutrition and income, improving the environment, full participation, training and education, spirituality, gender and family focus, and genuine need and justice. These cornerstones result in a model that seems to work in diverse settings for people with various levels of education. It is highly participatory and emphasizes local owners in the decision-making process; commitment of local resources; participation of people regardless of gender, ethnicity, or religion; and the use of traditional knowledge. Heifer International® positions itself on the cusp of capitalism using three communication strategies: perspective by incongruity, dissoi logoi/kairos, and the art of illusion (Clair and Anderson 2013). *Best Practice*: Build resilience within the local communities your organization touches.

I interviewed Steve Denne, Heifer International's® Chief Operating Officer, who described how they had recently made a strategic shift so that they could operate at scale. Despite the good they have done, he said:

> our success has been fragmented and small scale at the village level involving 200 families. In a 45 day period a few years ago we saw a shift in food prices that moved 100's of thousands of people in the wrong direction. We saw that at the rate we were going it would take an unimaginable long time to end poverty.

Now when Heifer International® undertakes a project, they work with communities of at least 1,000 families, build partnerships, link farmers to markets, and build a critical mass of stakeholders. They are scaling up to increase their impact by partnering with the private sector. Creating partnerships is a *Best Practice*. They work with organizations like Danone, Elanco, Green Mountain Coffee Roasters, and the Bill & Melinda Gates Foundation to "develop high impact partnerships that strengthen value chains, engage employees and enhance brand value while building the knowledge, business skills and social capital of our project participants" (www.heifer.org/partners). The farmers they work with, both in the USA and overseas, are in marginal communities often operating outside the market, so "We work to get the farmers ready for market. They learn to save, manage money, organize into coops and farmers associations, trust in relationships that can withstand market forces, and get a strong value chain underway." Empowering others is a *Best Practice*.

 City and County of Denver, CO This metropolitan region ranks among the top ten US cities for having the most energy-efficient buildings, best public transportation system, and best overall green living performance (Bushwick 2011). The City and County of Denver governmental organization has 11,000 employees and serves almost 650,000 people living in a 155 square mile area. It includes 21 agencies working under a mayor. As of 2014, it was a member of ICLEI.

In 2013, their Office of Sustainability was created and their 2020 sustainability goals were set. They have multiple specific and measurable goals addressing air quality, climate change, energy use, food, health, housing, land use, materials, mobility, water quality, water quantity, and workforce development. They gathered baseline data related to each of the sustainability goals and identified current initiatives associated with each goal. From there, they evaluated current initiatives, developed strategies for reaching the goals, and measured progress. They display *Best Practices* by measuring existing resource use and practices to develop a benchmark, researching effective change alternatives, implementing the selected change, measuring resource use and practices against the initial benchmark, and adjusting as necessary.

Jerry Tinianow, Chief Sustainability Officer for the City and County of Denver, told me:

> The model we are trying to foster here, we call it sustainability at scale. You have to produce numbers that are big enough to address the real challenges as best as you can project them for your city or your region. So that is the first challenge we are dealing with, getting people to think in terms of that scale.... We can't afford to compare ourselves to other cities, we have to compare ourselves to the uncertainties of the future that are peculiar to our own city and region.

 The Arbor Day Foundation Deforestation is a serious global problem with implications for global warming, agriculture, and freshwater availability. In 2013, a global map of deforestation was developed by the University of Maryland, NASA, and Google, a map which revealed that 888,000 square miles of forest vanished since 2000. Deforestation causes approximately 15 % of global greenhouse gas emissions. Carbon sequestration—the process of absorbing carbon into living things so that it stays out of the atmosphere—is a powerful tool against global warming. According to the Environmental Protection Agency, a 50-year-old oak forest could sequester 30,000 pounds of carbon dioxide per acre. It would take 40 acres to counter the amount of carbon dioxide emitted by 109 average cars based on 2007 emissions. Trees are part of the solution and the Arbor Day Foundation exists to help us enact this solution.

The Arbor Day Foundation was founded in 1973 to inspire people to plant, nurture, and celebrate trees. They have 70 employees at their home office and 200–300 seasonal workers at the Arbor Day Farm in Nebraska City, NE. The foundation runs ten programs including Trees for America, Tree City USA, Replanting Our National Forests, Nature Explore, Rain Forest Rescue, Tree Campus USA, Celebrating Arbor Day, Conservation Trees, Energy-Saving Trees, and Arbor Day Farms. Woodrow Nelson, Vice President of Marketing Communication, described one of the Arbor Day Foundation's core values saying:

> We build programs that are high-impact, and we define high impact programs as being life changing, large scale, partner-engaging and sustainable. And the way we define sustainability in that context is we want to build programs that create environmental impact that happens by itself. It doesn't need an annual budget, it doesn't need an annual influx of grants or donations. We want to build programs that manage themselves, that are sustainable unto themselves.... Our vision currently is that we will be a leader in creating worldwide recognition and use of trees as a solution to global issues.

Designing high-impact, self-managing systems is a *Best Practice*. Woodrow used their Tree City USA program to illustrate how the Arbor Day Foundation can reach 3,500 communities, with more than 140 million residents. The foundation established four core standards for a city to become a Tree City USA. Communities take ownership of their urban forest management and receive Arbor Day Foundation recognition when they achieve these standards. Approximately 3,500 towns and cities in the USA are doing forestry work, and the Arbor Day Foundation is "inspiring them to get onboard, but once they are onboard, they are in it and they believe in it," Woodrow said. Partnerships are important to the Arbor Day Foundation. In 2012 alone, the foundation planted nearly five million trees in America's forests in partnership with the US Forest Service and the National Association of State Foresters. Tree Campus USA with financial support from Toyota reached 152 colleges and universities in 2012. Enterprise Rent-A-Car pledged to plant 50 million trees over the next 50 years, a gift totaling more than $50 million. The Arbor Day Foundation is like other organizations you will read about. They provide road maps forward by providing standards and certification processes for others to follow. *Best Practice:* Develop, identify, and/or deploy a road map.

 ClearSky Climate Solutions   This professional services company located in Missoula, MT, provides consulting services to develop land-use plans to protect or develop forests. Services include forest carbon project design and development, strategic climate change consulting, carbon footprint assessment and reduction, and carbon credit brokering. Organizations concerned with offsetting their greenhouse gas emissions can purchase forest carbon offset credits. Despite its small business status, ClearSky has supported projects in more than 40 countries. A member of the UN Global Compact and UN Environment Programme's Caring for Climate Initiative, Keegan Eisenstadt, CEO and owner of ClearSky Climate Solutions, has pledged to fully offset his company's greenhouse gas emissions each year. The company was named the 2009 Sustainable Business of the Year by the Montana Sustainable Business Council. Keegan said:

> For us to really talk about sustainability and have a meaningful and significant impact in the next short term, 40–50 years, the only real choice we have is to embrace the fact that the capital markets are the deciders of natural resource use. . . .. There is not enough public will frankly to address this problem on a global scale.

As we sat together in the Mustard Seed Asian Cafe in Missoula, MO, Keegan went on to discuss the current global warming challenges humanity is facing. He said:

> The classic problem of capital is that the externalities are not internalized. One of the externalities that we did not know about is greenhouse gas emissions. It's my parents' legacy. They did not know. I cannot fault them. But now we know. We are wasting time. It's the challenge of my generation to try to begin to address the problem former generations caused without knowing.

In economics, an externality is a cost or benefit which affects a party who did not choose to incur that cost or benefit. Externalities can include environmental problems resulting from how something is produced which are not reflected in the price of goods, materials, or services. *Best Practice:* Be aware of the externalities associated with your organization and act now to offset them.

So now, you have met five individuals whose organizations recognize the need to adapt and get to scale so as to better manage impending challenges due to global warming. They represent groups working in the posh ski town of Aspen, empowering the poorest of the poor, providing services to the citizens of towns and cities, promoting the planting of trees, and working with greenhouse gas emissions and carbon offset credits. In the following pages, you will learn more about how each uses communication to enact sustainability-focused initiatives.

1.3   Why Communication?

Given the challenges associated with global warming:

> Effective communication is absolutely essential for the purpose of mobilization; achieving buy-in and agreeing through consensus over priorities. This communication is necessary especially because to a greater or lesser degree all climate change response measures involve trade-offs along with their benefits. Hence, a measure of consensus and synergy is required across the board; from the board room to the boiler room; and from the federal government to municipal courts (Okereke et al. 2012, p. 26).

[handwritten margin note: mobilize via comm.]

Although Okereke et al. (2012) spoke specifically of climate change responses, communication plays a critical role within organizations generally. Organizing is first and foremost a communication activity. Karl Weick, Professor Emeritus of organizational behavior and psychology, writes, "The communication activity is the organization" (Weick 1995, p. 75). Some scholars take this assertion seriously and acknowledge communication's constitutive role in creating organizations.

Communication occurs when sustainability-related issues are conceived, defined, discussed, planned, initiated within and between organizations, modified, and, perhaps, terminated. It is present when various stakeholders encounter and react to the initiatives.

A useful guiding theoretical framework for understanding communication's role rests in social constructionism, or the social construction of reality. This theory of knowledge emerged from sociology and communication to examine the processes underlying the development of our jointly constructed understandings of the world. The theory was introduced in Peter Berger and Thomas Luckmann's book *The Social Construction of Reality* (Berger and Luckmann 1966). Two of their points are especially relevant to our focus on communication: people construct a model of how the social world works which helps them make sense of their experiences, and language is the most important system through which reality is created. Whatever exists in the social world is the product of human communication. Through communication, social constructions (e.g., what sustainability means in an organization) are created, maintained, repaired, and changed. People communicate to create meanings for the physical world. For example, scientists say that 350 parts per million CO_2 in the atmosphere is the safe limit for humanity. Regardless of how they know this, an environmental action group sprang up around that number (see www.350.org). "Social actors use language to make things happen. Naming something gives them substance and makes them real" (Leeds-Hurwitz 2009, p. 893). Symbols (e.g., the number 350) and conversations function as critical tools in reality maintenance, to create a particular cultural (e.g., organizational) identity and to mobilize actions.

Communicators use mediated and nonmediated verbal and nonverbal channels to manage the ambiguity surrounding the term *sustainability* and to create and enact shared sustainability initiatives within dyads, small groups, and organizations and between organizations. Knowledge of theory and best practices, coupled with our personal experiences and trial and error allow us to communicate strategically toward our personal and organizational sustainability-related goals. Although this definition appears very functional, within any organization, the norms about what sustainability is and how it is symbolized and discussed are continually being created, challenged, and recreated through human interaction. Other definitions certainly exist. For example, Cox (2013) defines environmental communication as the pragmatic and constitutive vehicle used to shape our understanding of the environment as well as our relationships to the natural world. He writes that "it is a symbolic medium that we use in constructing environmental problems and in negotiating society's different response to them" (p. 19). This construction and response duality is addressed in Domenec's (2012) essay describing how Exxon, Chevron, and British Petroleum (BP) talked about environmental issues in their companies' annual letters. Domenec (p. 296) wrote:

> The recent BP oil spill evidenced that today, "communicating green" is as important as "acting green": in April 2010, BP was not only criticized for polluting the Gulf of Mexico, but Tony Hayward's awkward statements also caused a general uproar, further discrediting the company.

Like Cox, I view communication as pragmatic in that it educates, alerts, persuades, and helps us enact sustainability initiatives within and between organizations. It is constitutive in that it orients our consciousness by inviting a particular perspective, evoking certain values and not others, and creating referents for our attention and understanding. Frandsen and Johansen (2011) analyzed the constitutive and pragmatic nature of how automobile manufacturers discuss climate change. Recognizing that climate change is a social construction, they define it as a set of ideas that materialize through communication in (at least) three ways: (a) as representations or symbolic constructions of nature, the environment, and/or the climate (What and how do the car producers communicate about climate change? Has a new vocabulary been institutionalized?), (b) as strategic actions (What and how do the car producers communicate about their strategic-level climate-friendly initiatives?), and (c) as tactical actions (What and how do the car producers communicate about their tactical-level climate-friendly products and/or production processes?). *Action Plan*: Consider similar ways ideas materialize and are discussed in your organization or industry.

1.3.1   Communication Challenges Organizations Face

Organizations face multiple challenges (e.g., political disagreements as to what should be done internally or communicated externally; managing the potential distrust of key stakeholders) in terms of their sustainability-related messages. Here, I focus on only two: fear of speaking out and messages that are not processed.

1.3.1.1   Fear of Speaking Out

Many organizational leaders hesitate to communicate about their sustainability-related initiatives and accomplishments fearing a backlash if stakeholders perceive their motives are self-serving or think the organization is green-washing. Others worry that the promotion of sustainability will only invite increased stakeholder scrutiny and a cycle of rising expectations over time (Peloza et al. 2012). For years before Walmart spokespeople began speaking publicly about their sustainability initiatives, I'd heard about various initiatives from employees at their Bentonville, AR, office. Walmart remained publicly silent due to their concerns over the potential charge of green-washing or backlash from activists. Yet, Auden Schendler, Vice President of Sustainability at Aspen Skiing Company, urges leaders to speak out saying:

> Corporations are universally fearful of talking about their environmental work, whether its policy or operational greening. They are scared because they don't want to be called hypocrites. They want to do their work and be humble and not crow about it. But my advice to those people is that we are not going to solve the problem of climate change if you are quiet. All you have to be is honest when you talk about whatever you are doing. If you

are working on policy, you can say, 'we are working on policy, we have a lot to do cleaning up our own house but we think this is really important'. If you are working on changing light bulbs or operational greening, you can say 'we have done this. There is a long way to go and we understand that but this is where we are starting and we are going to do more'. Many corporations, if they communicate to the public, will roll out what they did as if they think they are saving the world and that is a mistake. It should simply be, 'hey look, we want you to know about what we are doing, we think this makes sense. We have a long way to go and this is the first step.'

Certainly, speaking out can be difficult. Auden spoke with me about his experience with transparency saying:

Probably the biggest communication issue we had in our history was my belief that we need to be a corporation trying to pursue sustainability. . .. radically transparent and honest and not focused on spin or marketing. And our management for a while. . . did not get that. . .. It [transparency] was seen as subversive instead of open.

About a year after Auden and his top management had a conflict over his radical transparency, Auden found a *Fast Company* magazine article on his desk describing how Patagonia was pushing for transparency in their product line, along with a note from his CEO which said, "Auden, should we be more, quote, transparent?" He described going into his CEO's office angry because this had been what he had talked about. "When I walked in there waving the article, he was kicked back in his chair laughing, meaning he got it." But sustainability coordinators and champions need to be prepared to educate their management team about open and transparent communication. During our interview, Auden, whose background is as a writer, shared with me his philosophy on communication:

When I came into this job I decided it was not enough to do the work. We had to talk about the work and the reason we wanted to talk about the work was we wanted to use Aspen as a stage to share our stories with the world and drive change in the business world and beyond. And so the work we have done has always been coupled with a really aggressive level of communication. It has taken multiple forms. One is talking to the press, even soliciting stories from the press. It has been applying for awards. It has been writing essays and papers across the spectrum of media. The public speaking is another piece of that. Reaching whoever we can: business, Wall Street groups, and corporations with the message. Groups like BICEP and POW are focused on furthering that message and getting a focus from the business community and others like athletes or government on policy change.

1.3.1.2 Messages Not Processed

An industry exists to rate and rank companies based on their sustainability efforts. Communications with rating agencies are very detailed, specific, and transparent. But this specific, and often technical, information isn't widely accessed or understood by stakeholders (e.g., customers, employees). Stakeholders are often unaware of specific company sustainability efforts and ratings either because they did not receive the information or they are not able and/or motivated to interpret it. In one study, Peloza et al. (2012) surveyed 2,400 individuals representing three key stakeholder groups—investment professionals, purchasing managers, and graduating students (representing potential employees) drawn from six counties (i.e., China, Germany, India, Japan, the UK, and the USA). They plotted the respondents' perceptions against third-party rankings of actual firm performance on sustainability metrics. Although 88 % of their respondents said good corporate citizenship was either somewhat or extremely important, most lacked the information they needed to integrate sustainability into their decision making. Stakeholders could not correctly identify companies that had better than average sustainability records.

In trying to identify how stakeholders decode and interpret sustainability messages, Peloza et al. turned to an important communication theory, the elaboration likelihood model of persuasion (Petty and Cacioppo 1986). Elaboration refers to the amount of critical thinking that a person gives to a persuasive message (Vaughan 2009). The persuasive impact of a message is influenced by an individual's level of elaboration (degree of cognitive processing).

> In the case of high elaboration, the receiver typically engages in careful consideration of the message [central route]... In the case of low elaboration, the receiver relies primarily on inferences about the message based on relatively simple cues from the communication, such as a famous spokesperson. This is known as the peripheral route to persuasion. A third possibility is when the individual is neither motivated nor able to process the communication. Known as passive processing, communication success relies on repeated exposure over time to create attitude change and simple connections between the brand and message elements such as a catchy jingle (Peloza et al. 2012, p. 81).

Central processing involves questioning and researching sustainability claims and then drawing conclusions and making judgments. People must be motivated and have the ability to interpret the message, but in order to be motivating, a message must be personally meaningful. Prior research suggests consumers lack the motivation to process sustainability-related messages. Also, they may lack the frames necessary to effectively understand and act upon the information (Lakoff 2010). Although they may indicate positive attitudes toward corporate sustainability, traditional product quality issues are more important for most customers. This is a frame they have been socialized to understand. Information regarding sustainability initiatives may impact perceptions of a company's reputation but not stakeholder behaviors as much as do issues related to quality and service. Also, individuals tend to ignore information about ethical attributes in their decision-making process in order to reduce the stress associated with the decision.

Information processed through the peripheral route receives less critical consideration. Decisions made through this route aren't based on the message itself but by the activation of simple decision rules (heuristics) or guiding principles which come to mind in response to the message such as source credibility, evaluation of the message's style and format, or the receiver's mood. Because most stakeholders' knowledge of and interest in sustainability is low, they use heuristics (Peloza et al. 2012). For example, people use knowledge of one sustainability initiative (e.g., an organization's recycling program) to make inferences about its performance across a broad range of sustainability issues (e.g., pollution control, workplace policies). This halo effect leads people to inaccurately estimate an organization's true investments in sustainability. Peloza et al. discussed four important heuristics: sustainability initiative form, category biases, brand biases, and senior management image. Initiatives that were self-oriented (e.g., fuel-efficient cars save money) are more closely related to the traditional product performance cues that dominate customer behavior. So if a stakeholder knows a company leads its industry in fuel efficiency, they are likely to believe it also leads to other forms of sustainability. Stakeholders negatively view some industries or product categories. For example, because of the financial industry's reputation crisis in the mid-2000s, stakeholders are more likely to rate the sustainability performance of financial companies lower than is actually true. Stakeholders have inherent positive or negative perceptions of some brands. For example, although Walt Disney and Walmart were rated similarly by rating agencies regarding sustainability, the Disney brand is seen as wholesome, clean, and caring, while the Walmart brand is associated with operational efficiency, cost, and some degree of ruthlessness. Finally, if the senior management team is seen as wasteful and out of touch based on media coverage, this influences perceptions of the organization's sustainability performance.

1.4 What Are We to Do?

The elaboration likelihood model of persuasion helps us realize that repetition can be helpful in getting the message out. Also, thoughtful message design is important. Mike Johnson, Associate Vice Chancellor for Facilities, at the University of Arkansas, Fayetteville, and a retired Rear Admiral in the US Navy's Civil Engineer Corps, talked to me about the difficulty getting the word out about sustainability-related initiatives. Although he was talking about intraorganizational communication, the same holds true for external communication. In discussing messaging, Mike said:

> Part of what we are talking about is, you can't be too complex. So what you need to do is pick some highlights, talk about 'we saved 5,000 kW h, that's 5 tons of coal not used to generate the electricity, that would power a classroom for 2 months.' And then go on with it, don't belabor it, come at it from a different direction and then retransmit. I used to tell my troops, if you don't transmit it 50 times, you probably have not gotten to everybody and by the time you get to the 50th time you need to start over, because the ones that originally understood it have forgotten about it. So you just need to keep cycling.

In this example, we see the use of repetition, but we also see the message designer seeking to frame the importance of one particular number.

I asked David Driskell, Boulder's Executive Director of Community Planning and Sustainability, to define *good communication*. He said:

> Thinking critically about the audience, where they are at in their understanding. How to present the information in an accessible way. Having different tiers of communication. So I can give the 30-section version or the 3-minute version or the 3-hour version, depending on the audience's ability and interest.

In 2013, effective communication about energy-related sustainability issues was especially important in Boulder, CO. Boulder was one of the first US cities to adopt the Kyoto Protocol. City leadership set a goal to reduce Boulder's greenhouse gas emissions to 7 % below 1990 levels by 2012. They worked hard to reach this goal. In 2006, voters passed the Climate Action Plan Tax, the nation's first carbon tax. Despite engaging in multiple efficiencies, the city realized they could not reach their goal unless they shifted their dependency away from the carbon-intensive coal their energy provider, Xcel, used. They asked Xcel to change the city's energy source from coal to alternative energy to help them meet their carbon goals. Xcel refused. In 2011, voters directed the city to explore different options for providing clean, reliable, low-cost, local energy to their community. One possibility included creating a local power utility. In November 2013, voters passed a measure supporting the creation of a local power utility if a number of conditions were met, rather than the countermeasure Xcel placed on the ballot to block further efforts in that direction. The energy future project involved "navigating an incredibly complex, political environment," David told me.

The city pulled together a communication team from various parts of the organization to think strategically about providing clear messages about complex technical information involving the city's energy supply, energy rate structure, and energy distribution system. David explained:

> You can get lost in the weeds [of the complex information] and the communication team has such a great way of looking at all that and figuring out what is important, what do people need to know? How do we present it in a way that they can hear it?

David discussed the need for the messaging to be informative rather than persuasive saying "it is a fine line to walk." He also talked about message framing:

> It has been a huge learning curve for us as the city and for the community about where energy comes from and what's happening in the energy industry and the choices that we have. And that combined with our climate commitment work, we've really tried to reframe it from doom and gloom, the planet is going to die if we don't do something, to we have an opportunity to be a leader in the future of energy in terms of creating a low-carbon economy. Invent it here and export it everywhere in the world and support our companies who are doing innovative, greener things, and I think it is a pretty exciting time.

Throughout this book, you will learn ways to more strategically and effectively communicate about your organization's sustainability initiatives with internal and external stakeholders. You will also see how other organizations are preparing for and embracing the opportunity to change the ways they do business.

1.5 Concluding Thoughts

The precautionary principle specifies that scientific uncertainty is no excuse for inaction on an environmental problem. Instead of reacting to environmental problems after the fact, this principle suggests we must try to avoid environmental

problems initially. Organizational forces aligned against the precautionary principle in the USA seek to create uncertainty. They use the trope of uncertainty to
nurture doubt in the public's perceptions of scientific claims and delay calls for
action (Cox 2013). Doubt strips people of the motivation to act and weakens our
public and political will to solve problems. Awareness of how the created uncertainty is a communication strategy used to delay action helps us better understand
why climate change is being debated long after a scientific consensus has emerged.
Understanding precedes action. And action must occur even in the face of uncertainty. The Union of Concerned Scientists (UCS), an alliance of more than 400,000
citizens and scientists, argue that scientific analysis—not political calculations or
market-based concerns—should guide the formation of our government policy,
corporate practices, and consumer choices. They believe thoughtful action based
on the best available science can help safeguard our future and the future of our
planet. Now we must quickly work within and through organizations to adapt to
global warming. I encourage each reader to use communication strategically to
move the global debate and action agenda forward. You can help your organization
be a leader in its sector. Many of the organizations you will read about in this book
are already leading the way.

References

Berger, P. L., & Luckmann, T. (1966). *The social construction of reality: A treatise in the sociology of knowledge*. Garden City, NY: Anchor Books.
BICEP. (2013). *Tackling climate change is one of America's greatest economic opportunities of the 21st century*. http://www.ceres.org/bicep/climate-declaration. Accessed 25 June 2013.
Blackburn, W. R. (2007). *The sustainability handbook: The complete management guide to achieving social, economic and environmental responsibility*. London: Earthscan.
Bushwick, S. (2011). *Top 10 cities for green living*. Scientific American. http://www.scientificamerican.com/article.cfm?id=top-10-cities-green-living. Accessed 23 Dec 2013.
Christensen, L. T., Morsing, M., & Thyssen, O. (2015). Discursive closure and discursive openings in sustainability. *Management Communication Quarterly, 29*, 135–144.
Clair, R. P., & Anderson, L. B. (2013). Portrayals of the poor on the cusp of capitalism: Promotional materials in the case of Heifer International. *Management Communication Quarterly, 27*, 537–567.
Cox, R. (2013). *Environmental communication and the public sphere* (3rd ed.). Washington, DC: Sage.
Domenec, F. (2012). The "greening" of the annual letters published by Exxon, Chevron and BP between 2003 and 2009. *Journal of Communication Management, 16*, 296–311.
Dryzek, J. S. (2005). *The politics of the earth: Environmental discourses* (2nd ed.). Oxford: Oxford University Press.
Eisenberg, E. (2007). *Strategic ambiguities: Essays on communication, organization, and identity*. Thousand Oaks: Sage.
President Obama's Plan to Fight Climate Change. (2013). http://www.whitehouse.gov/share/climate-action-plan. Accessed 20 Dec 2013.
Frandsen, F., & Johansen, W. (2011). Rhetoric, climate change, and corporate identity management. *Management Communication Quarterly, 25*, 511–530.

Intergovernmental Panel on Climate Change. (2013). *Climate change 2013: The physical science basis.* http://www.climatechange2013.org/report/. Accessed 20 Dec 2013.

International Energy Agency. (2011). http://www.iea.org/. Accessed 23 Dec 2013.

Lakoff, G. (2010). Why it matters how we frame the environment. *Environmental Communication: A Journal of Nature and Culture, 4*, 70–81.

Leeds-Hurwitz, W. (2009). Social construction of reality. In S. W. Littlejohn & K. A. Foss (Eds.), *Encyclopedia of communication theory* (Vol. 2, pp. 891–894). Los Angeles, CA: Sage.

Leitch, S., & Davenport, S. (2007). Strategic ambiguity as a discourse practice: The role of keywords in the discourse on 'sustainable' biotechnology. *Discourse Studies, 9*, 43–61.

National League of Cities. (2013). *Climate change.* http://www.nlc.org/find-city-solutions/city-solutions-and-applied-research/sustainability-as-section-index/climate-change. Accessed 15 Nov 2013.

Okereke, C., Wittneben, B., & Bowen, F. (2012). Climate change: Challenging business, transforming politics. *Business and Society, 51*, 7–30.

Peloza, J., Loock, M., Cerruti, J., & Muyot, M. (2012). Sustainability: How stakeholder perceptions differ from corporate reality. *California Management Review, 55*, 74–97.

Petty, R. E., & Cacioppo, J. T. (1986). *Communication and persuasion: Central and peripheral routes to attitude change.* New York: Springer-Verlag.

Schendler, A. (2009). *Getting green done: Hard truths from the front line of the sustainability revolution.* New York: PublicAffairs.

Seelye, K. Q. (2012). Rising temperatures threaten fundamental change for ski slopes. *New York Times.* http://www.nytimes.com/2012/12/13/us/climate-change-threatens-ski-industrys-livelihood.html?pagewanted=alland_r=0. Accessed Oct 2013.

United Nations Environmental Programme. (2013). *Climate change.* http://www.unep.org/climatechange/NewHome/tabid/794594/Default.aspx. Accessed 12 Jan 2014.

Vaughan, D. R. (2009). Elaboration likelihood theory. In S. W. Littlejohn & K. A. Foss (Eds.), *Encyclopedia of communication theory* (Vol. 1, pp. 330–322). Los Angeles, CA: Sage.

Wals, A. E. J., & Schwarzin, L. (2012). Fostering organizational sustainability through dialogic interaction. *The Learning Organization, 19*, 11–27.

Weick, K. E. (1995). *Sensemaking in organizations: Foundations for organizational science.* Thousand Oaks: Sage.

Wexler, M. N. (2009). Strategic ambiguity in emergent coalitions: The triple bottom line. *Corporate Communications: An International Journal, 14*, 62–77.

World Commission on Environment and Development. (1987). *Our common future.* Oxford: Oxford University Press.

5 RS reduce reuse recycle repair reimagine
Com. to Stakeholders = clear, inform, strategic
heuristic – category | brand | sustain form | senior mangmt

Chapter 2
Changing Paradigms, Shifting Societal Discourses, and Organizational Responses

Abstract This chapter begins with a description of the transformational changes undertaken at the Portland Trail Blazers' basketball arena campus when messages within their external environment changed to highlight environmental sustainability. Two paradigms generally used to describe the belief systems of people and people groups regarding our relationship with the natural environment are described: the dominant social paradigm and the new ecological paradigm. The societal shift toward the growing importance of the sustainable development *Discourse* within businesses, cities, and universities is described. Paradigms are differentiated from *Discourses* and ten environment-related *Discourses* are identified (i.e., the industrialism *Discourse*, survivalism, the Promethean response, administrative rationalism, democratic pragmatism, economic rationalism, green politics, green consciousness, ecological modernization, and sustainability). Criticisms of the ecological modernism and sustainability *Discourses* are reviewed. Communication's role in reinforcing and challenging paradigms is discussed. The chapter ends by discussing forces which influence how environmentally related issues are framed, contested, and reframed. In addition to paradigms, *Discourses*, and ideology, theories or theoretical concepts highlighted include discursive closure, critical theory and the neo-Marxian perspective on sustainable development, framing, schemata of interpretation, systematically distorted communication, and social judgment theory. Throughout the chapter, interview data gathered from small businesses, an activist organization (the Natural Resources Defense Council), a nongovernmental organization, South Dakota's state government, multiple cities (e.g., the City and County of Denver), two sports organizations (the Portland Trail Blazers, Aspen Skiing Company), a university, and two multinational organizations (Tyson Foods, Sam's Club) is integrated.

Lewis and Clark and the Natural Environment When Lewis and Clark set out on their journey, the dominant view of the human–environment interface in America was a tradition of repugnance. Settlers loathed the wilderness and/or saw it as something to be exploited (Cox 2013). Shortly after their return, an idea started growing about America's Manifest Destiny. In the July–August 1845 issue of the *United States Magazine and Democratic Review*, an anonymous author proclaimed "our manifest destiny to overspread the continent allotted by

© Springer International Publishing Switzerland 2016

M. Allen, *Strategic Communication for Sustainable Organizations*, CSR, Sustainability, Ethics & Governance, DOI 10.1007/978-3-319-18005-2_2

Providence for the free development of our multiplying millions." There were at least three basic themes to Manifest Destiny: the American people and their institutions had special virtues; it was their mission to remake the West into an agrarian America; and the nation had an irresistible destiny to accomplish this duty. Westward expansion and efforts to tame the wilderness were rooted in our collective belief in Manifest Destiny. However, within 25 years after Lewis and Clark returned, voices rose in art, in literature, and on the lecture circuit challenging the view of nature as alien or exploitable. Almost 80 years after their return, America had its first national park—Yosemite National Park. This example illustrates how societal discourses and worldviews change over time—the focus of this chapter.

2.1 The Trail Blazers: An Example of Organizational Change

In Oregon, over 30 large and small businesses including ski resorts and investor groups signed the BICEP Climate Declaration. Three major brands were represented in that list: Adidas, Nike, and the Portland Trail Blazers. I wondered what prompted these highly visible sports-related organizations to make such a public and unified statement. Although the Trail Blazers employ only about 300 people, tens of thousands of volunteers and part-timers work in their arena. In 2012, the team had the second highest attendance in the National Basketball Association (NBA) with an average of 20,496 fans per game. Their website suggested they had an interesting story to share, so I arranged to speak with Justin Zeulner, their former Senior Director of Sustainability and Public Affairs.

One August afternoon, I sat down with Justin in a building adjacent to the Moda Center, the world's first professional sports arena to achieve LEED (Leadership in Energy and Environmental Design) Gold certification under the US Green Building Council's (USGBC) Existing Buildings standard. LEED is a voluntary, consensus-based, market-driven program that provides third-party verification of buildings. It was developed in 2000 by the USGBC, a nonprofit committed to the construction of cost-efficient and resource-saving (e.g., energy, water) buildings. New and renovated buildings which meet LEED-developed criteria can earn points which result in Certified, Silver, Gold, or Platinum level rating. In 2014, 1.7 million square feet of building space was being certified *per day* around the world. Like the Arbor Day Foundation, the USGBC provides a road map used by others seeking to become more environmentally sustainable.

Justin described the evolution of thought regarding sustainability for him, his organization, and the surrounding business community. In the early to mid-2000s, Justin was working as operations leader and his executive management team asked him "What is this sustainability?" Simultaneously, frontline employees asked him "Why aren't we recycling?"

> I remember going to a sustainability summit at the Nike headquarters here in Portland. The topic was financial sustainability and how can corporations stay viable over time because very few companies do survive for three decades. And when I went back the next year (to the summit) all of a sudden they were talking about environmental stewardship. It changed from more of a corporate sustainability and finance focus to this different path.

Within 5 years, the Trail Blazers had renovated the Moda Center. In 2010 the Trail Blazers invested $560,000 in operations improvements around the arena. By 2011 they had recouped $411,000 in energy savings, $165,000 in water savings, and $260,000 in waste diversion savings, with a total savings of $836,000. They projected that their savings would reach over $1 million by the end of 2012. They joined forces with northwest-based teams from six professional sports leagues to launch the Green Sports Alliance. Today, they are among the most progressive teams in the world of sports in terms of their environmental initiatives. They helped moved the NBA toward a deeper engagement in environmental issues (Henly et al. 2012).

This brief glimpse into changes occurring with the Trail Blazers illustrates how societal *Discourses* are changing for organizations. Executives, employees, and summit planners heard things which caused them to ask questions, to seek and provide new information, and to ultimately implement change.

2.2 Changing *Discourses*: The Organization-Environment Interface

In his book *The Structure of Scientific Revolutions* (1962), Thomas Kuhn, an American physicist, historian, and philosopher of science, talked about paradigms and paradigm shifts. Kuhn defined a scientific paradigm as "universally recognized scientific achievements that, for a time, provide model problems and solutions for a community of practitioners" (p. 10). A paradigm shapes what is observed, the questions asked in relationship to the subject, how these questions are structured, how answers are sought, how results are interpreted, and what solutions are preferred. Paradigm shifts can occur swiftly or incrementally. It is common to see a mix of old and new paradigms as old ideas are challenged.

In the USA, over time people challenged the prevailing view of the human–environment interface resulting in four antagonisms (Cox 2013). These antagonisms involved preservation or conservation (newer paradigm) vs. human exploitation of nature (older paradigm); human health (newer paradigm) vs. unregulated business and pollution of our air, water, and soil (older paradigm); environmental justice (newer paradigm) vs. nature as separate from where we live and work (older paradigm); and sustainability and climate justice (newer paradigm) vs. unsustainable social and economic systems (older paradigm).

The belief systems reflected within paradigms are built up, reinforced, and changed by the *Discourses* flowing in a given society. A *Discourse* is a system of statements made about aspects of our world which carry a set of assumptions, prejudices, and insights—all of which are historically based and limit the consideration of other alternatively valid statements (Ashcroft et al. 1998). Messages occurring at meta-levels (e.g., internationally, nationally, professionally) reflecting these *Discourses* influence how people communicate about sustainability within and between organizations (Olausson 2011). Talk and text appearing in corporate annual reports, public speeches, news articles, and advertisements can be analyzed to show the larger societal *Discourses*. In the next section, after discussing paradigms briefly, the discussion turns to some of the major *Discourses* taking place in industrialized societies about the organization–environment interface. Although the sustainability *Discourse* is a dominant and growing *Discourse*, sometimes *Discourses* clash. Criticisms of the sustainability *Discourse* are introduced.

2.2.1 Paradigms and Discourses in Societies

Although various scholars have described multiple paradigms when seeking to explain the human–nature interface (e.g., Cohen 1976; Colby 1991), I focus on the dominant social paradigm (DSP) and the new ecological paradigm (also called the new environmental paradigm) (NEP). Sustainable development emerged as part of a paradigm shift from the DSP and displays elements of both the DSP and the NEP.

The DSP has been the main belief system in most parts of the Western world for at least two centuries with one of the earliest references to it in Pirages and Ehrlich (1974). This paradigm involves political, economic, and technological dimensions (Shafer 2006). Politically, limited government intervention, private property rights, liberty, and economic individualism are stressed. Unlimited economic growth is seen as achievable and free enterprise promoted. Supporters feel that science and technology can solve any human problems including those resulting from environmental degradation. Nature is valued as a resource for humans who have domination and humans are assumed to be exempt from the laws of nature. There is strong support for the status quo and faith in future material abundance and prosperity. Traditionally, organizations embracing this orientation often wasted resources, did not seriously consider their environmental impact, protected the environment only to the extent outlined by regulatory requirements, and emphasized technological solutions to environmental problems. They responded to environmental issues mainly for compliance reasons, rarely had a comprehensive environmental policy, and had environmental goals unlinked with other management goals.

Over the past few decades, the NEP emerged as an outgrowth of the fundamental conflict between continued economic growth and ecological sustainability (Shafer 2006). The NEP was supported by the growing US environmental movement in the 1970s and reinforced by several large environmental disasters. For example, in one of the world's largest industrial disasters, the 1984 gas leak at the Union Carbide India Limited pesticide plant in Bhopal, India, resulted in thousands of deaths and hundreds of thousands of injuries and forced the chemical industry to use risk communication to regain its legitimacy (Chess 2001). NEP supporters think unlimited growth within a finite ecological system is impossible, feel we are approaching the limits of ecological sustainability, and believe we must learn to live in ways redesigned to avoid ecological catastrophe. They challenge the belief that we humans have the right to modify the environment as we desire and do not believe we can solve all environmental problems even if we develop new technologies. They support greater restraints on free enterprise, private property rights, and the pursuit of economic individualism.

Dunlap et al. (2000) developed a widely used scale used for measuring individuals' values in terms of these paradigms. *Action Plan*: Seek out and complete this scale. Scale items illustrating the DSP include "People have the right to modify the natural environment to suit their needs," "The balance of nature is strong enough to cope with the impacts of modern industrial nations," and "People were meant to rule over the rest of nature." Scale items illustrating the NEP include "Plants and animals have as much right as people to exist," "The balance of nature is very delicate and easily upset," and "Despite our special abilities, people are still subject to the laws of nature." Five dimensions are measured including the realization of limits to growth, antianthropocentrism, the fragility of nature's balance, the rejection of exceptionalism, and the possibility of an ecocrisis.

Paradigms are reinforced, challenged, and changed at the societal level through communication including media coverage of environmental, economic, and political issues, political activity within governments (e.g., lobbying, speeches on the House and Senate floor, the wording of legislative bills), education, and dyadic communication (e.g., conversations, arguments, debates, or questioning). We see the paradigms clashing (historically in favor of the DSP) in media accounts partly because conflict is one of the common media framing devices for environmental issues (Cox 2013). Often a false dichotomy is created. *Key Point*: Be aware of false dichotomies; new perspectives can bring new opportunities.

2.2.1.1 The False Dichotomy Between Jobs and the Environment

We often see the false dichotomy between jobs and the environment appearing in media coverage. But over time, the media debate is shifting to include a green-job theme. A green job, also called a green-collar job, is, according to the United Nations Environment Program (UNEP) (2008):

> Work in agricultural, manufacturing, research and development (R & D), administrative, and service activities that contribute(s) substantially to preserving or restoring environmental quality. Specifically, but not exclusively, this includes jobs that help to protect ecosystems and biodiversity; reduce energy, materials, and water consumption through high efficiency strategies; de-carbonize the economy; and minimize or altogether avoid generation of all forms of waste and pollution.

In 2007 the UNEP, the International Labor Organization, and the International Trade Union Confederation jointly launched the Green Jobs Initiative. Simultaneously, the symbolic institutionalization of green jobs occurred within one of the dominant institutions of our culture—the government. The 2009 American Recovery and Reinvestment Act (ARRA) included provisions for new jobs in industries such as energy, utilities, construction, and manufacturing with a focus toward energy efficiency and more environmentally friendly practices. Worldwide, an estimated eight million green jobs have evolved in sustainable organizing and changes are occurring in fields including consultancies, public relations, marketing, design, manufacturing, and engineering (Mitra and Buzzanell 2015).

But still, a false dichotomy may or may not be reinforced by the interpersonal communication occurring within social groups:

> The manner in which we see our environment depends largely on what we are looking for in it. But what we look for is not just an individual or idiosyncratic matter—it depends on our cultural conditioning, our accustomed social roles, and our definition of the situation from which we relate to the environment. (Cohen 1976, p. 49)

Boulder, CO, is a town which has intentionally sought to negate the false dichotomy between jobs and the environment according to David Driskell, Boulder's Executive Director of Community Planning and Sustainability:

> [We] really stay away from any kind of attitude that you have got to choose between the environment and jobs. I think Boulder is a great example of a place that has refused to frame things that way. And as a result it has a very successful economy because it offers a great quality of life. When people created the Open Space program [a local land conservation program] they weren't doing it as an economic development tool—they were trying to preserve the landscape. But that's our signature. That's why people want to be here.

2.2.1.2 Difficult Conversations and Discursive Closure

Communicating across paradigms is difficult, yet necessary. When we communicate, tensions emerge between people who embrace one paradigm over the other. Often our fear of the tensions works to silence the discussion. A profile of George Marshall, Cofounder of the Climate Outreach and Information Network, COIN, and climatedenial.org, a blog about the psychology of climate change denial, shares how Marshall will ask complete strangers what they think about the changing climate (Woodside 2013). His personal passion is to talk to those who disagree that global warming is a serious man-made problem. Woodside quotes Marshall as saying, "My core contention is that climate change is a contested narration that is shaped by social negotiation. . . . In society as a whole—and I think this applies at a micro level in individuals and peer groups and institutions—I think there is a deliberately negotiated silence on climate change." Marshall promotes bringing up the topic even if you feel uncomfortable:

> I think any time you have a conversation with anyone about the weather you should bring climate change into the conversation, not in a hectoring, judgmental, on-your-soapbox way but just drop it in there every single time. 'Weird weather we're having . . . . Yeah well, personally, I believe its climate change and that something weird has been going on. And it's been getting weirder.' Just put it out there.

Often when he raises such issues with strangers, they simply stop talking. But mentioning the issue helps "establish staging points in that void where it is acceptable to talk about it." One of Marshall's favorite examples involves a man at a dinner party with retired professionals. The man raised the issue of how could they possibly fly so much, and everyone there was completely silent until someone said, "What a delicious spinach tart." And then they spent the next 10 minutes talking about spinach tart—in obsessive detail. It wasn't just a random thing. A spinach tart was a more comfortable topic to discuss than climate change.

This negotiated silence illustrates discursive closure, a concept developed by communication scholar Stanley Deetz (1992). Discursive closure involves the unobtrusive strategies used by the proponents of a particular *Discourse* to suppress potential conflict and prevent alternative views from being freely expressed. Strategies are difficult to notice but include disqualification, naturalization, neutralization, topical avoidance, subjectification of experience, meaning denial and plausible deniability, legitimation, and pacification. Disqualification occurs when people are denied access to speaking opportunities because they lack the needed expertise or skills. Naturalization occurs when one view is seen as the way things are while overlooking the social historical processes which created that normative view. For example, in the USA, few people rely on passenger trains to get from city to city; however, that has not always been the case. In the 1950s and 1960s, protesters fought to stop or reroute the interstate construction. Neutralization refers to the process by which value positions are hidden and value-laden activities are treated as value-free. Today, our unconscious reliance on our private automobiles for such trips reinforces our cultural value of autonomy, burdens people with the expense of automobile purchase and maintenance, and results in environmental damage and pollution. Topical avoidance is self-explanatory. Subjectification of experience occurs when someone says something is just a matter of opinion. Meaning denial and plausible deniability occurs when someone makes a statement in a way that allows them to deny it later, depending on the audience response. For example, I might say, "We should shut down Interstate X because it costs the U.S. too much to repair all the crumbling bridges," believing this assertion is worthy of discussion but wanting the flexibility to say *just kidding* depending on my audience's reaction. Legitimation invokes higher-order explanatory devices (e.g., corporate goal statements exist to make decisions appear acceptable rather than to guide decisions and actions). Pacification acknowledges the conflict but discounts the issue's significance, its solvability, or the ability of the participants to do anything about it. On the US Department of Transportation Federal Highway Administration website (2015), I read:

> Because of concerns about what highways, dams, and other public works projects were doing to the environment, the Congress passed the National Environmental Policy Act (NEPA) of 1969. . .. NEPA was a major turning point for the Interstate highway program. It gave us a framework for studying proposed projects—for changing them where needed and making sure the public has an opportunity to comment. Thanks to NEPA, we have learned a

lot about how to build highways, including Interstates, that protects the environment or make it better. We still face controversies because it is sometimes difficult to improve our transportation network without affecting the environment. But overall, the highways being built today are much better suited to the Nation's needs than would otherwise have been the case.

This quote illustrates legitimation (i.e., NEPA) and pacification. The discursive closure concept and the eight strategies can be applied to the conversations occurring between people who differ in whether or not they embrace the DSP or the NEP. *Key Point*: Recognizing these strategies is the first step to managing the communication challenge being imposed (e.g., talk about climate change rather than the spinach tart).

Although standardization and quantification are important to many sustainability initiatives, a reliance on predefined matrices and indices suggests that sustainability is to be performed, not explored, discussed, and created through interaction (Christensen et al. 2015). Standardization and quantification act as a form of discursive closure. "When discussion is thwarted, a particular view of reality is maintained at the expense of equally plausible ones" (Deetz 1992, p. 188). Micropractices stabilize social relations, suppress creative solutions, and limit engagement in solving sustainability problems. Christensen et al. (2015) offer a way to challenge discursive closure which they call a license to critique. This approach draws on stakeholders' experiences, ideas, and enactments and encourages them to detect and report discrepancies between organizational talk and action. Communication about sustainability focuses on formulating definitions, challenging the status quo, articulating ideals, laying down principles, contesting standards, publicizing visions, and putting forward plans. Good communication involves allowing and cultivating a variety of perspectives to ensure that established positions are continuously challenged. Creative solutions and commitment among involved parties emerge and continuously evolve through input and challenge. Given the range of paradigms and discourses discussed in the next section, clashing views over sustainability exist within many organizations.

2.2.2 Paradigms and Discourses in Organizations

Moving to the organizational level, paradigms exist that include the collective values and beliefs of organizational members which influence how issues are interpreted and acted upon (Andersson and Bateman 2000). Some organizations reflect the DSP. However, increasingly organizational paradigms are changing. Management practices such as recycling and waste management, incorporating environmental criteria onto the balance sheet, and recognizing the environment as a source of competitive advantage convey a moderately strong pro-environmental paradigm. In those organizations with a strong pro-environmental paradigm, there are strong pro-environmental attitudes among top management, rewards for environmental performance, support for sustainability-oriented innovation, and the initiation of and involvement in environmental partnerships. These qualities were evident in the Neil Kelly Company.

The Neil Kelly Company and the Emerging Paradigm As we drove along the Columbia River toward Portland, I reached out to Andy Giegerich, editor of *Sustainable Business Oregon*, a publication of the *Portland Business Journal*, asking for suggestions as to whom I should interview. He recommended Tom Kelly, President of the Neil Kelly Company, a family-owned business with locations in Portland, Lake Oswego, Bend, Eugene, and Seattle. Neil Kelly is the largest residential contractor in Oregon with approximately 170 employees. The company provides remodeling and energy services and handymen and builds custom homes. Like all contractors, Tom's company was challenged by the Great Recession of 2009. Tom admitted, "When you are really challenged financially it is hard to apply some of your values in a way that you would like." However, their core focus on sustainability survived.

One afternoon, I sat down with Tom and Julia Spence, Vice President of Human Resources. Under Tom's leadership, their South Portland showroom became the first LEED-certified commercial building on the West Coast and the fourth in the county. Tom's mountain home was the first LEED-certified single-family home on the West Coast. Active in fundraising, he helped raise money for a LEED-certified sleeping village (kid's camp), LEED-certified Catholic grade school, and LEED Platinum-certified grade school. He has worked on several green Habitat for Humanity houses and is on the board of a nonprofit purveyor of Forest Stewardship Council (FSC) certified wood called Sustainable Northwest Wood. "So a lot of the stuff we do in the community ends up reflecting our values," Tom said. Forests receive FSC certification after documenting that they utilize environmentally appropriate forest management techniques, reward and encourage local people to sustain the forest resources, and do not generate financial profit at the expense of the forest resource, the ecosystem, or communities.

Early on, as company president, Tom was trying to find a way to implement his values regarding sustainability and the environment into his business. When he was invited to a half-day training by Natural Step, he went. Although impressed, he wouldn't adopt it unless his management team agreed. All of the management team attended the next training, the approach resonated with them, and they became a Natural Step company. *Best Practice*: If you learn of training about sustainability, investigate it. If you like what you see, invite others to attend.

What is Natural Step? Developed in the late 1980s by Dr. Karl-Henrik Robèrt, a Swedish doctor and cancer scientist, Natural Step provides advisory services, certificate courses, and e-learning globally to organizations interested in becoming more sustainable. The approach articulates four system conditions. In a sustainable society, nature is not subject to systematically increasing concentrations of substances extracted from the Earth's crust, increasing concentrations of substances produced by society, or degradation by physical means; and people are not subject to conditions that systemically undermine their capacity to meet their needs. Their five operational strategies involve fostering sustainability-focused mindsets, providing advisory services, launching or participating in system change initiatives, incubating and spreading innovations and new ventures, and building a movement.

Reflecting on his experience with Natural Step, Tom said:

> Certainly, it was a key juncture in how I look at business and how it can influence the community, my customers, and my employees in a way that stands by my environmental values. That is when we really started taking initiatives. . .. We have worked really hard to implement sustainability as a core-company value. . . So that is pretty good for waking up in the morning and saying 'I am making a business but I am also making a triple bottom-line difference, I am also making a contribution to society'. . .. The words for me that are important in the scheme of what sustainability is are balance, fairness, and justice, but most importantly keeping nature in balance because if we don't keep nature in balance we will all be done. That is what it comes down to. I believe that. I think what is going on now is just the edges of where we will land with global warming.

So now we have read about paradigms and how the DSP is shifting to include elements of the NEP. That transition occurred as the major *Discourses* of our culture expanded. In the next section, we focus on the broader sociopolitical *Discourses* surrounding the human–environment interface.

2.3 What Are the Major *Discourses*?

Discourses reflect the general and enduring systems of thought which, in turn, influence the formation and expression of ideas within a historically situated time (Grant et al. 2004). Communication scholars Gail Fairhurst and Linda Putnam (2004) discuss how *Discourses* order and naturalize the world, are culturally standardized and shared ways of understanding the world, and establish power/knowledge relations. Bound up in political power, each *Discourse* advances the interests of some and suppresses the interests of others. Understanding the range of environmental *Discourses* is important because they shape how environmental debates are framed and what is considered to be a reasonable option (Dryzek 2005) when environmentally related choices are being made. Embedded in language, *Discourses* allow us to put bits of information together into coherent stories and construct meanings for what is common sense and legitimate knowledge (see Fairhurst and Putnam 2014, for a discussion of organizational discourse analysis).

For example, Al Gore's 2006 documentary, *An Inconvenient Truth*, entered the public discourse and helped change our perception of climate change and reshape the role of environmental protection in our lives from an infrequent conversation to a moral obligation (Walker and Wan 2012). Because many of the film's narratives fit within our long-standing cultural perspectives of nature (Rosteck and Frentz 2009), viewers could identify with the arguments being made. The documentary included actions viewers might take to minimize global warming. The Climate Project which was launched with the documentary has trained over 3,500 people to give Gore's presentation in their communities worldwide.

Although multiple typologies exist to identify environmentally related *Discourses* existing in Western societies (e.g., Prasad and Elmes 2005), in this section I focus on the work of Dryzek (2005), a professor of social and political theory, who described ten environmentally related *Discourses* in hopes of promoting "critical comparative scrutiny of competing discourses" (p. 20). Because each *Discourse* is based on different "assumptions, judgments, and contentions that provide the basic terms for analysis, debates, agreements, and disagreements" (p. 8), they often conflict.

Industrialism has been "the long-dominant discourse of industrial society" (Dryzek 2005, p. 13) with its commitment to growth in the quantity of goods and services and the hope of material well-being such growth might bring. The industrialism *Discourse* reflects the DSP and is the point of contrast for the other nine *Discourses* (i.e., survivalism, the Promethean response, administrative rationalism, democratic pragmatism, economic rationalism, green politics, green consciousness, ecological modernization, sustainability). Over the last 30–40 years, movement away from the industrialism *Discourse* has occurred. I interviewed Allen Hershkowitz, senior scientist with the Natural Resources Defense Council (NRDC), about his work with business organizations. The NRDC is a New York City-based nonprofit international advocacy group with offices in Washington, San Francisco, Los Angeles, Chicago, and Beijing. It consists of more than 350 lawyers, scientists, and other professionals and has grassroots support from 1.4 million members and online activists. NRDC priorities include curbing global warming, creating a clean energy future, reviving the world's oceans, defending endangered wildlife and wild places, preventing pollution, ensuring safe and sufficient water, and fostering sustainable communities.

Allen discussed the legacy of our reliance on the industrialism *Discourse* saying:

> There are a lot of challenges in advancing our issue [environmental stewardship] because you know we have built-up our existing industrial system based on environmentally ignorant practices which is why we are in the mess we are in. Whether it is global-climate disruption or biodiversity loss or the proliferation of waste or water scarcity or deforestation in ecologically rare places, I mean, there is such a diverse amount of pressures happening around the world and that's in large part because of our past environmental behavior and production techniques.

2.3.1 Do We Face Global Limits?

Two of the *Discourses* Dryzek (2005) identified deal with global limits: survivalism and the Promethean response. The survivalism *Discourse* is concerned with the Earth's carrying capacity, which is the maximum population of a species that an ecosystem can support in perpetuity. Populations crash when their environment's carrying capacity is exceeded. In *Collapse: How Societies Choose to Fail or Succeed* (2005), Diamond argued this happened to the inhabitants of Easter Island, the Anasazi of southwestern North America, and the Maya of Central America. We see a similar warning in Al Gore's documentary, *An Inconvenient Truth*. Carrying capacity was addressed in Garrett Hardin's (1968) influential essay "The Tragedy of the Commons." Hardin, a professor of human ecology, described that when faced with a decision about whether or not to put an extra cow to graze on the village commons [shared green space], a villager will rationalize that if he or she puts a cow on the commons he or she will benefit, whereas the costs [environmental stress due to overgrazing] will be shared with the other villagers. Each individual villager quickly places another cow on the commons, which is subsequently destroyed by overgrazing. Computer simulations sponsored by the Club of Rome in the early 1970s and published in *The Limits to Growth* (Meadows et al. 1972) established that exponential growth cannot go on forever in a finite system. Concern for limits permeated the Brundtland Commission's report which sparked growth in the sustainable development *Discourse*.

Many organizations recognize that resource limits exist and proceed accordingly. For several years Sam's Club periodically hosted private sustainability lectures at their Bentonville, AR, headquarters where representatives of their vendor companies spoke about their sustainability initiatives. I attended when Bruce Karas, Vice President for Environment and Sustainability at Coca-Cola, spoke. He described his company's 2020 goal to replenish 100 % of all the water it uses to make its products and to reduce water consumption used in manufacturing by 25 % (from 2010 levels). Coca-Cola is working with the US Department of Agriculture to restore damaged watersheds in our national forests, watersheds which supply drinking water to over 60 million people. They worked with DEKA to develop a low-energy water purification system called a slingshot to deploy in poor Third World communities so people have access to clean water. One slingshot unit can purify up to 300,000 L of water each year—enough daily drinking water for roughly 300 people—producing 10 gallons of clean water an hour while consuming less power than it takes to run a handheld hair dryer. Karas said:

> Coca-Cola has been in business for more than 125 years and we want to make sure we are here for the next 125, and strike a balance between business and being sustainable. An essential part of business is that, if sustainability is not a part of your business model, you won't be around. If our community is healthy, then our business will be healthy, representing a key balance. (Crognale 2012)

In contrast, those who adopt the Promethean response deny that there are limits to resources and to growth. They argue that market pricing and technology will stimulate creativity which will result in the identification or creation of alternative raw materials once one set of resources is depleted. Pollution is "just matter in the wrong place in the wrong form, and with enough skilled application of energy, that can be corrected" (Dryzek 2005, p. 57). The business-as-usual approach taken by many organizations suggests this *Discourse*, which is closely aligned with the DSP, remains strong in Western organizations.

2.3.2   How Can We Solve the Problems We Face?

Dryzek (2005) identified three *Discourses* which focus on environmental problem solving: administrative rationalism, democratic pragmatism, and economic rationalism. Although each recognizes ecological problems do exist, these problems are treated as manageable within our basic industrial society framework. Each embodies the fundamental belief of the NEP that unregulated growth and pollution is impossible within a finite system. Administrative rationalism relies on experts and the use of institutional and policy responses including professional resource-management bureaucracies, pollution control agencies, regulatory policy instruments, environmental impact assessments, and policy analysis techniques. When we think of the regulatory compliance issues many organizations face as well as their use of experts to improve eco-efficiencies, we see this *Discourse* shapes a great deal of what takes place in Western for-profit organizations. Democratic pragmatism involves interactive problem solving within the basic structure of a liberal capitalist democracy. We see it played out when organizations consult with the public, through the use of alternative dispute resolution mechanisms, policy dialogue, lay citizen deliberation, and right-to-know legislation. US laws mandate public forums where citizens can gain information and communicate about environmental concerns (Cox 2013). Economic rationalism involves the use of market mechanisms to achieve public ends. Resource privatization argues people care more for what they own privately rather than what they share in common with others. Another is the idea of managed markets and we have seen heated and politicized discussions in the USA about voluntary markets for trading carbon and/or pollution credits.

ClearSky Climate Solutions and Economic Rationalism   During my interview with Keegan Eisenstadt, of ClearSky Climate Solutions, he noted:

> The [stock] market has deeply internalized the need to satisfy demands at the end of the fiscal quarter, at the end of the day's trading close deadline, at the end of the stock price premerger. The demands of the market have no linkage to the long-term demands of our planet and our societies.

That said, he remained optimistic about the role of the voluntary greenhouse gas and carbon emission markets. He described how the European Compliance Market is doing well and how, in 2006, California passed the Global Warming Solutions Act, which set the 2020 greenhouse gas emissions reduction goal into law. In 2011, they adopted cap-and-trade regulation. Keegan also shared the results of a survey sponsored by the *Ecosystem* Marketplace, a leading source of news, data, and analytics on markets and payments for ecosystem services (e.g., water quality, carbon sequestration, and biodiversity). The study addressed why companies are purchasing forest carbon offset credits on the greenhouse gas market:

> The first one you would hope is because of their values, either the value of the board of directors, the founder or CEO, or employees.… The second is pre-compliance learning. Businesses say carbon is going to cost us something in the future, let's invest now to learn

what the problem is before we are whacked over the head with the compliance stick. . . . Another important piece of the puzzle that I think is more important in Europe [than the U. S.] is reputational risk management. . . . Another is co-branding. . . . The last one is that carbon markets provide an opportunity for people to diversity their business model. . . . These are the big reasons people are joining forest carbon, and that does not sound like a tree-hugger, right?

Critics contend that carbon trading overlooks the crucial point that trading does not reduce emissions, but rather redistributes the allowance of further emissions. Growth, production, consumption, and emissions continue. By creating a false sense of relief, public concern is shifted away from dealing with the global problems of increasing CO_2 emissions and natural resource scarcity (Christensen et al. 2015). Their point is very valid.

2.3.3 What About Something a Bit More Radical?

Two of the *Discourses* Dryzek (2005) identifies as radical and imaginative: green politics and green consciousness. These two come closest to embodying the NEP. Green politics includes green parties, social ecology, environmental justice, environmentalism and the global poor, and anti-globalization and global justice. Green consciousness includes deep ecology, ecofeminism, bioregionalism, ecological citizenship, lifestyle greens, and eco-theology. Bioregionalism is a political, cultural, and ecological system or set of views based on naturally defined areas called bioregions, defined through physical and environmental features (e.g., watershed boundaries). The determination of a bioregion is also a cultural phenomenon influenced by local populations, knowledge, and solutions.

Ecotrust and Bioregionalism Ecotrust, based in Portland, OR, is an economic development and conservation incubator for social enterprise. The organization seeks to identify and test innovations between public, private, for-profit, and nonprofit organizations around issues related to fisheries, food and farms, forests and ecosystem services, knowledge systems, marine consulting initiatives, indigenous affairs, natural capital, and watershed restoration. On their website (www. ecotrust.org), it reads:

The challenges of the twenty-first century are immense, but at Ecotrust we believe that true wellbeing is possible. To get there, we need radical transformation of current institutions—those ways of living from banking to building, from transportation to tree harvesting, that dominate our lives. We see urgency in building up an economy that restores nature and invests in people. And we believe the way to build that economy is through bold experimentation in the bioregion we call home—the Pacific Northwest. We are continually creating and supporting new businesses, nonprofits, alliances, networks and programs that build wellbeing in nature and community and deliver economic prosperity. The industrial economic model that's reigned for the last 300 years simply can't last. It's time for a more natural model of development. At Ecotrust, we believe the new economy starts here—in our backyards, communities, cities and regions.

I interviewed Oakley Brooks, their senior media manager, who described how their digital magazine *Commonplace* takes a place-based look at sustainability-related issues. Their first issue focused on the Skeena River Basin in British Columbia. Residents there wanted to think of new ways to grow and live in the twenty-first century while still protecting one of the best salmon rivers in North America. They were receiving external pressures to develop. "It is part of that ongoing question of how do you grow and continue to be part of the global economy in the twenty-first century but also keep your self-determination and desire for a viable place and an intact ecosystem relevant," Oakley said.

Canyouhave
econot sustain?

2.3.4 Ecological Modernization and Sustainable Development

This book mainly focuses on the two *Discourses* which Dryzek identifies as imaginative yet non-radical ways to "dissolve the conflicts between environmental and economic values" (2005, p. 14). Both argue our capitalistic goals and processes can be modified to be ecologically sustainable and that externalities such as social and political problems can be managed. Both promise that dramatic and inconvenient changes in our lifestyles (e.g., a reduced standard of living, reduced consumption patterns) and basic institutions are unnecessary. But meanwhile the capitalist version of the good life based on consumption and acquisition continues to outpace our shared concern for the common good. Will these *Discourses* stimulate enough solutions to equip us for the impending climate changes? Probably not, but that doesn't mean we shouldn't try.

Sam's Club and Communities Brian Sheehan, former Sustainability Manager at Sam's Club, appeared reflective when we met over a cup of coffee one late afternoon at the University of Arkansas campus. Brian had worked in sustainability for a city before being hired by Sam's Club in 2011. When asked to discuss sustainability, he said:

> Walmart has its three aspirational sustainability goals. Personally, I think those are great and worthy goals to achieve, but that does not get you to sustainability.... Some people would be surprised that even if you are powered 100 % by renewable energy and diverting 100 % of your waste from landfills and you are selling products that help sustain people and the environment, there is always more that you could do. So how are you helping communities thrive? If Walmart has limited its impact to almost zero, is that still enough for all the communities that it operates in to be sustainable?

Sam's Club is a US chain of membership-only retail warehouse clubs owned and operated by Wal-Mart Stores, Inc., founded in 1983 and named after Walmart founder Sam Walton. As of 2012, the Sam's Club chain served 47 million US and Puerto Rican members and was the 8th largest US retailer. Over 2.2 million associates globally work at Walmart and Sam's Club combined. The aspirational goals for both organizations combined are to divert 100 % of store waste from landfills, utilize 100 % renewable energy, and sell products to sustain people and the environment. Their sustainability-related mission statement states, "We're working to improve the quality of life now and for generations to come by operating our business in environmentally responsible ways and by offering Members sustainable products."

2.3.4.1 The Promise of Ecological Modernization and Limited Political Will

The ecological modernization *Discourse* addresses how environmental damage can be addressed through foresight, planning, and economic regulation. The *Discourse* is that it pays to promote efficiencies, develop new technologies, promote green consumerism, and invest in ecologically sound practices. It is more cost effective to redesign environmentally benign processes today than to pay for higher cleanup costs later. Businesses can make money by improving efficiencies and by making and selling green products and services. Keegan Eisenstadt, of ClearSky Climate Solutions, described how, when he returns to companies he has previously worked with, management says that the greatest benefit they received from entering into a sustainability process is they found many other ways to increase their company efficiency:

> The process started with a review of their greenhouse gas footprint and then along the way they recognized, 'Oh my gosh, we can save money on water, we can save money on power, we can save money on heating and cooling, we can save money by telecommuting or having video conferences,' and they have found huge financial savings from efficiency gains just because somebody said, 'Let's look'. … They are saving tens of thousands to millions of dollars, depending on the client, from efficiency gains. From thinking about the way they schedule employees, the way they schedule shifts, the way they bring in inputs, the way they manage supply chains and inventory, the way they market. It may not be that in the corporate suite they have a cleaner conscience, it's that they are amazed that being smart throughout their business could save them so much money.

Cities such as Denver and Portland are working to develop technologies and processes they can export. Susan Anderson, Director of Portland's Bureau of Planning and Sustainability, discussed their *We Build Green Cities* program. She said:

> We have found that by trying to do the right thing here for our own reasons, we actually ended up growing this industry [the LEED building industry]. … We have really keyed into that. For example, we look at what else is out there around storm water mitigation, around all sorts of different environmental quality issues where we think there is going to be job growth. [We ask ourselves] what if we put in standards here? So people are coming up with ways to do it here. Then they have this expertise to export to everyplace else.

Ecological modernization requires business and government working together. As Portland illustrates, a government can stimulate and reinforce environmentally preferable private sector actions (i.e., resource use reduction, new benign product development). In Europe, ecological modernization appears to have moved beyond the perceived conflict between economic development and environmental quality. The relationship between government, business, and environmental groups is more cooperative (Schlosberg and Rinfret 2008). However, European-style ecological modernization is unlikely to work at the federal level in the USA at this point in time given the partisan gridlock in Washington. Bipartisan political consensus was possible in the USA in the 1970s as illustrated by the large number of federal laws passed (e.g., the Clean Water Act, the Clean Air Act, the National Environmental Policy Act, the Endangered Species Act, and the Safe Water Act) and by the creation of the Environmental Protection Agency. Party lines emerged over environmental issues in the 1980s when an anti-regulatory, anti-environmental stance characterized the Republican agenda. Later the Republican majority in the House and Senate worked against Clinton's environmental agenda. In the 2000s, Bush's environmental agenda focused on economic interests, rather than on environmental protection. More recently, Obama increased the use of renewable energy technology through the use of tax credits, launched a new grant program to fund a residential solar and wind project, and raised fuel standards for cars and light trucks. See Kraft and Vig (2000) and Vig (2000) for a discussion of presidential leadership on the environment. However, by 2013, political gridlock surrounded the issue of climate change and little was being done to incorporate ecological modernization ideas into federal policy.

Lacking systematic incentives from the US federal government, organizations still engage in eco-efficiencies in terms of designing or redesigning buildings, products, and processes to utilize less energy, water, or natural resources and/or minimize or repurpose waste materials. Eco-efficiency is one of the most popular industrial strategies for environmental change, establishing a direct link between efficiency and environmental preservation. Eco-efficiency arguments resonate with organizations because they fit within a familiar corporate mindset regarding efficiency and cost management (Bullis and Ie 1997). Many for-profit organizations have had environmental management systems and/or continual improvement systems in place since the 1990s, if not before. Yet often eco-efficiency changes are limited to incremental modifications in production processes and the initiation of green consumerism.

Hydropower, Pierre, SD, and State Government Eco-Efficiency As we followed the Lewis and Clark Trail along the Missouri River, we learned how a chain of six main multipurpose dams in the upper Missouri River Basin supplies hydroelectric power to North Dakota, South Dakota, Nebraska, Minnesota, and Montana communities. In 2008 hydroelectric power accounted for almost 67 % of the USA's renewable energy and over 6 % of its total electricity. At least 34 states have hydroelectric plants. A large hydroelectric plant was located just outside our destination of Pierre, SD. Pierre is the South Dakota state capital and a town of almost 14,000.

In response to record rains in June 2011, the Army Corps of Engineers opened the dams along the Missouri flooding many parts of Pierre. Jody Farhat, Chief of the Missouri River Basin water management in the corps' Omaha District, said, "We had this incredible rainfall event. That was a rainfall event in May, and that was the game-changer in terms of system operations." High-magnitude water events and droughts are part of the very high-probability scenarios projected in the Intergovernmental Panel on Climate Change's (2013) report, *Climate Change 2013: The Physical Science Basis*. The rivers the US Army Corps of Engineers "tamed and harnessed" surprise people during high-magnitude water events.

After we discussed the flood, Mike Mueller, Sustainability Coordinator for the South Dakota Bureau of Administration, and I talked about how the state primarily defines sustainability in terms of achieving eco-efficiencies:

> The focus of my job, by design, is less on environmental impacts than it is on operational cost savings. . ... And that's because so many more people are receptive to saving money than saving the planet. So the view here has always been if we can be more efficient we'll reduce costs and therefore be able to keep taxes down for taxpayers, which is a very laudable primary goal.

Mike's job as the state's first sustainability coordinator is to pay attention to energy and water conservation and waste reduction in all state-owned buildings. He also focuses on fleet travel and fuel use. The South Dakota state government received over $1.3 million dollars from the American Recovery and Reinvestment Act of 2009 (ARRA). They used ARRA funds to pay for an energy assessment of all state-owned buildings. Then they designed a spreadsheet prioritizing the buildings which needed to be retrofitted. Almost $23 million was spent on projects such as installing new windows, buying new boilers and chillers, and installing new lighting systems in state-owned buildings. In addition, in 2008 the South Dakota State Legislature passed a bill requiring that all new state government buildings be built to LEED Silver certification. A few years later, the state's procurement laws changed to require environmentally preferred purchasing by all state agencies. Anything that uses electricity must be Energy Star certified and water fixtures must have WaterSense certification. *Key Point:* Ecological modernization partnerships and government leadership are critical in the face of global climate change.

2.3.4.2 The Shift Toward the Sustainability *Discourse*

The Shift in Business

Following the Rio de Janeiro Earth Summit of 1992, the World Business Council for Sustainable Development (WBCSD), a group consisting of 162 of the world's largest corporations, mainly in the manufacturing, mining, and energy sectors, and some with less than an exemplary environmental record, coauthored *Changing Course: A Global Business Perspective on Development and the Environment*, along with Stephan Schmidheiny (1992). Between 1992 and 2002, when the World Summit on Sustainable Development was held in Johannesburg, a number of important books were published discussing the role of business including *The Ecology of Commerce: A Declaration of Sustainability* (Hawken 1994),

The Hungry Spirit: Beyond Capitalism: The Quest for Purpose in the Modern World (Handy 1999), *Cannibals with Forks: The Triple Bottom Line of 21st Century Business* (Elkington 1999), *Open Society: Reforming Global Capitalism* (Soros 2000), *The Mystery of Capital: Why Capitalism Triumphs in the West and Fails Everywhere Else* (De Soto 2000), *Natural Capitalism: Creating the Next Industrial Revolution* (Hawken et al. 2000), *The Civil Corporation: The New Economy of Corporate Citizenship* (Zadek 2001), and *When Corporations Rule the World* (Korten 2001). In 2002, at the World Summit on Sustainable Development, the world's largest-ever international conference, the corporate leaders present articulated that business must be a major participant in sustainable development, not a source of problems to be overcome. Members of the Chamber of Commerce and the WBCSD argued that economic growth produced by free trade was the only hope for the world's poor and succeeded in positioning business as the dominant tool for pursuing sustainable development.

Such publications and the influence of USA-based groups, such as the WBCSD and the Chamber of Commerce, brought an awareness of sustainability into the boardrooms and management suites in organizations across the USA. For example, the late Ray Anderson, CEO of Interface, the world's largest manufacturer of commercial carpets and floor coverings, which is based in Atlanta, GA, was one of the first to adopt greener business practices after reading Paul Hawken's *The Ecology of Commerce* (1994). Hawkens wrote about how economics should be designed so that business and natural systems can coexist in balance. Anderson said in a TED talk which aired in February 2009:

> In his book Paul charges business and industry as (1) the major culprit in causing the decline of the biosphere, and (2) the only institution that is large enough and pervasive enough and powerful enough to really lead humankind out of this mess. And by the way he convicted me as being a plunderer of the earth. And then I challenged my people at Interface to lead our company and the entire industrialized world to sustainability which we defined as eventually operating a petroleum intensive company in such a way as to take from the earth only what can be renewed by the earth naturally and rapidly. Not another fresh drop of oil, and to do no harm to the biosphere.

At one point, Anderson was described as America's greenest CEO. He said, "From plunderer, to recovering plunderer, to America's greenest CEO in 5 years. That frankly is a sad commentary on American CEOs." Anderson, who died in August 2011, estimated that after 2001 he gave more than 1,000 speeches making the business case for sustainability.

Ideally, sustainability carries with it a set of assumptions different from the DSP and industrialism as illustrated in the shift at Interface. Increasingly, a wide variety of organizations are finding environmental considerations to be integral to their strategic positioning and are including environmental considerations into their mission and vision statements; integrating environmental concerns into their operations, decision making, and measurement systems; seeking to position their organization as an industry leader in environmental concern; developing innovative services and products for evolving markets; and considering their relationships with external stakeholders.

Elkington (1999) linked corporate sustainability with the idea of the triple bottom line which involves profits, planets, and people. Ten years after Elkington coined the term, Google Scholar, a search engine dedicated to scholarly writing, showed 57,000 references to the term. A year later, there were 206,000 entries (Wexler 2009). The triple bottom line refers to the idea that an organization's economic performance (profits), social performance (valuing and caring for employees, consumers, and communities), and environmental performance (the organization's ecological footprint) are interconnected. The environmental dimension involves activities that do not erode natural resources due to prudent corporate environmental management efforts. The social dimension encourages organizations to consider their impact on society and addresses issues such as community relations, support for education, and charitable contributions. Finally, the economic dimension centers on the value creation and enhanced financial performance of an organization's sustainability-related activities. The triple bottom line is also called the three pillars of sustainability: social, environmental, and economic. Sustainable organizations create goals and initiatives focused around each pillar. Economic value for shareholders is no longer the dominant purpose but rather a balance is sought as the organization admits that it has a role in promoting an ecologically sustainable world. Partnerships are formed with stakeholders (e.g., NGOs) also concerned with sustainability and justice. Products and production pressures are reconsidered, not just with an eye toward eco-efficiency but also toward ecosystem protection. Environmental and human costs are incorporated into accounting systems (i.e., full-cost accounting is utilized).

Tyson Foods and the Bottom Line   The week before we started our journey on the Lewis and Clark Trail, I met Kevin Igli, Senior Vice President and Chief Environmental Health and Safety Officer at Tyson Foods, Inc., in his office at their corporate headquarters in Springdale, AR. Tyson Foods is one of the world's largest processors and marketers of chicken, beef, and pork. Their merger with the Hillshire Brands company in August 2014 created a company with more than $40 billion in annual sales and a portfolio that includes brands such as Tyson®, Wright®, Jimmy Dean®, Ball Park®, State Fair®, and Hillshire Farm®. Following the merger, Tyson Foods employed 124,000 people worldwide. They provide products and services to customers in the U.S. and 130 other countries.

Their sustainability mission statement, which is a direct quote from John Tyson, their Chairman of the Board, reads:

> We recognize the importance of being a responsible corporate citizen. Our Core Values— which define who we are, what we do, and how we do it—are the foundation of corporate sustainability at Tyson. We are committed to making our company sustainable—economically, environmentally, and socially. Our progress in this endeavor will be measured by how we develop and market our products, how we care for the animals, land and environment entrusted to us, and how we treat people, including our Team Members, consumers, suppliers, and the communities in which we live and operate.

Kevin, who has worked with environment, health and safety issues throughout his career, explained:

> At Tyson Foods we decided a couple of years ago to follow that triple bottom line model. But in so doing we decided to add a fourth p which is products. Because that is really what we do, we make products. So we talk about sustainability as people, planet, profit and products. And we look at it as all four of those have to be managed in a balance in order for your business to be sustainable for the long term.

Responding to Pressures and the Lure of the Business Case for Sustainability

The MIT Sloan Management Review published the findings of a 2010 sustainability and innovation global executive study and research report conducted in conjunction with the Boston Consulting Group entitled *Sustainability: The 'Embracers' Seize Advantage* (Hannaes et al. 2011). More than 3,000 business executives and managers from organizations located across the world were surveyed. Participant organizations ranged from 500 to 500,000 employees. Many respondents indicated external pressures led them to adopt sustainability-driven management processes. The external pressures came from public policy changes at the local, national, and global levels, customer and potential employee preferences, trends for more measurement and accountability, and pressures from investors and pension funds (e.g., the Carbon Disclosure Project).

Organizations often follow a three-stage model in their sustainability efforts (Jabbour and Santos 2006). First, an organization reacts to pressures such as environmental legislation and product requirements either by complying or by seeking to change the regulatory environment (e.g., lobbying, making political donations, or bringing lawsuits). If an organization decides to augment or abandon those reactions, then it will focus on preventing harm to the environment (e.g., preventing pollution). The final stage some organizations take involves voluntary proactive actions to ensure long-term sustainability. Organizations adopt self-regulation, technological innovation, industry-wide codes, and/or certifications. Among big companies, tools such as triple-bottom-line accounting, a sustainability balanced scorecard, life-cycle assessments, eco-efficiencies, and environmental information and management systems are used. Common sustainability initiatives involve recycling; reducing the use of energy, water, and other natural resources; and switching to the use of more environmentally responsible products and services.

Organizations vary in why they respond. Some are reactive and compliance oriented, while others are more proactive either because their management team believes it is the right thing to do, they see it as presenting a strategic advantage, or both. The business case for sustainability is a set of popular arguments that often appear in the business literature (both scholarly and practitioner oriented). The main argument is there can be a positive relationship between environmental, social, and financial performance. A variety of reasons are given including reduced operating costs, competitive parity, regulatory advantages associated with environmental performance (e.g., lower compliance costs, greater flexibility when adapting to legislative changes, the ability to influence environmental laws and regulations), reduced waste, improved efficiency and productivity, product differentiation, international competitive advantage, greater appeal to consumers, strengthened firm reputation, the ability to sell pollution control technology, creation of entry barriers, development of new market opportunities, better access to markets, and reducing or avoiding legal liabilities (Walker and Wan 2012). Blackburn (2007) provides those who would like to make a case for sustainability with support for seven business case arguments involving increased reputation and brand strength; more competitive, effective, and desirable products and services; new markets; productivity; lessened operational burden and interference; lower supply chain costs; lower cost of capital; and less legal liability. *Best Practice*: Read Blackburn before pitching sustainability to your organization's leaders.

However, strategies that protect the Earth do not always result in corporate savings nor do markets always embrace green products. Not all scholars agree there is a strong positive causal relationship between financial and environmental performance. Salzmann et al. (2005) review the theoretical frameworks, instrumental studies testing the hypothesized positive relationship, descriptive studies examining managers' perceptions, and tools used to test the relationships. Some of the theoretical arguments (e.g., the trade-off hypothesis, managerial opportunism hypothesis, and negative synergy) argue for a negative link, whereas others (e.g., social impact hypothesis, slack resources hypothesis, and the idea of positive synergy or a virtuous cycle) argue for a positive relationship. They conclude that the research fails to show a strong causal relationship between the variables, but they provided no information supporting the basis for this conclusion. A lack of unified findings is not surprising since so much depends on organizational dynamics, industry sector, environmental challenges faced, and tools used to measure environmental, social, and financial performance.

The Shift in Cities

Concerned with the implications of environmental issues such as global warming on their communities, local governments are developing their own climate action plans. Some cities joined ICLEI which was formed in 1990 at UN headquarters. ICLEI is regarded as a paragon of Agenda 21 implementation. Five years after the Brundtland Commission's report, at the 1992 United Nations Conference on Environment and Development's Earth Summit in Rio de Janeiro, the 171 national delegates endorsed Agenda 21 and it was reaffirmed at the 2012 United Nations Conference on Sustainable Development. Agenda 21 is a nonbinding voluntarily implemented action plan developed by the UN to guide sustainable development.

I provide an overview here so you can see what it covers because, as you will read later, it is part of an artificially created controversy. The report consists of 300 pages with 40 chapters grouped into four sections: Section I: Social and Economic Dimensions—combat poverty, especially in developing countries, change consumption patterns, promote health, achieve a more sustainable

population, and utilize sustainable settlements in decision making. Section II: Conservation and Management of Resources for Development—provide atmospheric protection, combat deforestation, protect fragile environments, conserve biodiversity, control pollution, and manage biotechnology and radioactive wastes. Section III: Strengthen the Role of Major Groups—protect children and youth; empower men, women, NGOs, local authorities, business, and workers; and strengthen the role of indigenous peoples, their communities, and farmers. Section IV: Means of Implementation—use science, technology transfer, education, international institutions, and financial mechanisms.

Other groups exist to help towns and cities manage resources wisely while creating more livable cities. The Urban Sustainability Director's Network (USDN) is a peer-to-peer professional network which develops and shares solutions to challenges faced when designing and constructing buildings and neighborhoods, transporting people and goods, managing resource use, and fueling growth and development. The Sustainable City Network is a business-to-government media and publishing operation based in Dubuque, IA, that provides information on sustainability products, services, and best practices. They provide excellent webinars, many of which are archived, informative online content, and useful collaboration tools. Their primary audiences are city and county government professionals, elected officials, academicians, business leaders, and federal officials. *Action Plan*: If you are interested in working on city- and regional-level sustainability initiatives, check out their webinar archives. Cities seeking assistance with implementing sustainability initiatives will find ample resources to assist them. Blackburn (2007) provides a checklist for action for those considering implementing sustainability within their city.

Creating Sustainable Cities: Boulder and Portland For cities, the triple bottom line focuses more on people than profits. As I drove to an interview in Boulder, CO, I passed through some of Boulder County's 97,000 acres of open space. For decades, area citizens have worked to conserve the area's natural, cultural, and agricultural resources, provide land for public use, and design a good quality of life within their city. Meeting in his second story office overlooking the tree canopy sheltering a public gathering space, David Driskell, Boulder's Executive Director of Community Planning and Sustainability, told me:

> About 25 years ago there was a lot of analysis done and they said if we keep widening the roads, it isn't going to solve the problem. People are just going to drive more and the roads will be just as congested. So you need to find ways for people to get out of their cars and on their bikes or the bus or walking. So we've have 25 years of creating the multi-mobile transportation network. But if you look at it, it has had huge economic benefits for our community. Companies are here because their employees love to bike to work. And we know that there is value in being close to the bike path system. Every employee who works downtown gets a free bus pass. They love it. So there are economic benefits, there's obviously environmental benefits, and there are social benefits. We know it makes people happier. They are happier. They are more productive. They are physically fit. It is a more active lifestyle. All these things have come together in terms of what was driven initially by reducing congestion. But if you think about it holistically, about all the different things that you can advance that are around social, economic and environmental sustainability, it all leads you to it is worth investing in the multi-mobile transportation network.

Agenda 21

In 2012, Boulder's population was 101,808, while the Boulder Metropolitan Statistical Area population was over 310,000. The Boulder Valley Comprehensive Plan is a joint plan between the City of Boulder and Boulder County which informs and guides the shared planning and development of Boulder Valley. Sustainability is a unifying framework as they seek to create a welcoming and inclusive community, a culture of creativity and innovation, strong city and county cooperation, a unique community identity and sense of place, compact and contiguous development and infill to create a more sustainable urban form, open space preservation, great neighborhoods and public spaces, environmental stewardship and climate action, a vibrant economy based on quality of life and economic strengths, a diversity of housing in multiple price ranges, an all-mode transportation system, and physical health and well-being among residents. It has been identified as one of the Five Happiest Cities in America, Top 100 Best Places to Live, America's Most Productive Metros, Best Urban Green Spaces in North America, Most Popular City for Tech Startups, The Best Places for Business and Careers, and Best Places for Work-Life Balance. David commented:

> I am not a big fan of the three-legged stool [the triple bottom line]. I feel like it implies that if you have too much environmental sustainability you have to cut off the leg to make it equal with the others. They are not separate. My whole thing here has been we need to think about them [people, planet, profit] but sustainability is at the core where they intersect. So we now draw more interlocking circles and talk a lot about the sweet spot in the middle. That is sustainability. There is environmental quality, there is economic vitality, there is social equity but the sweet spot is sustainability where you are advancing all of those at the same time. … We talk about how do we challenge ourselves regardless of the initiative we are working on to think really comprehensively, strategically, about how we advance all areas at the same time.

Portland is located is near the confluence of the Willamette and Columbia rivers. In 2012, it had 600,106 residents, making it the 28th most populous city in the USA. With 2,289,800 people in the Portland metropolitan area, it is the 19th most populous metropolitan statistical area in the USA. The city has a commission-based government headed by a mayor and four commissioners as well as Metro, their regional government. The city is noted for its superior land-use planning and investment in light rail. Because of its public transportation networks and efficient land-use planning, it has been referred to as one of the greenest cities in the USA.

Like Boulder, Portland offers lessons useful in other cities. Susan Anderson, Director of Portland's Bureau of Planning and Sustainability, described how cities have a large footprint, internal purchasing power, numerous city employees, and influence through zoning and building codes. She said, "I think the impact of purchasing power is something that cities can use better, as a tool for change." She described a city ordinance Portland passed more than 10 years ago that said if developers want to build affordable housing projects or commercial projects that include city funds or receive a city tax incentive, they must build to LEED standards. Today they must build to LEED Silver and all city facilities are LEED Gold. This one action shifted everything. Now Portland has 300 LEED-certified buildings and they are exporting the expertise of their planners, architects, builders, and product developers across the globe.

In planning for their future, Portland integrated more than 20,000 resident comments into their *Portland Plan*, a guiding text which articulates their 25-year goals and 5-year action plans for community prosperity, business success, and equity; education and skill development; sustainability and the natural environment; human health, food, and public safety; design, planning, and public spaces; neighborhoods and housing; transportation, technology, and access; quality of life and civic engagement; and arts, culture, and innovation. In a document summarizing the highlights of the *Portland Plan*, the authors write:

> Sustainability means more than environmental stewardship; it is also about caring for our economy and for each other. It means recognizing that our actions matter and that each individual choice makes a difference to our health and to the health of our community. When pursuing our vision for a sustainable city, equity matters. If we are going to thrive, we need to ensure all Portlanders have access to the jobs, quality housing, education, art, nature, recreation and other services and amenities we need to live full and enriching lives. We value our diverse communities, so it is important to ensure that we have the social networks and built environment that helps us stay connected.

Aware of the importance of strategically communicating about sustainability, Susan said:

> We really worked at making sure the message was, it's not about the environment. It's about the environment and its connections to all these other things. Sustainability is not just about the environmental movement. It's much broader. It's a jobs movement. It's a social movement. It's a personal health movement and one of the things that came out of it [the community input sessions] was the issue of equity.

Movement Across Universities

The operations side of a college or university is similar to that of a town or city. There are streets to maintain, buildings to heat, grounds to keep, and utilities to pay. Some have large staffs or own substantial acreage. All impact the economic, social, and environmental aspects of the communities within which they are located. In 2006 the American College and University Presidents' Climate Commitment (ACUPCC) was introduced. This is a long-term commitment at the presidential and chancellor level to achieve carbon neutrality and reduce greenhouse gas emissions by making changes related to electricity, heating, and travel. As of October 2013, over 677 institutions representing over 30 % of the total US higher education enrollment were participating.

In 2010, the Association for the Advancement of Sustainability in Higher Education (AASHE) launched STARS, a program which rates organizations in terms of the campus operations, education, research, and administrative components of campus sustainability. By 2011, 289 institutions from the USA and Canada were participating. Participants can create a climate action plan, conduct a greenhouse gas inventory, develop a climate neutrality plan, take greenhouse gas reduction measures, and/or incorporate sustainability into the curriculum. Collectively, in 2013, ACUPCC institutions offered 12,482 sustainability-focused courses and 358 - sustainability-related undergraduate or graduate degrees. Although I discuss

AASHE and the ACUPCC here, other organizations exist to help colleges and universities pursue sustainability. Blackburn (2007) devotes a chapter to colleges and universities and provides a checklist for action for those considering implementing sustainability at their college or university.

Higher education enrollment accounts for nearly 2 % of the total annual US greenhouse gas emissions and ACUPCC signatories represent approximately 0.6 % of the US total, equivalent to about a quarter of the State of California's emissions. Thus, the long-term commitment to carbon neutrality by ACUPCC signatories may result in measurable reduction emissions (Sinha et al. 2010). Between 2007 and 2012, signatories had already achieved a 25 % (10.2 million MtCO2e) reduction in greenhouse gas emissions. MtCO2e stands for million metric tons of carbon dioxide equivalent. Based on current projections, they are expected to reach a 93 % reduction by 2050 (Mulla 2013).

Sustainability Efforts at the University of Arkansas My university, the University of Arkansas, Fayetteville, was one of the first 100 institutions to sign the ACUPCC in 2007. In 2007, we established our aspirational goal to achieve climate neutrality by 2040 and set two strategic goals: reduce greenhouse gas emissions to 2005 levels by 2014 and scale down to 1990 levels (approximately 121,000 MtCO2e) by 2021. Short-term plans involve increasing energy conservation and efficiency measures in our building and transportation systems while creating campus policies to facilitate energy and water savings and promote solid waste recycling. By 2013, the short-term goals were achieved by a 3.31 % decrease in greenhouse gas emissions despite an increase in students, staff, and faculty of 33 % and a gross square feet building increase of 21 % since 2002. But we need to reduce our current emissions by 121,000 MtCO2e in 19 years if we are to meet our aspirational 2040 goal. Increasing efficiencies is the low-hanging fruit and more transformational changes are needed. Many other organizations of all types, having increased efficiencies, are now facing the need to make transformational changes.

But eco-efficiencies are not the only things happening on my campus. The ACUPCC provided campuses with other actions to consider. Our campus hired its first Director of Campus Sustainability in 2007 and formed a Sustainability Council which includes campus *and* community representatives. Our Applied Sustainability Center, the precursor of the Sustainability Consortium, was developed following a grant from Walmart. We have a sustainability minor, a Graduate Certificate in Sustainability, and a BS in Sustainability is developed but awaiting funding. Numerous student groups and researchers are thinking about, researching, and/or working toward sustainability. Our organizational structure promotes closer dialogue among top administrators about sustainability-related issues. Mike Johnson, Associate Vice Chancellor for Facilities, reflected on our campus' journey saying:

> Are we at the point where we have reached critical mass, and it's starting to be self-sustaining for lack of a better word? I don't know. I think we are, but I think we are still up on that crest and we need to keep pushing it and getting it on to the other side.

Sustainability is a process rather than an end point. Different things spark the process for different organizations.

2.3.4.3 Critics of the Ecological Modernization and Sustainability *Discourses*

A number of criticisms have been leveled against these two *Discourses* including their failure to seriously challenge the capitalistic production and consumption relationship, their emphasis on practicality, and the way sustainability facilitates corporate posturing. Critical theorists and those supporting a neo-Marxist perspective target the capitalist means of production and consumption as the root causes of the global environmental problems (e.g., Aras and Crowther 2008; Prasad and Elmes 2005; Springett 2003). Economic growth is based on growth in production, which consumes natural resources and generates waste, and growth in production requires continual consumption. True environmental sustainability is incompatible with the traditional economic paradigm. Critics reject the Brundtland definition as promoting a dangerous liaison between growth, environmental integrity, and social justice (Springett 2003) and argue that the terms sustainability and the triple bottom line act as political cover for irreconcilable ideologies (e.g., continual growth, resource exploitation). Sustainable development that focuses on green business as usual fails to address how human and natural resources are exploited and how consumer needs are manufactured to increase an organization's profit margin. Much of the current green rhetoric used by industry really is about actions which do little to protect the natural environment (Prasad and Elmes 2005). Rather than adjusting the market to its ecological limitations, political systems and environmental groups, at least in the USA, have been co-opted to fit within market limitations to ensure economic growth. Issues such as how much is enough both for consumers and company profits, how do we want humanity to live, and why do consumers have unlimited desires are overlooked. The looming future environmental challenges we face due to global warming are not seriously addressed.

A second criticism involves the privileged positioning of practicality. As college courses, university degrees, and professional associations focusing on environmental management and sustainability proliferate in the USA, a core ideology being taught is practicality. The strong focus on practical problem solving shields the ideology of practicality from serious critique, limits subsequent swift and creative action (see Sect. 7.1.3 for how to stimulate group creativity), and inhibits alternative environmental discourses (e.g., anti-consumerism movements, ecological feminism) (Prasad and Elmes 2005). Any incompatibility between ecological and economic thinking is minimized, if not completely overlooked, in favor of short-term economic thinking. Practicality repackages ecological issues into economic, technical, and managerial issues and problems. Holistic systems-level problems are subdivided. Specific roles are assigned to individual issues and individuals are provided with specific scripts for solving their small piece of the problem. Although innovations do occur, true creativity is narrowed and channeled. Changes to meaningfully address the ecological imperatives associated with global warming are not made. Using critical discourse analysis, Prasad and Elmes "examine the hegemonic dimensions of the language of pragmatism, showing how the discourse of practical relevance continues to limit rather than enhance a more ecologically viable condition" (p. 846).

How does an idea such as practicality maintain its dominant hold? Practicality is aligned with the concepts of economic utilitarianism, compromise, and interorganizational collaboration which already have widespread sociocultural appeal (Prasad and Elmes 2005). Economic utilitarianism arguments stress that going green makes practical sense because it makes a for-profit organization more competitive and enhances its bottom line. However, economic utilitarianism can be far from practical in human and ecological terms. Compromise between economic growth and biospheric conservation ostensibly can avoid unnecessary conflict between competing positions and resolve pressing environmental problems. Yet the environment cannot speak for itself, compromises occur between different organizational interests, and compromises may not solve the actual environmental problems or halt environmental deterioration. Finally, collaboration involves working within the system and involving all existing stakeholders in the generation of visions and strategies. But different stakeholders may hold incompatible visions regarding our planet's ecological future and accommodations are rarely equal. In terms of environmental organizations, these interorganizational collaborations with business interests may be practically advantageous (e.g., financial backing, image enhancement) but may not seriously address environmental degradation at any meaningful scale. Prasad and Elmes argue that what is called practical should better be called convenient because it emphasizes minimal socioeconomic disruption and maximum conflict avoidance.

Prasad and Elmes' work helps us understand why the argument for practicality resonates with people. I met Jim Ekins in a bustling coffee shop in downtown Coeur d'Alene, ID, to discuss his work with the University of Idaho Sustainability Center in Moscow. The Sustainability Center is a student-led and student-funded group which focuses on education and small campus projects. On their website I read:

> Sustainability involves reorganizing our life support systems: climate, energy, biodiversity, food, consumerism and consumption, waste, transportation and built environment. We do this through projects that reduce our environmental footprint and increase participation and collaboration among students, faculty, staff, and community members in addressing sustainability-related issues.

Jim serves on their advisory board, although his real job is area water educator for the University of Idaho Extension Service. He described how the Sustainability Center has a quote, "Some people call it sustainability. We just call it common sense," saying how that approach was successful in a state where there was the potential for conflicts over the environment. "It was not difficult to get people on board with what they [the students] are doing, and what they are doing is very benign.... If they had been an activist organization they would have gotten a lot more push back," Jim explained. But instead, student teams do visible projects to make their campus more sustainable. These projects function as proof of concept and conflict is avoided.

On the other hand, Auden Schendler, Vice President of Sustainability at Aspen Skiing Company, shared his concern with me regarding collaboration. I asked him what message frames he thinks work best with the various groups he communicates with. He said:

> I don't know that I have an answer to that but I can tell you that in the industry talking about catastrophe has not been helpful. We think that talking about opportunity is a good frame but it is not even clear to me that it gets people's attention. Recently we did a project to convert methane leaking from a coal mine into electricity. We partnered with a coal mine and its owner who politically is very different from us. That story was unbelievably resonate and remains resonate. It's a story about collaboration and bipartisanship and taking advantage of wasted resources. People like that positive frame but I am skeptical because I have seen other people [who used] a relentlessly positive frame and it is not working. We are not seeing the solutions play out at the speed and scale that we need them to.

Finally, Ihlen (2015) described several rhetorical strategies businesses use when talking about sustainability which allow them to maintain the status quo while making only minor changes and promoting trust in technological solutions. Most often, sustainability rhetoric promotes the balance metaphor. Indeed, through this book, you will find researchers and interviewees alike discussing the need for balance. Its widespread acceptance is akin to the argument for practicality just mentioned. Generally balance involves weighing profit against environmental concerns. Businesses argue that economic stability is critical to their ability to make pro-environment changes. But at the end of the day, it is environmental concerns brought on by global climate change that humans are responding and must respond to. Many businesses have shifted from discussing their journey toward sustainability to arguing they are now sustainable. Others argue they are truly sustainable because they strive for sustainability. Because there is no agreed definition of what sustainability is, these arguments are difficult to challenge.

In this section, we discussed some of the competing macro-level *Discourses* surrounding the environment and sustainability. These competing *Discourses* make for a confusing array, leaving open the opportunity for an organization's management to create and communicate their own version of sustainability to their internal and external stakeholders through business communication channels (e.g., training materials, sustainability reports) (Allen et al. 2012). Organizations are discursive constructions (Fairhurst and Putnam 2014). *Key Point*: It is important to understand these *Discourses* because individuals bring ideas drawn from these various *Discourses* into their lives as citizens and organizational members. "Organizational life" is the sum of the socially shared belief systems of the organizational members (van Dijk 1998).

2.4   We Must Reframe the Issue

Societal discourses influence how individuals think about environmental issues because they create frames for us. Frames are unconscious structures in our brains that include semantic roles, relations between roles, and relations to other frames (Lakoff 2010). Frames are "schemata of interpretation that help actors reduce socio-cultural complexity in order to perceive, interpret and act in ways that are socially efficacious" (Goffman 1974, p. 21). These structures are physically realized in neural circuits in our brain. What this means is that when we hear a certain word, that word opens up an entire set of related ideas and the relationship between ideas, as well as often directly connecting to the emotional regions of our brain. A for-profit organizational frame might include top management, employees, goals, quarterly profits, products, and stockholders. So, given this frame, when we hear goals, this might lead us to think in terms of quarterly profits, and if there are limited profits, this may stimulate emotional stress for those charged with showing profits. Sustainability is a powerful framing word because it opens up a set of connected frames.

People trained in public policy, science, economics, and law often appear to believe that if you just tell people the facts, they will reason through to the right conclusion, but if the facts don't make sense in terms of the listeners' frames, they will be ignored (Lakoff 2010). Complex facts must be communicated strategically in order to activate frames people understand. It is through shared frames that thinking takes place, coordinated discussions can occur, and problems can be addressed. These shared frames define what is or is not an appropriate action. But with the impending climate and resource threats we face, few people have a system of cognitive frames in place that will help them take action. If a hearer doesn't have a frame in place, communicators must carefully build up those frames. Some new frames are being built such as the regulated commons (e.g., cap-and-trade efforts, carbon offset credits) and the economics of well-being (e.g., the happiness index).

What we need is a reframing of the human–organization–environment interface. We need a new vocabulary and series of discourses spoken by a complex array of actors, including governments, local authorities and communities, citizens, corporations, and those who speak on behalf of nature. This is climate communication in the global public sphere (Bortree 2011). Symbolic systems (e.g., language) play a major role in the maintenance and change of social order. Frames can be charged by critical communities and social movements that create and advocate for alternative field frames (Brulle 2010). Critical communities are small groups of critical thinkers who help each other develop a set of cultural values different from the larger society. These alternative frames allow us to analyze problems differently, offer different solutions, and can create an alternative map for action around which individuals can mobilize. Social movements spread alternative frames and seek to generate political pressure to implement institutional changes based on the alternative worldview they support. For example, Al Gore's 2006 documentary, *An Inconvenient Truth*, helped shift the public perception of climate change. Public awareness resulted in pressure on corporations since they were portrayed as a key cause of climate change and environmental problems (Walker and Wan 2012).

Working to Reframe the Issue: Protect Our Winters, the Green Sports Alliance, and the NRDC   With the support of his management team, Auden Schendler at Aspen/Snowmass worked to help create a movement in 2007 around Protect Our Winters (POW). He said:

> What we are doing mostly focuses around the work to protect our winters where we are trying to say, 'Here is a constituency that could be the core of a social movement. It is 21 million people who are avid, engaged, influential, sometimes famous, often wealthy, that could become the core of a social movement on climate. So we are working very hard on that effort and engaging others for support, not just individuals but corporations, brands, trade groups—that is the big push.

Justin Zeulner and the Portland Trail Blazers were active in helping create the Green Sports Alliance. Justin suggested I interview Allen Hershkowitz, Senior Scientist with the NRDC. These three men are among the many who are working to spark new discussions and actions around the organization–environment interface. Allen told me:

> There has been a cultural shift that is ongoing. The objective of my work is to instigate a cultural shift in expectations and behaviors as it concerns our relationship to the planet. We need a cultural shift in the way that people think about their relationship to the planet. That's why I am working with sports, for example, or the entertainment industry, with cultural elites. That is the most important thing we can do—get a cultural shift in the way people think about the planet.

Allen had been working with the sports industry since 2004 and that work appears to be paying off. In the NRDC report *Game Changer: How the Sports Industry Is Saving the Environment* (Henly et al. 2012), the authors describe how all the professional sports league commissioners have made commitments to environmental stewardship and are actively encouraging their league teams to incorporate sustainability-related measures into their operations. In 2013, 15 professional North American stadiums or arenas had achieved LEED green building design certifications, 18 had installed onsite solar arrays, and almost all had or were developing recycling and/or composting programs. Of the 126 professional sports teams, 38 had shifted to using some renewable energy, and 68 had energy efficiency programs. All of the large sports concessionaires had developed environmentally preferable menus for at least some of their offerings. All Jewel events, including the World Series, the Super Bowl, the Stanley Cup Playoffs, the NBA Playoffs and Finals, the MLS Cup, the US Open Tennis Championships, and all of the league All-Star Games, had incorporated greening initiatives into their planning and operations. All leagues were educating their fans about environmental issues, especially recycling and the need to reduce energy and water use. League efforts had resulted in millions of pounds of carbon emissions being avoided, millions of gallons of water being saved, and millions of pounds of paper products shifted toward recycled content or eliminated altogether. In light of this activity, it was interesting to read Ciletti et al.'s (2010) investigation of how 126 professional sports teams across four different leagues communicate about sustainability on their websites. Most downplay economic issues and highlight social issues. Although some teams communicate about sustainability on their website, others do not, and communication about environmental factors varied by league. Speaking of sports, Allen told me:

> These are the mainstream cultural arbiters of our society. Sports is a trusted network. It provides a trusted network where people have refuge from political debates. The (playing) court says yes climate change is happening and it is real and we have to do something about it. That information is not offered in a political context but more from the context of a trusted network—sports, non-partisan, non-political. Overall, 13 % of Americans follow science, 63 % follow sports. And it is not Democrat or Republican, it is not male or female, or Black or White. You know it is everybody. So you have professional sports doing things that address global climate change and biodiversity loss and water scarcity and recycling. The change is in the best interest of the marketplace. The things that the environmental community have been asking for so long that now are becoming mainstream.

2.4.1 Changing Frames Is Contested

All frames are political. Antonio Gramsci, Italian political theorist and linguist, expanded on how Marxist thinkers of the twentieth century discussed hegemony. He described how the ruling elite use the institutions of civil society (e.g., church, media, schools) to disseminate messages that contend the ruling elite's interests and the interests of the masses are the same. Over time, the interests of the ruling elites become taken for granted as the right and proper course of events. In other words, ideologies develop which play a critical role in establishing and maintaining societal unity. The repetition of ideological language strengthens the circuits for that ideology in a hearer's brain (Lakoff 2010). If repeated often enough that language becomes normally used language which continues to unconsciously activate that ideology in the hearer's brain. The dominant frame regarding the environment has been that it is a resource for short-term private enrichment, that it exists to be used by people, and that we should leave it up to market forces to manage the resources the environment provides for people. These frames have become reified by our institutions, industries, and cultural practices. Once reified, they are difficult to change. Yet they must and are being changed.

2.4.1.1 Efforts to Control the Discussion

Those who control the frames work hard to maintain them. A powerful way to silence a group is to own the language of the debate (Shafer 2006). Although the term sustainable development grew out of ecology, international development, and the environmental movement, as early as 1992, powerful political forces (e.g., the WBCSD) began taking charge of how sustainable development was conceptualized and discussed.

In the face of ecological crises, defenders of the status quo often try to discredit or marginalize threats posed by ecological issues. They engage in systematically distorted communication designed to marginalize the perceived severity of global warming's threat so as to protect traditional or large economic interests. Systematically distorted communication refers to the instrumental manipulation of language by powerful interests which corrupts our everyday networks of communication practice. The work of the German political philosopher, Jürgen Habermas, was foundational to our understanding of how systemically distorted communication occurs at the societal level. A number of strategies associated with systematically distorted communication include universalization (treating sectional interests as universal); disqualification (treating a subject or source as trivial), and neutralization (treating topics as if they have no political significance) (Ganesh 2009). There is a long and well-documented history of business interests and their allies in many countries who seek to create scientific uncertainty about the existence and causes of global warming.

Climate change deniers and those who seek to limit or block environmental regulations utilize a variety of other communication and legal strategies including intentionally disseminating disinformation, using intimidation and threats, supporting public and media outreach from conservative think tanks, funding sympathetic political candidates and scientists, legally challenging the right of citizens to speak for the environment (i.e., the right of standing), weakening environmental regulations, bringing strategic litigation against public participation (SLAPP) lawsuits, using fear appeals (e.g., jobs versus the environment, government takeover of individual liberties) to mobilize citizens, and contesting the credibility of scientists (Cox 2013). One of my interviewees, Keegan Eisenstadt, CEO and owner of ClearSky Climate Solutions, discussed how scientists face:

> Special interests that have hired very good communicators. Whether or not they talk about the truth or fact, is irrespective of the fact that they communicate well. And their communication is about fear, and the fear is the pocketbook of the family, it is not about future hypothetical grandchildren.

A clear effort to control the language of the debate is being played out today as towns and cities consider membership in groups such as ICLEI. In 2012 Glenn Beck, a politically conservative television and radio host, published *Agenda 21*, a ghost-written, dystopian novel based around the UN's Agenda 21. Beck, who appears not to believe climate change is real, sought to discredit Agenda 21, as have other conservative commentators. Ideologically motivated articles are appearing on the Internet about the horrors of Agenda 21 and ICLEI. In light of the disinformation being circulated, some citizens are opposing anything connected to Agenda 21. In response, ICLEI placed information on their website "Setting the Record Straight about ICLEI" and took the open list of ICLEI members off the site. Some cities are not renewing their membership in an organization designed to help them plan for future resource access by their residents and how to reduce their greenhouse gas emissions.

The battle for control is ongoing. NRDC Senior Scientist Allen Hershkowitz said:

> I really do think that the work we have done with sports has played a very influential role in helping instigate a shift. The deniers of climate change can attack the National Academies of Science or the Environmental Protection Agency, but they can't attack the NFL, Major League Baseball, and the National Hockey League. And NASCAR, we have a partnership with NASCAR on energy efficiency and climate change issues. You can't get more mainstream than NASCAR.

2.4.1.2 Changing Language Use

Realizing the term sustainability carries baggage with it either because it has been co-opted by business interests, is a target for political pressure, or is simply not descriptive of their effort, some organizations have consciously decided to use different terms. For example, Heifer International® and Ecotrust both prefer the term resilience. Oakley Brooks from Ecotrust said:

> Ecotrust is very much in the center in grappling with all the issues of sustainability but we deliberately try not to use that word. . ... We have tried to come up with our own terms to push the debate in different directions. . ... The framework we are using now is resilience and that is related to a school of thought, the Stockholm Resilience Center and others that are thinking of all the big changes that are going on in the world and the idea that the best work you can do as an organization. . .is really to focus on how you make the environmental, social, economic systems stronger and resilient—able to withstand big shocks, be they financial or climatic or social. . ... I think the international development community has been ahead of the local and even international green community, conservation community, using that term and thinking about resilience. . ... We would like to expand that to thinking about how does that translate into everyday wellbeing. . ... How does that make my life better today? We think that there are benefits today and benefits when those big shocks come.

Increasingly, organizations are talking about resilience, especially climate resilience. Researchers, policy makers, and community organizers find the term attractive (Polk and Servases 2015). Resilience occurs when a system, enterprise, or a person has the capacity to maintain its core purpose and integrity during dramatically changed circumstances. A truly resilient system can reorganize both the ways it achieves its purpose and redesign its operational scale. Polk and Servaes describe the role of participatory communication within one Transition Town. The Transition Town Network is a global social movement focused on developing community resiliency and sustainability. The first initiative began in 2006. By 2014, there were more than 1,170 groups in 47 countries, plus 11 official and 14 developing National Hubs (Hopkins 2014). Transition Towns start projects to increase sustainability in the areas of food, transport, energy, education, housing, and waste. Three communication processes are key: the provision of a framework that allows community members to respond collectively to challenges, resources to raise awareness of problems and to share community-developed solutions, and opportunities to share best practices through the Transition Network's central online hub. The group's philosophy is that a strategic community-led approach to sustainable social change must meet the need of residents already in crisis as well as those who have the resources needed to prepare for crises.

Although not a Transition Town, resilience is being discussed in Fayetteville, AR. Fayetteville is the third-largest city in Arkansas with a population of almost 75,000 in 2012. It was the first city in Arkansas to join ICLEI in 2006 but did not renew its membership in 2012. It operates with the mayor-city council form of government. The city created the first property-assessed clean energy (PACE) district in the State of Arkansas for commercial building. PACE is an innovative way to finance energy efficiency and renewable energy upgrades to buildings. A newly constructed city building over 5,000 square feet must meet a minimum LEED Silver standard. In 2012 the City began requiring all new homes to be energy rated and receive a RESNET HERS index score. This code emphasizes improvements in building thermal envelopes, including insulation improvements, improved window U-factor requirements, the testing of air leakage from HVAC ducts, and increased lighting efficiency requirements. It was recognized as a Bronze-level Bicycle Friendly Community, ranked sixth in the 2009 Natural Resources Defense Council Smarter Cities Project, and won the 2013 National Wildlife Federation Rex Hancock Wildlife Conservationist of the Year Award. Fayetteville's 2030 comprehensive land-use plan was adopted unanimously in 2011. Their goals include making appropriate infill and revitalization their highest priority, discouraging suburban sprawl, making the traditional town form the standard, growing a livable transportation network, assembling an enduring green network, and creating opportunities for attainable housing.

Peter Nierengarten, Director of Sustainability and Resilience for Fayetteville, AR, told me how he participated in an hour-long discussion about the use of the term *sustainability* at the 2012 USDN meeting in Portland, OR. Cities are using terms such as livable (Milwaukee), homegrown (Minneapolis), and greenovator (Boston). Austin, TX, is challenging their citizens to "rethink"—rethink water, rethink energy, and rethink housing. Susan Anderson, Director of Portland's Bureau of Planning and Sustainability, admitted she rarely uses the word sustainability but said that the topic was woven throughout the myriad of conversations she has with other department directors. Peter said:

> I don't use sustainability because it turns off skeptics. On our website it says 'The City of Fayetteville is focused on becoming a resource efficient community of livable neighborhoods that meets present needs without compromising opportunities for health, well-being and the prosperity of future generations.' Basically that is the definition of sustainability without using the word. Livable to me is a nice replacement word.

Peter's department is focused on a triple-bottom-line approach (i.e., economic, environmental, and social) to policy development and project management.

Given that the definition of what is sustainable is often political and contested (Jacobs 1999), social judgment theory, developed by Sherif and his associates, helps us understand the relationship between attitude change and communication involving statements with challenged connotative meanings. The theory conceptualizes an attitude as a range of possible positions on an issue, some extreme and some moderate, which fall along a continuum (Seiter 2009). Any given individual's attitude falls within a particulate range on this continuum and each of us has an anchor point which is our preferred position. Some statements fall within our latitude of acceptance, others fall within our latitude of rejection, and finally others fall within our latitude of noncommitment (i.e., we are ambivalent). Messages falling closer to our anchor point are distorted favorably and seen as closer to our attitude than they may be. Those falling further from our anchor point are distorted negatively. More persuasive communicators engage in audience analysis and create messages which fall within an individual's latitude of noncommitment. Ego-involved people have wider attitudes of rejection. For example, as someone who grew up in eastern Kentucky and knew many miners who suffered from coal miners' pneumoconiosis (i.e., black lung disease), I automatically reject any messages utilizing the term *clean coal*, as do many environmentalists. Social judgment theory helps me understand why some of my interviewees sought to use words other than sustainability in their discussions. They realized that if they use words that were uncontested, listeners will be more likely to consider the content of the message rather than immediately reject it. Also, audiences may respond more positively to ambiguous messages delivered by credible sources. *Best Practice*: If the term sustainability doesn't resonate in your organization, find one that does which addresses sustainability-related concerns.

2.5 Concluding Thoughts

Throughout this book, most of the discussion surrounding communication and sustainability focuses on public, interpersonal and group communication. In this chapter, the discussion focuses on paradigms and *Discourses*. It is through discourse that ideas are developed, spread, and changed. "How we 'talk' about and represent the natural environment have serious ramifications for how we will conceptualize and enact our future relationship with it" (Prasad and Elmes 2005, p. 853). Each *Discourse* focuses us on different aspects related to how humanity will view and respond to global warming challenges. "Language matters. . . the way we construct, interpret, discuss, and analyze environmental problems has all kinds of consequences" (Dryzek 2005, p. 10). Climate change sounds like part of the natural cycle, while global warming has connotations that it is the outcome of human action. According to the National Aeronautics and Space Administration (2014), 97 % of climate scientists agree that climate-warming trends very likely are due to human activities, and worldwide the leading scientific organizations have issued public statements to that effect. Yet people in the USA feel more comfortable discussing climate change than global warming. Why? In 2002 Frank Luntz, a Republican political strategist, recommended Republican politicians promote the term *climate change*. Based on his focus group data, the general public finds it less frightening than the term global warming (see Luntz Memorandum 2002). Subsequently, Republican congressional and executive leaders increasingly used the term climate change, and the media and general public followed suit. Today political affiliation is one of the strongest correlates with individual uncertainty about climate change, not scientific knowledge (McCright and Dunlap 2011). So, *Discourse*, ideology, and language do matter. They shape awareness, understanding, beliefs, attitudes, and action.

References

Allen, M. W., Walker, K. L., & Brady, R. (2012). Sustainability discourse within a supply chain relationship: Mapping convergence and divergence. *Journal of Business Communication, 49*, 210–236.

Andersson, L. M., & Bateman, T. S. (2000). Individual environmental initiative: Championing natural environmental issues in U.S. business organizations. *Academy of Management Journal, 43*, 548–570.

Aras, G., & Crowther, D. (2008). Governance and sustainability: An investigation into the relationship between corporate governance and corporate sustainability. *Management Decision, 46*, 433–448.

Ashcroft, B., Griffiths, G., & Tiffin, H. (1998). *Key concepts in post-colonial studies*. London: Routledge.

Blackburn, W. R. (2007). *The sustainability handbook: The complete management guide to achieving social, economic and environmental responsibility*. London: Earthscan.

Bortree, D. S. (2011). The state of environmental communication: A survey of PRSA members. *Public Relations Journal, 4*, 1–17.

Brulle, R. J. (2010). From environmental campaigns to advancing the public dialog: Environmental communication for civic engagement. *Environmental Communication, 4*, 82–98.

Bullis, C., & Ie, F. (1997). Corporate environmentalism. In S. May, G. Cheney, & I. Roper (Eds.), *The debate over corporate social responsibility*. New York: Oxford University.

Chess, C. (2001). Organizational theory and the stages of risk communication. *Risk Analysis, 21*, 179–188.

Christensen, L. T., Morsing, M., & Thyssen, O. (2015). Discursive closure and discursive openings in sustainability. *Management Communication Quarterly, 29*, 135–144.

Ciletti, D., Lanasa, J., Ramos, D., Luchs, R., & Lou, J. (2010). Sustainability communication in North American Professional Sports Leagues: Insights from web-site self-presentations. *International Journal of Sport Communication, 3*, 64–91.

Cohen, E. (1976). Environmental orientations: A multidimensional approach to social ecology. *Current Anthropology, 17*, 49–70.

Colby, M. (1991). Environmental management in development: The evolution of paradigms. *Ecological Economics, 3*, 193–213.

Cox, R. (2013). *Environmental communication and the public sphere* (3rd ed.). Washington, DC: Sage.

Crognale, G. (2012). *Coca-Cola builds sustainability from the watershed up*. http://www.sustainableplant.com/2012/10/coca-cola-builds-sustainability-from-the-watershed-up/?show=all. Accessed 12 Nov 2013.

De Soto, H. (2000). *The mystery of capital: Why capitalism triumphs in the west and fails everywhere else*. New York: Basic Books.

Deetz, S. A. (1992). Systematically distorted communication and discursive closure. In S. A. Deetz (Ed.), *Democracy in an age of corporate colonization: Developments in communication and the politics of everyday life* (pp. 173–198). Albany: SUNY Press.

Diamond, J. M. (2005). *Collapse: How societies choose to fail or succeed*. New York: Viking.

Dryzek, J. S. (2005). *The politics of the earth: Environmental discourses* (2nd ed.). Oxford: Oxford University Press.

Dunlap, R. E., Van Liere, K. D., Mertig, A. G., & Jones, R. E. (2000). Measuring endorsement of the new ecological paradigm: A revised NEP scale. *Journal of Social Issues, 56*, 425–442.

Elkington, J. (1999). *Cannibals with forks: The triple bottom line of 21st century business*. Gabriola Island: New Society.

Fairhurst, G. T., & Putnam, L. (2004). Organizations as discursive constructions. *Communication Theory, 14*, 5–26.

Fairhurst, G. T., & Putnam, L. L. (2014). Organizational discourse analysis. In L. L. Putnam & D. K. Mumby (Eds.), *The SAGE handbook of organizational communication* (pp. 271–296). Thousand Oaks: Sage.

Ganesh, S. (2009). Critical organizational communication. In S. W. Littlejohn & K. A. Foss (Eds.), *Encyclopedia of communication theory* (Vol. 1, pp. 226–231). Los Angeles: Sage.

Goffman, E. (1974). *Frame analysis: An essay on the organization of experience*. London: Harper and Row.

Grant, D., Hardy, C., Oswick, C., & Putnam, L. L. (2004). Introduction: Organizational discourse: Exploring the field. In D. Grant, C. Hardy, C. Oswick, & L. L. Putnam (Eds.), *The Sage handbook of organizational discourse* (pp. 1–36). Thousand Oaks: Sage.

Handy, C. (1999). *The hungry spirit: Beyond capitalism: The quest for purpose in the modern world*. New York: Broadway Books.

Hannaes, K., Arthur, D., Balagopal, B., Kong, M. T., Reeves, M., Velken, I., et al. (2011). *Sustainability: The 'embracers' seize advantage*. MIT Sloan Management Review and the Boston Consulting Group Research Report. http://sloanreview.mit.edu/reports/sustainability-advantage/. Accessed 20 Dec 2013.

Hardin, G. (1968). The tragedy of the commons. *Science, 162*, 1243–1248.

Hawken, P. (1994). *The ecology of commerce: A declaration of sustainability*. New York: HarperCollins.

Hawken, P., Lovins, A., & Lovins, L. H. (2000). *Natural capitalism: Creating the next industrial revolution*. Boston: Little, Brown and Company.

Henly, A., Hershkowitz, A., & Hoover, D. (2012). *Game changer: How the sports industry is saving the environment*. NRDC Report R:12-08-A. http://www.nrdc.org/greenbusiness/guides/sports/game-changer.asp. Accessed 20 Dec 2013.

Hopkins, R. (2014). *Transition network's annual report 2014*. http://www.transitionnetwork.org/news. Accessed 12 Jan 2015.

Ihlen, O. (2015). "It is five minutes to midnight and all is quiet": Corporate rhetoric and sustainability. *Management Communication Quarterly, 29*, 145–152.

Intergovernmental Panel on Climate Change. (2013). *Climate change 2013: The physical science basis*. http://www.climatechange2013.org/report/. Accessed 20 Dec 2013.

Jabbour, C. J. C., & Santos, F. C. A. (2006). The evolution of environmental management within organizations: Toward a common taxonomy. *Environment Quality Management, 16*, 43–59.

Jacobs, M. (1999). Sustainable development as a contested concept. In A. Dobson (Ed.), *Fairness and futurity: Essays on environmental sustainability and social justice* (pp. 21–45). Oxford: Oxford University Press.

Korten, D. C. (2001). *When corporations rule the world*. Bloomfield: Kumarian Press.

Kraft, M., & Vig, N. (2000). Environmental policy in Congress: From consensus to gridlock. In N. J. Vig & M. E. Kraft (Eds.), *Environmental policy: New directions for the twenty-first century* (5th ed., pp. 121–144). Washington, DC: CQ Press.

Kuhn, T. S. (1962). *The structure of scientific revolutions*. Chicago: University of Chicago Press.

Lakoff, G. (2010). Why it matters how we frame the environment. *Environmental Communication: A Journal of Nature and Culture, 4*, 70–81.

Luntz Memorandum to the Bush White House. (2002). *A cleaner safer, healthier America*. https://www2.bc.edu/~plater/Newpublicsite06/suppmats/02.6.pdf. Accessed 17 May 2014.

McCright, A., & Dunlap, R. (2011). The politicization of climate change and polarization in the American public's views of global warming, 2001–2010. *The Sociological Quarterly, 52*, 155–194.

Meadows, D. H., Meadows, D. L., Randers, J., & Behrens, W. W. (1972). *The limits to growth.* New York: Universe Books.

Mitra, R., & Buzzanell, P. M. (2015). Introduction: Organizing/communicating sustainably. *Management Communication Quarterly, 29,* 130–134.

Mulla, R. (2013). *Ten institutions receive prestigious climate leadership awards. Second nature: Education for sustainability.* http://secondnature.org/news/ten-institutions-receive-presti gious-climate-leadership-awards-0. Accessed 23 Oct 2014.

National Aeronautics and Space Administration. (2014). *Consensus: 97% of climate scientists agree.* http://climate.nasa.gov/scientific-consensus. Accessed 27 May 2014.

Olausson, U. (2011). "We're the Ones to Blame": Citizens' representations of climate change and the role of the media. *Environmental Communication, 5,* 281–299.

Pirages, D. C., & Ehrlich, P. R. (1974). *Ark II: Social response to environmental imperatives.* New York: Viking Press.

Polk, E., & Servaes, J. (2015). Sustainability and participatory communication: A case study of the Transition Town Amherst, Massachusetts. *Management Communication Quarterly, 29,* 160–167.

Prasad, P., & Elmes, M. (2005). In the name of the practical: Unearthing hegemony of pragmatics in the discourse of environmental management. *Journal of Management Studies, 42,* 845–867.

Rosteck, T., & Frentz, T. S. (2009). Myth and multiple readings in environmental rhetoric: The case of An Inconvenient Truth. *Quarterly Journal of Speech, 95,* 1–10.

Salzmann, O., Ionescu-Somers, A., & Steger, U. (2005). The business case for corporate sustainability: Literature review and research options. *European Management Journal, 23,* 27–36.

Schlosberg, D., & Rinfret, S. (2008). Ecological modernization, American style. *Environmental Politics, 17,* 254–275.

Schmidheiny, S. (1992). *Changing course: A global business perspective on development and the environment.* Cambridge: MIT.

Seiter, J. S. (2009). Social judgment theory. In S. W. Littlejohn & K. A. Foss (Eds.), *Encyclopedia of communication theory* (Vol. 2, pp. 905–908). Los Angeles, CA: Sage.

Shafer, W. E. (2006). Social paradigms and attitudes toward environmental accountability. *Journal of Business Ethics, 65,* 121–147.

Sinha, P., Schew, W. A., Swant, A., Kowaite, K. J., & Strode, S. A. (2010). Greenhouse gas emissions from U.S. institutions of higher education. *Journal of the Air & Waste Management Association, 60,* 568–573.

Soros, G. (2000). *Open society: Reforming global capitalism.* New York: PublicAffairs.

Springett, D. (2003). Business conceptions of sustainable development: A perspective from critical theory. *Business Strategy and the Environment, 12,* 71–86.

U.S. Department of Transportation Federal Highway Administration. (2015). http://www.fhwa. dot.gov/interstate/brainiacs/environmental.htm. Accessed 28 Jan 2015.

United Nations Environmental Programme. (2008). *Green jobs: Towards decent work in a sustainable, low-carbon world.* www.unep.org. Accessed 20 Dec 2013.

van Dijk, T. A. (1998). *Ideology: A multidisciplinary approach.* London: Sage.

Vig, N. J. (2000). Presidential leadership and the environment. In N. J. Vig & M. E. Kraft (Eds.), *Environmental policy: New directions for the twenty-first century* (5th ed., pp. 98–120). Washington, DC: CQ Press.

Walker, K., & Wan, F. (2012). The harm of symbolic actions and green-washing: Corporate actions and communications on environmental performance and their financial implications. *Journal of Business Ethics, 109,* 227–242.

Wexler, M. N. (2009). Strategic ambiguity in emergent coalitions: The triple bottom line. *Corporate Communications: An International Journal, 14,* 62–77.

Woodside, C. (2013). *What makes climate communicator George Marshall tick?* http://www. yaleclimatemediaforum.org/2013/05/what-makes-climate-communicator-george-marshall-tick/. Accessed 18 Nov 2013.

Zadek, S. (2001). *The civil corporation: The new economy of corporate citizenship.* London: Earthscan.

Chapter 3
Legitimacy, Stakeholders, and Strategic Communication Efforts

Abstract When organizational leaders undertake sustainability-related initiatives, they are concerned with whether or not their key stakeholders perceive their organization and its actions to be legitimate. In this chapter, legitimacy is defined, the types of legitimacy are described, the benefits of being seen as legitimate and having a positive reputation are discussed, and actions used to manage legitimacy are reviewed. Communication helps manage perceptions of the legitimacy of industries as well as individual organizations. Message credibility is important. Standardized communication directed simultaneously toward multiple internal and external stakeholders is discussed (i.e., annual public meetings, websites and sustainability reports, reporting frameworks, certifications, and architecture such as LEED-certified office buildings). Best practices are provided. The chapter reviews stakeholder theory and discusses the need to adapt messages to engage different stakeholders. In addition, theories or theoretical concepts highlighted include population ecology theory, resource dependency theory, institutional isomorphism, agency theory, normative discourse, the elaboration likelihood model of persuasion, actor–network theory, speech act theory, attribution theory, visual rhetoric, narrative theory, and organizational co-orientation theory. Interview data drawn from a variety of organizational types illustrates the applicability of theory and research. Organizations represented include Tyson Foods; WasteCap Nebraska; Missoula Sustainability Council; Ecotrust; the Neil Kelly Company; the Arbor Day Foundation; Aspen Skiing Company; the University of Arkansas, Fayetteville; the City of Boulder; Sam's Club; Assurity Life Insurance; Heifer International®; the Natural Resources Defense Council; the University of Colorado, Boulder; and the University of Colorado, Denver.

Lewis and Clark, Legitimacy, and Stakeholders Jefferson had wanted to explore the West since the 1790s. He and Meriwether Lewis began preparing in 1802, but told no one. Lewis submitted a $2,500 budget which he kept small so as to limit Congressional criticisms and Indian fears of invasion. Jefferson omitted the request from his proposed annual budget. In 1803, he sent a secret message to Congress requesting the money, arguing it would aid American commerce by opening up trade with native peoples and result in a river route to the ocean—actions which would allow the Americans, rather than the British, to control the fur trade. Congress received the secret message after they approved paying for the Louisiana Purchase. Why the delay? Opposing forces existed. The Federalists saw no point in spending money on western exploration and feared expansion would dilute their political power base on the eastern seaboard. They ridiculed the stories being told about the unknown interior and condemned the acquisition of a wilderness. Aware of the opposition and because he feared the Louisiana Purchase would be deemed unconstitutional, Jefferson drafted a constitutional amendment authorizing the national government to purchase new lands and promote settlement of the new territory. Then, he decided to submit a treaty instead of an amendment. The treaty was quickly ratified. In this example, issues emerge which we cover in this chapter including legitimacy (e.g., the questionable constitutionality of the Louisiana Purchase) and stakeholders (e.g., the Federalists, the Congress, the explorers, and the native peoples). Broadly defined, stakeholders are "any group or individual who can affect or is affected by the achievement of an organization's objectives" (Freeman 1984, p. 46). More narrowly defined, they are those groups which are vital to an organization's survival and success.

3.1 Legitimacy

A range of potentially competing *Discourses* exist as organizations decide how to act (or not) around issues such as resource degradation, pollution, carbon emissions, and climate change. How an organization decides to position itself strategically around such issues has implications for that organization's legitimacy in the eyes of society and its key stakeholders. Sometimes organizations develop sustainability initiatives to gain or maintain legitimacy (Emtairah and Mont 2008).

In his discussion of how 30 of the world's largest corporations address climate change, Ihlen (2009) reviewed the legitimacy literature. Legitimacy is defined as "a generalized perception or assumption that the actions of an entity are desirable, proper, or appropriate within some socially constructed system of norms, values, beliefs, and definitions" (Suchman 1995, p. 574). Three broad types of legitimacy exist: pragmatic legitimacy, moral legitimacy, and cognitive legitimacy. Stakeholders accord organizations with pragmatic legitimacy when they envision an organization's actions will benefit them personally. Moral legitimacy involves assessments of whether or not an organization's actions are the right thing to do and contribute to societal welfare. Cognitive legitimacy focuses on whether or not an organization's actions are seen as understandable. Is it utilizing socially accepted techniques and processes? How is the organization structured? Are its leaders credible and charismatic?

Organizations rely on an implied social contract that if they will undertake socially desired actions, key stakeholders will support their goals and continued existence. A legitimacy gap exists when society expects something from an organization that it can or will not deliver (Ihlen 2009), and this can present a threat for the organization (Dowling and Pfeffer 1975) (e.g., a loss in clients, customers, or donors, government sanctions, citizen protests, difficulty attracting employees). If an organization can manage the legitimacy gap, it can maintain maximum discretionary control over its internal decision-making and external practices. Some organizations depart from societal expectations without suffering the loss of legitimacy if their actions go unnoticed or if there is no public disapproval. But to do so is a risk. Sethi (1979) proposed legitimacy-gap theory to explain this gap, generating two scenarios. In the first, the organization's behavior is knowingly inappropriate. Perhaps its behavior changed or the problematic behavior had been hidden. In the second, society's norms or expectations changed.

Societal expectations for a business organization's behaviors have changed over time. Initially, maximizing shareholder profits was of primary value. Later, as laws (e.g., environmental, occupational safety, and health) were passed, the expectations changed. In the early 1970s, with the emergence of environmental concern among citizens of industrialized societies, business organizations faced an identity crisis. Many went from being seen as the provider of jobs, income, and prosperity to being seen as the destroyers of the natural environment and of peoples' health (Emtairah and Mont 2008). Today, the public generally expects Western for-profit and public organizations to proactively manage their impact on the natural world and treat their employees and surrounding communities better than is legally prescribed (Domenec 2012). In order to maintain the social license to operate whereby organizations have access to public support and resources, and in the case of for-profit organizations, access to markets, management engages in and seeks to communicate their organization's engagement in corporate social responsibility (CSR) activities (Emtairah and Mont 2008).

3.1.1 Legitimacy, Reputation, and Influence

Legitimacy is a critical resource associated with improved reputation, which can in turn result in increased benefits and reduced risks. A reputation is a type of social assessment-based memory connected with what a certain organization should be. It represents the attractiveness of the organization. Corporate reputations are built up over time in a dynamic and lengthy process and can be hard to manage directly (Biloslavo and Trnavcevic 2009). Reputations are associated with stakeholder trust that the organization will proactively protect its stakeholders. Strong positive reputations act as insurance partially protecting an organization against crises. Indeed, some executives see reputation insurance as a primary benefit of sustainability initiatives. Organizations seen as more caring avoid costly and dangerous negative scrutiny (Hunter and Bansal 2007). If something negative occurs, prior sustainability initiatives can be used to counter charges of unethical behavior allowing market value to be sustained (Peloza et al. 2012). Negative reputations are likely to prompt criticism and regulation in the event of a sustainability-focused crisis. One of the main challenges of reputation management for extraction-based (e.g., oil) companies involves communicating about environmental issues (Domenec 2012).

In terms of reputation, companies can be divided into three groups: green companies, companies that are on their way to becoming green, and others (Biloslavo and Trnavcevic 2009). The ideal green company fully promotes financial, social, and environmental sustainability. Its employees communicate ethically among themselves and with the company's stakeholders. Few organizations achieve this ideal, even environmental NGOs and nonprofits. However, the green company reputation is achieved when most of its stakeholders (both internal and external) believe the company is fully committed to the long-term, ideal goal of zero: zero emissions, zero waste, and zero environmental impact. Such a reputation helps an organization differentiate itself from competitors, adds value to the brand, and improves stakeholder relationships. Legitimacy can do more than protect an organization and provide it with access to resources. It can serve as a source of influence. Those with high legitimacy may have the opportunity to lead debates about appropriate environmental practices (Hunter and Bansal 2007). The formation of a green company reputation is enabled by two interrelated processes: perception and communication. Perceptions are influenced by an individual's own experiences and everything he or she has heard or read about the organization which carries information about its operations and activities. Different communication activities are used to create a green reputation (e.g., promotion, sponsorship). Often, the communication surrounds a proactive change in the company's operation (e.g., a change of business policy, new or modified products, or a change in the production process). A green reputation is more likely when both the production process and the product itself change or were pro-environmental initially.

3.1.2 Discussing Sustainability Within and Between Institutional Fields

Traditionally, legitimacy research fits into one of two distinct groups: the strategic approach or the institutional approach (Suchman 1995). The institutional approach is interested in entire fields or sectors of organizations (e.g., all professional sports leagues). Studies based on the institutional approach investigate how sector-wide structural dynamics generate cultural pressures that go beyond a single organization's purposive actions. This approach downplays managerial agency and manager–stakeholder conflicts, focusing instead on societal or sector-level changes. The pattern of values and priorities in a particular institutional field influences the behaviors of managers and employees (Lammers 2009). This approach directs us to investigate how organizations address the environmentally related legitimacy challenges most relevant to their institutional field and how the institutional field's focus may shift over time.

WasteCap Nebraska and Institutional Change WasteCap Nebraska was developed by the Lincoln, NE, business community when, in the early 1990s, a federal law required communities to have an integrated solid waste management plan. WasteCap served as a liaison between businesses and recyclers to ensure fair and equitable services. Over time, business waste management plans became institutionalized and organizations with WasteCap's original goals were no longer needed. Today, the organization's services include providing education, assessment, and certification services to about 100 paying business members. Carrie Hakenkamp, the organization's executive director, explained how her organization was reinventing itself:

> We've really been trying to define our role and who we are and what role we are going to play in sustainability. We determined that really what we do is look at sustainability through the lens of waste—wasting resources, wasting time, wasting energy, wasting water, wasting. Looking at it through the lens of waste gave us our starting point on what we intend to address.

Several theories can help us understand the WasteCap situation. Population ecology theory (Hannan and Freeman 1977) and resource dependence theory (Pfeffer and Salancik 1978) both focus us on how organizations must adapt to changing conditions in their institutional fields as well as in the broader institutional environment. Both have been important theories in guiding our understanding of macro-organizational phenomena since the 1970s. Population ecology scholarship investigates issues related to an organization's life cycle (i.e., birth, growth, and mortality) and the power of external forces. Salimath and Jones (2011) review the population ecology literature from 1996 to 2010 and explain the theory's implications for sustainability. They argue that population ecology theory suggests a focus on sustainability (in business practice and strategy) may lead to the formation and survival of new companies and industries, while unsustainability may lead to the demise of companies, if not industries. Both theories are appropriate when looking at institutional fields and legitimacy. The population ecology theory informs us that regardless of how legitimate a single organization is perceived to be, if its industry does not adapt to changes in the broader environment, that organization will fail unless it moves away from industry norms. Resource dependency theory helps us understand how critically important it is that an organization effectively communicates its legitimacy to powerful stakeholders and how organizations working together can shape the normative discourse.

Institutional beliefs and processes penetrate and transcend particular organizations, often through the process of institutional isomorphism. They become part of the taken-for-granted normative systems people rely on (Lammers 2009). Much of an organization's structure is a symbolic reflection of external forces (e.g., laws governing pollution, market forces promoting more sustainable products, reactions to reporting guidelines) (Meyer and Rowan 1977). Organizations incorporate institutionalized rules or rationalized myths which represent what is taken for granted as appropriate conduct or which symbolize rationality and progress either within their organizational field (i.e., same industry, same sector) or in the broader society. For example, we see this occurring as all the professional sports league commissioners made commitments to environmental stewardship and began actively encouraging their league teams to incorporate eco-efficiencies and other sustainability-focused initiatives into their operations (Henly et al. 2012).

Institutional isomorphism includes three types: coercive, normative, and mimetic. Coercive isomorphism occurs when the change is due to new laws or rules. For example, in 2005, the Arkansas legislature passed a law that directed all new or major renovations of state-owned buildings to be built to the USGBC LEED standards or the Green Building Initiative Green Globes system—a law which dramatically influenced building construction on my campus. Illustrating how coercive isomorphism spreads, in 2008, a similar law was passed in South Dakota. With normative and mimetic isomorphism, organizations have agency or the ability to self-direct their actions, meaning that they may decide not to adopt sustainability initiatives similar to other organizations in their institutional field. Normative isomorphism occurs when the change is in response to new norms associated with a professionalization process (e.g., professional accreditation). For example, since the LEED certification procedure began in 1998, as of June 2013, 187,428 people had attained LEED professional accreditation and currently over 4.3 million people live and work in LEED-certified buildings, numbers which are steadily growing. Finally, mimetic isomorphism occurs when organizations imitate influential and successful organizations. It can occur unintentionally (e.g., through employee transfer) or explicitly (e.g., through trade associations or consulting firms). Modeling can lead to further innovations as organizations imperfectly imitate others.

Tyson Foods and Mimetic Isomorphism As Tyson Foods prepared to write their first sustainability report, they benchmarked against Procter and Gamble, Weyerhaeuser, Dow Chemical, and Johnson & Johnson. "We looked at different entities in American business. We went to Cincinnati and met with Procter and Gamble's sustainability team back in 2004 to learn 'How did you get into this?' Because they had already been into [sustainability reporting] 12 or 14 years by then," Kevin Igli, Senior Vice President and Chief Environmental Health and Safety Officer at Tyson Foods, Inc., told me. Now companies contact Tyson Foods viewing them as a role model in sustainability reporting. Kevin said:

> Sustainability has to be unique to your organization and you should not allow too much outside influence on what it means to you. . .and your culture. You are going to get plenty of input and plenty of criticism from the outside. You want to invite outside thought at some point but when you are originally in the incubator stage you have got to tie it to the core values of who you are. If you don't you are going to make a big mistake. It really has to be about your organization.

Given the pressures toward isomorphism, researchers have investigated how some organizations can transcend regulative and normative expectations to establish innovative practices which ultimately result in changes at the level of the institutional field. Walls and Hoffman (2013) investigated influential factors including the skills and the experience of an organization's board of directors and the pressure on an organization to conform to institutional norms. Innovation was more likely when the board members had varied backgrounds, which included dealing with environmental issues, and the organization was on the periphery of its institutional network (i.e., not tightly connected with similar or related organizations).

3.2 Actions Organizations Take to Influence Legitimacy

Organizations use a range of actions, only a few of which will be discussed here. In their quest for legitimacy, organizations engage in various actions including changing their business performance, lobbying to change the normative environment, seeking to shape industry norms, and using communication strategically to frame a situation (Ihlen 2009).

3.2.1 Changing Business Performance

An organization which seeks to create a reputation as a green organization can change its business performance (e.g., engage in eco-efficiencies, create alliances with other organizations, develop new products and services, change working conditions and labor practices). Others monitor internal operations (e.g., use formal environmental management systems) to prevent any potential misconduct that would undermine long-term legitimation efforts (Emtairah and Mont 2008). Several of my interviewees who worked for nonprofits discussed how their organization followed elements of the for-profit business model. Resource dependency theory suggests this choice may be resource driven (e.g., to gain increased legitimacy or funding).

Oakley Brooks, Senior Media Manager of Ecotrust in Portland, OR, discussed the development of Ecotrust Forest Managements, their for-profit subsidiary. Growth in green building in the Northwest increased the need for more Forest Stewardship Council (FSC) certified wood. Ecotrust staff researched why the supply was limited and decided to create their own business model for simultaneously performing forest management and getting sustainable forests to the mill. They sought investors, raised more than $50 million, and bought forest land around the Northwest. In 2013, they owned and managed approximately 30,000 acres. They found ways to layer different activities into their forest management which resulted in different revenue streams (i.e., they sell carbon offset credits and habitat conservation easements). When I sat with Oakley on the green roof of their LEED-certified home office, he said:

> That follows the capital arc that we like to think of as the way we do business. We've got this grant-funded stage and research and development. If it's viable you move it to the market stage. Then you have, just like in the for-profit world, your early investors. These are people who are often conservation-minded. They have a sensibility for doing things in a

different way. They are willing to invest in somewhat risky propositions. And then we push it up to the mainstream banking and investing environment to fund things. We are not quite to the mainstream investment arena for Ecotrust Forests' funds but it is getting more and more mainstream.

Ecotrust Forests is two investment funds that buy land on behalf of investors. Ecotrust Forest Management, the for-profit subsidary of Ecotrust, manages the land on behalf of investors. Investment funds and foundations are investing and, thereby, conveying legitimacy and resources to the enterprise. Ecotrust designed its business performance in a creative way to become more fiscally sustainable while simultaneously working to promote environmental sustainability.

3.2.2 Changing the Normative Environment

Organizations engage in normative discourse (e.g., education, lobbying, and public affairs) to shape societies' perceptions of which behaviors are legitimate (Ihlen 2009). Normative discourse occurs when institutions make claims about how socially valued issues or things should be, how to value them, which things are good or bad, and which actions are right or wrong. Industry lobbying during an environmental or human rights crisis is an archetypical legitimation-seeking strategy. But industry lobbying can be used to proactively benefit human and environmental conditions. Tom Kelly, President of Neil Kelly Company, Portland, OR, described how his desire to see some pro-environmental lobbying going on in the Oregon legislature led him to enlist others in the business community and they started the Oregon Business Association (OBA). "The association's values are centered on a good community period: good environment, good school funding, good transportation system. The OBA is now the most powerful business lobby in Oregon," Tom said. When the societal-level moral discourses are changing, organizations often band together to shape the normative discourse. For example the WBCSD was formed to be a voice for business on sustainability issues, to provide research on emerging outlooks and future prospects to develop global sustainability scenarios, and to help companies learn how to successfully frame environmental issues. Similarly, the Green Sports Alliance is working with the NRDC to change the normative discourse in the USA.

3.2.3 Creating New Standards

When organizations lack established models or standards, they will seek to formalize new activities by developing new structures and procedures. We saw this when the USGBC developed the LEED standards. Another example involves when Walmart sponsored the creation of The Sustainability Consortium (TSC) in 2008. The TSC is a group of academics and others charged with establishing scientific standards for measuring the sustainability of consumer products. Walmart realized their greatest environmental impact would not occur through modifying their internal operations but rather from working with their supply chain to influence resource use on a global scale. Makower (2009) quoted Dr. Jay S. Golden, from the School of Sustainability at Arizona State University, as saying:

> The idea of quantifying the sustainability of products arose. ... It became very clear to us that no one researcher, no one institution, could do that because you're dealing with geographies around the world and with various sciences—physical, life, and engineering—and that required a multidisciplinary approach. So we outlined a proposal to Walmart to develop a consortium of academic researchers from institutions to think through the process and try to bring it to life based on the best available sound science and engineering principles available.

The group's activities concentrate on conducting foundational research. "We're trying to put the right science, the right data, and the right metrics in there, and then place the data into an open-source system for all to use," Golden was quoted as saying. As TSC continues to do their research and work with Walmart and its suppliers, the discussions which occur within a variety of institutional fields are likely to shift dramatically. The promise this process holds is what led *Scientific American* magazine to identify TSC as among the top ten World Changing Ideas of 2012. Brian Sheehan, former Sustainability Manager at Sam's Club, thinks the group is having an important impact:

> For a long time Walmart was talking about the goal of creating this tool that buyers and suppliers could use to understand what the true performance of their products and categories are. There are just so many claims that have been made that it's tough for any one buyer to fully comprehend or vet. Of course to vet every single claim was not possible nor would it be efficient to do. So partnering with TSC, helping to support the creation of that group, actively supporting the growth of the tools that TSC is creating and contributing to the knowledge base, being an active member in that network and also pulling in new members. I think that is just one example of how Walmart is contributing to putting more products on shelves that are more sustainable.

In 2015, TSC had over 90 members employing over 8.5 million people with combined revenues totaling over $2.4 trillion. Today, TSC works to develop methodologies, tools, and strategies to help drive producers and supply networks to address environmental, social, and economic imperatives. Their knowledge products include a category sustainability profile which is a summary of the "best available, credible and actionable knowledge about the sustainability aspects related to a product over its entire life" (The Sustainability Consortium 2013). Working groups provide key performance indicators (KPIs) which are "questions that retailers can use to assess and track the performance of brand manufacturers on critical sustainability issues." Through their KPIs, TSC provide a list of the issues buyers and suppliers should discuss when seeking to make informed sustainability-focused decisions. KPI questions are developed in collaboration with multiple stakeholder groups (i.e., companies, academics, civil-society organizations, and government agencies). "I think it is safe to say that a company that is engaged in the process and is sponsoring research that explores lifecycles that are relevant to that company—they are probably in a better position to learn from the process," Dr. Jon Johnson, from the Sam M. Walton College of Business at the University of Arkansas, was quoted as saying (Makower 2009).

However, developing KPIs can involve conflict. Kevin Igli, Senior Vice President and Chief Environmental Health and Safety Officer at Tyson Foods, Inc., described what occurred when the KPIs for beef were developed:

> [TSC] came under fire from the National Cattleman's Beef Association (NCBA) because they did not like what they [TSC] had done. And the reason is that they [TSC] did not engage them [NCBA]. In fact, they put out KPIs on beef and they had never set foot in a cow-calf operation or any of the parts of the beef supply chain. They just took a table top academic approach and they came out with these KPIs and the NCBA did not like that. The whole beef industry did not like that. So the good news is TSC said, 'OK we've got work to do.' They went out to these places and they engaged more and now they are coming back around to update the KPIs, but they are not done yet. The process of developing KPI's for chicken went more smoothly. I, and many others, went to TSC and they worked with us. I said 'Listen, you have got to survey all of the top poultry science universities in the U.S. You need to go to the U.S. Poultry and Egg Association and the National Chicken Council, go to the Sierra Club, go to all these other groups, survey them and get as much rich information as you can from everybody'. And they did. We are in a much better starting place on poultry than we were on beef because we learned some valuable lessons and that is what this is all about.

Kevin's example points to the importance of communication as TSC develops KPIs designed to guide future discussions around products.

3.2.4 Changing Descriptions and Absorbing Information

Organizations can change how they describe their performance (e.g., public relations, marketing, and branding) (Ihlen 2009). Corporate sustainability communication (CSC) is an evolving concept that refers to corporate communications about sustainability issues (Signitzer and Prexl 2008). Coming from a public relations perspective, CSC grew out of the reactive corporate social reports and environmental communication programs of the 1970s and 1980s when organizations in certain industries (e.g., chemical, oil) faced environmental scandals. Under pressure, they created communication programs which focused mainly on crisis communication and one-way reporting about environmental success stories. Social reports published in the 1970s and 1980s were often advertising instruments lacking honesty and transparency. By the end of the 1970s, although many companies stopped publishing such information, some were successful in their framing efforts. Royal Dutch Shell Group utilized discursive alliance (Livesey 2002) after encountering intense public pressure due to their environmental and social performance. Their inclusion of their consultant's, John Elkington, expository text in their first annual report to society in 1998 provided legitimacy and authority for Shell's own effort. Elkington, who wrote *Cannibals with Forks* (Elkington 1999), had credibility as an environmentalist, promoted a win–win environmental message, and sought to develop more comprehensive forms of environmental reporting. Through the use of discursive alliance in their annual report, the company moved into the sustainable development community and became identified as an exemplary case of progress.

Institutional theory offers practical implications for organizations. Some organizations align themselves via media relations with powerful or legitimate organizations (e.g., co-branding, joint sustainability initiatives). Others engage in boundary spanning by seeking institutional learning from shared institutions (e.g., trade associations, professional associations). Institutional theory also provides some exciting research opportunities for scholars interested in the role of communication (Lammers and Garcia 2014) in promoting sustainability (e.g., Frandsen and Johansen 2011) at the institutional level. At this level, our attention is directed to information crossing organizational boundaries and to formal communication or the medium of written codes (e.g., laws, law-like policies, professional codes, certifications). Industry- and profession-wide norms and regulations provide important clues for understanding actions within organizations. Institutional forces such as guidance from trade associations, peer expectations, and models of success are increasingly talking about sustainability. Many commonalities are likely to occur among organizations within sectors (e.g., the education sector). As sustainability initiatives become the norm, companies will lose external legitimacy if they don't develop and publicize their own sustainability initiatives.

3.3 Communication and the Strategic Approach to Legitimacy

In this section, I focus specifically on proactive and strategic communication as enacted by individual organizations (i.e., the strategic approach to legitimacy). Message credibility and greenwashing are discussed. Then, rather than discuss public relations, marketing, public affairs, and crisis management strategies in any depth, the focus is on communication directed simultaneously toward multiple internal and external stakeholders which reflects standardization efforts involving the use of signs and symbols in annual meetings, on websites, through certifications, and in the design of physical spaces.

Individual organizations communicate strategically with their stakeholders to gain environmental legitimacy (Allen and Caillouet 1994; Hunter and Bansal 2007). They might invite stakeholders into the decision-making process or respond to stakeholder concerns over organizational processes. Or they may use various tools, including symbols and texts, to communicate about their sustainability—although stakeholders increasingly are more interested in tangible indicators (e.g., reduced CO_2 emitted, reduced employee health claims, reduced energy or water use) (Suchman 1995):

> The process of managing an organization's relationship with its external environment of other organizations, regulators, competitors, customers, and diffuse media may be thought of as institutional positioning via the use of institutional rhetoric, or messages that offer specific interpretations of social issues of importance to the organization's survival and success. Examples of institutional rhetoric include corporate statements of social responsibility and policies on the environment. (Lammers 2009, p. 523)

Today, individual organizations engage in public disclosure practices, provide corporate donations and sponsorships, sign on to principled ideals from institutions of moral authority (e.g., sign onto the UN's Kyoto Protocol), and participate in conferences, public discussions, and open panels.

Their communication may be proactive or reactive. This chapter focuses primarily on proactive communication:

> Proactivity has come to refer to a more or less unspecified set of nondefensive or nonreactive practices through which organizations handle their relations with the external world. Instead of waiting for threats and opportunities to become manifest imperatives, the proactive organization attempts to influence and shape external developments in ways considered favorable in terms of its own aspirations. (Cheney and Christensen 2001, p. 253)

Some organizations position themselves as proactive when they are actually engaging in a discursive fiction (i.e., saying they are taking positive actions when they are not) (Zoller and Tener 2010). True proactivity involves working with multiple and varied stakeholders to anticipate potential harms and to adopt environmentally sustainable practices (Bullis and Ie 2007). True environmental leaders make sustainability initiatives and communication an integral part of their core business strategy; create alliances to foster progress on targeted sustainability issues; implement Global Reporting Initiative (GRI) reporting and fully and transparently meet the standards; integrate sustainability into the brand and client value propositions; integrate sustainability into corporate stories, mission, vision, and values; and direct varied, yet complementary, communication toward key stakeholder groups (Peloza et al. 2012).

3.3.1 Credible Communication and Greenwashing

An organization's environmental legitimacy is based on perceptions of its environmental performance, not necessarily its actual performance. Therefore, the credibility of an organization's environmental communication is an important determinant of its environmental legitimacy (Hunter and Bansal 2007). Credible communication provides detailed information on topics that stakeholders would expect to be discussed, supplemented with illustrative examples. Non-credible communication is more opaque and general; omits important facts, topics, and discussions; or presents more favorable information than would be expected. Credible communication is transparent and comprehensive. Transparent communication involves letting external (and internal) stakeholders know what the organization is "really" doing. For example, a transparent environmental report presents specific information about activities and goals, such as "reduce CO_2 emissions by 20 % by the year 2006." Opaque communication uses less specific terms such as "reduce emissions." Comprehensiveness is whether the "full story" is presented or if enough detail is provided to meet reasonable expectations. A comprehensive environmental report provides information about a range of practices such as the amount spent on an environmental initiative, the number of people executing it, and the type and amount of pollution to be reduced. Less comprehensive reports provide little if any information about activities or policies. *Best Practice*: Craft transparent and comprehensive messages.

Credible Communication at Tyson Foods and the Arbor Day Foundation Two of my interviewees shared their thoughts about credibility. In deciding what to include in Tyson Foods' sustainability reports, Kevin Igli, Senior Vice President and Chief Environmental Health and Safety Officer at Tyson Foods, Inc., said:

> As you begin to examine and write a report about your organization, you have to create a fact book. We go through a process that is similar to a massive financial audit. Any time we write one of these reports we have a room full of lawyers and others. We fact check everything we say. I mean we lock it down and button it up. You have to. And for us the result has been the challenges we have seen to our sustainability reports is that there have really not been challenges.

I asked Woodrow Nelson, Vice President of Marketing Communication, how the Arbor Day Foundation managed their reputation so that other organizations feel comfortable partnering with them. His comment pointed to a *Best Practice*: Be accurate and do what you promise:

> Number one, we are going to do what we say we are going to do, and we are not going to promise something that we are not going to be able to deliver. We have corporations who come and want to do the right thing—they want to give back, they want to plant trees... There are others who want to count carbon offset credits, and we are not going to do that because we are not able. We can't [we don't have the staff] verify that carbon still stands in a particular forest. If we can't verify it, we are not going to tell people we can.

Information not seen as credible may be labeled as greenwashing. Greenwashing is a strategy that companies adopt when they engage in symbolic communication about environmental issues without substantially addressing them through action (Walker and Wan 2012). It involves disseminating misleading information so as to present an environmentally responsible public image (Cox 2013). For example, some companies highlight a sideline of the business which is seen as more legitimate (e.g., British Petroleum's claim to be a world leader in solar energy while they derive 90 % of their income from hydrocarbons) or stress their operational efficiencies although it is their products that come into question (Emtairah and Mont 2008).

In one of the first studies to link greenwashing with the financial performance of a firm, Walker and Wan (2012) examined over 100 top performing Canadian firms in visibly polluting industries. They differentiated between greenwashing and green highlighting. Green highlighting is when an organization provides both symbolic and substantive action talking about what they are currently doing or have done (substantive action) or their green walk and what they plan to do in the future (symbolic action). They measured financial performance using return on assets (ROA) and found a negative relationship between financial performance and greenwashing and between financial performance and only communicating symbolic actions (e.g., goals and future actions). They did not find a relationship between financial performance and communicating about substantive actions or green highlighting. The strength of their article was in their distinction between greenwashing and green highlighting rather than their conclusions about the financial implications of the messaging. Obviously, an organization's ROA is a function of much more than the messaging strategy appearing on the corporate website.

3.3.2   Annual Meetings, Sustainability Reporting, Websites, and Architecture

In this section, I discuss five forms of communication about an organization's sustainability initiatives which are simultaneously directed toward a broad range of stakeholders: annual meetings, sustainability reports, websites, certifications, and architecture. All are the product of multiple communicators working together to shape sustainability-related messages. Similar to the work of Gatti (2011) and Zhang and O'Halloran (2013), semiotics and visual rhetoric play an important role in four of the five. Semiotic theories are based on the assumption that almost anything can be a sign or a symbol, standing in for and eliciting in the viewer's mind a concept separate from the sign itself. Signs can have a complex array of direct and indirect meanings. Images carry with them denotative and connotative messages. The connotative message includes all the values and emotions the image calls forth in the viewer (Hill 2009). Of the four, the least communication research has focused on architecture although it is an important tool for communicating about sustainability.

3.3.2.1   Annual Meetings

Top management uses various forums to communicate stewardship information to stakeholders including their annual report, Internet-based stakeholder forums, and investor meetings. But the annual general meeting is an often underutilized forum (Carrington and Johed 2007). It is a real-time event where stakeholders can ask and pursue questions, thereby presenting them with an opportunity to challenge what are often rehearsed and edited corporate messages. Annual general meetings are a time when a CEO can position him or herself as a good steward of both company resources and the environment.

Sam's Club, the Neil Kelly Company, and Annual Meetings Brian Sheehan, former Sustainability Manager at Sam's Club, thinks the annual meeting's big stage is a very important forum for discussing an organization's sustainability initiatives. Since 2007, Walmart has held sustainability summits which are televised into all Walmart and Sam's Club locations. These summits serve as a platform for announcing aspirational goals and reporting on progress made in relationship to sustainability initiatives. Brian said:

> Where we get the most impact is with the biggest scale. And for us the biggest scale happens at the big meetings on the big stage in front of a room of five thousand people. Hearing from our leadership, hearing from keynotes who are experts in sustainability and have a great personal passion for it and are good story tellers. ... Getting a celebrity that can help make sustainability desirable, and communicate in a way that is easy to understand and rewarding.

Annual meetings are not only important for large organizations. Julia Spence, Vice President of Human Resources at Neil Kelly Company, Portland, OR, talked about the annual meeting of that family-owned business. Before the annual meeting, everyone helps develop the strategic plan for the next year. Employees are invited to participate in open meetings to brainstorm options, opportunities, goals, changes, improvements, etc., and/or to submit ideas in writing. All are invited to think and contribute strategically. Then, the management team reviews the suggestions along with statistics, achievements, and client feedback. Plans for the coming year and beyond are drafted. At the annual meeting, the management team provides results and analysis of the previous year, recognizes employees, and presents the plan for the next year, a plan which makes strong use of employee planning contributions. Julia explained:

> [The annual meeting] is a time once a year where the management team gets up in front of everybody [employees] and says here is what we accomplished as a group and here is where we are headed. And it has come out of what everyone has contributed both to production for the past year and [for setting] the goals for the coming year. We do talk about our environmental goals and where we are headed. It is a time when we are accountable as a management team.

Coming from a financial accounting perspective, Carrington and Johed (2007) attended 36 annual general meetings of Swedish companies. Only a third of the questions the CEO answered dealt with financial accounting, and the other questions dealt with nonfinancial aspects of stewardship (i.e., company efforts regarding environmental, equality, and ethical issues). They based their study on actor–network theory, a theory initially developed by science and technology scholars interested in a performative view of the production of science that acknowledged both humans and nonhumans (e.g., machines, texts) accomplish things (Cooren 2009). Cooren and colleagues (e.g., Caster and Cooren 2006; Cooren et al. 2006) urge researchers to investigate forms of nonhuman agency (e.g., textual, technological). What distinguishes an actor is whether or not it influences other actors. Therefore, a written report can be an actor effecting an organization directly as well as creating the organization and making it visible. Actor–network theory is increasingly appearing in the communication scholarship. For example, Allen, Walker, and Brady (2012) discussed how organizations are partly discursive constructions created through the agency of both nonhuman (e.g., sustainability-related training materials) and human actors. Two key concepts of the theory are actor–network and translation. An actor (human or not) can act or speak on behalf of something or someone else which becomes part of a network being spoken for and therefore capable of influence. Translation involves the process by which actors make present the interests of the network they represent. The annual general meeting is a central and indispensable actor because reports are provided describing the state of the organization, shareholders can ask questions to hold management legally accountable, and management is elected or reelected. Unlike other forms of corporate communication (e.g., annual report, Internet forums), top management must answer face-to-face questions. Top management seeks to be identified as good stewards of the company by going through a translation process which involves four steps: problematization, interessement, designated enrolment, and mobilization (Carrington and Johed 2007). Problematization involves imposing one's own definition of a situation on others and convincing them that their interests are being served by what is being proposed. Interessement is when the actor locks others into assigned roles. Designated enrolment is when the locked actors are manipulated to act in a way benefiting the management. Finally, using mobilization methods, spokespeople represent and control their collectives. The general meeting results in a highly scripted and controlled event.

Sometimes organizational spokespeople encounter hostile questions when publicly discussing their organization's environmental performance or sustainability initiatives. This is less likely during an annual meeting and more likely in an environmental public meeting. A hostile question is one that contains a proposition that is both undesirable and erroneous. Campbell et al. (1998) suggest speech act theory can provide communicators with guidance on how to respond. Carrington and Johed (2007) also discuss speech act theory. This theory has had a major impact on communication research. Usually associated with the work of John Searle (Littlejohn 2009), it focuses on how people use a set of language games, complete with rules, to achieve goals. The listener understands the rules of the game, both constitutive and regulative, and can interpret the message accordingly (e.g., this is a promise, a question, or a command). A speech act is not completed until the listener responds. Every speech act does something in addition to saying something. If the goal of a public meeting is continued interaction between the audience and the speaker, then speech acts allow the spokesperson to accomplish two apparently contradictory goals: (a) satisfy the questioner by providing an answer and (b) satisfy the organization by providing an indirect answer. In response to hostile questions, at least three speech act strategies apply (i.e., desirability, agency, and timing). Although based on a small sample size, Campbell et al. recommend using either the timing or the desirability strategy. Timing strategies include claiming the questioner's request had already been addressed. The desirability strategy involves providing reasons why the questioner might not really want that particular request fulfilled.

3.3.2.2 Reporting About Sustainability

In an effort to gain legitimacy and to signal their organization's commitment to the triple bottom line, many organizations publish sustainability reports (Ferns et al. 2008; GRI 2013). These reports present a company's values and governance model and show how company strategy reflects its commitment to sustainability. The process of writing these reports can help companies set and measure goals, understand some of their social and environmental impacts, and communicate about their economic, environmental, social, and governance performance. Senior decision makers can use a report's information to shape their organization's strategy and policies and improve its performance. From a critical theory perspective, the argument is these reports seek to produce "statements of facts," tell a more or less passive audience that everything is fine, and discourage further questioning of the organization rather than functioning as part of a dialogic, problem-solving, educative process (Stiller and Daub 2007). Bowers (2010) argues they represent a genre organizations use to define the exigence (i.e., situation requiring urgent action) by making the business case for sustainability. But sustainability reports only approximate reality. Even the best only show how that organization would like to understand sustainability and have others see their sustainability efforts. To be of benefit, these reports must be fully auditable and address inherent conflicts and trade-offs between an organization's profit focus and its social and environmental responsibilities (Mitchell et al. 2012). Ideally, after reading a sustainability report, employees will engage in dialogue around triple-bottom-line issues, experience double-loop learning, and enact behavioral changes (see Sect. 7.1.2).

A Change over Time

Scholars have traced the evolution of sustainability reports (e.g., Bowers 2010; Livesey 2002; Maharaj and Herremans 2008). Some of the earliest corporate environmental reports were in response to public pressure following an environmental disaster. Increasingly, companies operating in sectors likely to experience environmental disasters (e.g., extraction-based companies, chemical manufacturers) began to report their environmental data. These early reports were technical accounts of environmental impacts and remedial efforts which included metrics collected for other business purposes, were required by regulatory agencies, or met voluntary guidelines created by industry associations and international bodies. Early reports stressed compliance. They often lacked significant quantitative data, explanations for trends, negative news, or proposed future actions to improve negative results. In 1991, Royal Dutch Shell Canada published one of the first reports labeled as sustainable development. That subsidiary's commitment to more transparent and broad sustainability reporting and communication influenced the parent company's reporting a few years later when its reputation was at its lowest level (Livesey 2002). Under pressure, Royal Dutch Shell published their first ethics report in 1998. By 1999, their reputation had improved, partly because of their stakeholder engagement efforts, especially their social and environmental reporting. The reports increased in sophistication and included more rigorous reporting of how environmental policies were being implemented and about the company's triple-bottom-line impact. Later reports included feedback mechanisms for their stakeholders, identified measurable annual targets and objectives, and incorporated the employee perspective on pro-environmental efforts.

Over time, the message frames appearing in the reports shifted. Exxon, Chevron, and BP changed the way they framed climate change in their annual letters to stockholders and stakeholders between 2003 and 2009 (Domenec 2012). Environmental values went from being portrayed as a potential threat to something that yields value, and oil spills were acknowledged. The emphasis was on how the companies behaved responsibly and proactively regarding environmental issues. Bowers (2010) found something similar in his rhetorical study of the sustainability reports of ten global companies at two points in time. The issues which led to the creation of the first sustainability reports remain unchanged (e.g., greenhouse gases, human rights, toxic emissions). What changed was why corporations responded and the way they defined the rhetorical situation from compliance to value generation. Bowers found a growing emphasis on economic value in the opening sections of the report and the CEO letters emerged over time. Increasingly reports provided information about measurable economic outcomes associated with various sustainability activities.

Developing a Reporting Framework

Due to the lack of standardization between reports, in 1997, the GRI was developed to provide a comprehensive sustainability reporting framework for measuring and reporting sustainability-related impacts and performance. The GRI's mission is to make sustainability reporting standard practice for all organizations. Their services include coaching and training, software certification, and reporting guidance for small- and medium-sized enterprises. The first framework was available in 2000, and in 2002, GRI was formally inaugurated as a United Nations Environmental Programme collaborating organization. Their framework is used worldwide. In their 2011/2012 annual report, GRI summarized a study showing that 95 % of the world's 250 biggest companies disclosed sustainability performance information in 2011, and 80 % of those used the GRI guidelines. In 2009, 1,000 organizations had used the framework, a 47 % increase from 2007, and the framework's use has grown steadily since. Most sustainability reports are prepared by business organizations (Dart and Hill 2010; Guthrie and Farneti 2008). That is slowly starting to change. The GRI has a sector supplement for public agencies. In 2012–2013, Fall River, MA, was the first US city to produce a GRI report; the Washington State Department of Ecology was the first state environmental agency to release a GRI report; and the San Diego International Airport was the first US airport to release a GRI report.

Tyson Foods and the GRI   Tyson Foods turned to the GRI when planning their first sustainability report. Kevin Igli, Senior Vice President and Chief Environmental Health and Safety Officer at Tyson Foods, Inc., explained:

> We wanted to make sure that we were holding ourselves accountable to some type of a standard with respect to reporting. . ... We decided that if we are going to compare ourselves, we need to compare ourselves to the gold standard. The reason is if you don't do that then someday you may decide to and then you have to completely switch formats. . . I think they do a great job. . ... We use a lot of ISOs' work around 1401 environmental issues to help guide the way we build programs. They have done wonderful work and I am not taking anything away from them. . ... GRI is so far out in front of everybody that they are the way to go. GRI is global and they do bring very broad perspectives. Most large, multinational companies that use sustainability reporting engage with GRI, and it just made sense for us.

Tyson Foods used the GRI reporting framework as a model for preparing their first two sustainability reports:

> We were fortunate that the very first outing [of preparing a report], while there were a lot of categories where we did not meet their expectation, there were so many where we did. . ... The title of our first report was *Living Our Core Values*. We chose that because we have core values as a corporation. So we decided that the best way to tell our story is to see *do we walk our talk*. If we lay out our core values against the principles of sustainability, how are we doing? So that was what our first report actually discovered and discussed. It discovered how we were doing. It was very interesting.

The company decided to have their third GRI report graded and they received a B. Kevin said, "We are a very competitive company, so the B resonated okay with senior management. It said, you are better than average. But what does it take to get an A [management asked]?" They earned an A for their fourth report in 2013. "To get an A plus you have to bring in an independent third party and it gets a lot more complicated. We have not decided if we will go after that. But if we could get another A, we would be happy."

Multiple entities besides the GRI exist to guide organizations interested in reporting on their sustainability efforts. Some are sector or industry specific. *Action Plan*: Identify which guidelines are most relevant to the organization and sector you are interested in. However, as long as sustainability reports lack standardization (Ferns et al. 2008), this makes it hard to benchmark performance or achieve a best practices status and it creates a vacuum filled by the report readers' personal expectations and interpretations. Many sustainability metrics are reported on a voluntary basis and measures are not standardized unlike financial reports' generally accepted accounting practices. *Best Practice:* Use the GRI to compare across organizations and sectors.

Deciding on Report Content

The GRI sustainability reporting framework has influenced report content. For example, a focus on materiality, a major consideration when writing a GRI report, gives business leaders and managers an opportunity to articulate how sustainable development issues can be integrated into specific business practices. Economic issues are one of the ways businesses decide if something is material. Some reports include statements describing how lessening environmental impacts and attending to social problems can reduce operating costs, generate revenue, open new markets, and provide a competitive advantage (Bowers 2010). Other report content seeks to identify and report on an organization's triple-bottom-line goals, targets, and initiatives, provide specific and measurable context and reference points (e.g., comparable data from previous years or cross-industry comparisons), show performance against benchmarks, share an organization's values, identify the processes organization's use to minimize any harm its activities produce, set specific goals for the future, and address the needs of various stakeholder groups (Maharaj and Herremans 2008). Others provide emission data, mechanisms used to measure the success of environmental practice implementation, internal communications systems, environmental policies and practices outside the USA, and environmental sustainability-related requirements for suppliers (Axelrod 2000). Realizing sustainability reports can influence corporate reputation, many include statements from a chief executive officer, chief financial officer, and/or chief sustainability officer. Ferns et al. (2008) concluded that a CEO's status and credibility is critical to building public trust. Some reports include information on regulatory problems, remediation liabilities, environmental challenges, and failures encountered when enacting sustainability-related initiatives (Axelrod 2000; Maharaj and Herremans 2008). Ihlen (2009) notes that most do not address the fundamental issue—how their industry and the lifestyles they support produce more environmental harms than benefits. Numerous scholars have researched the content of the corporate sustainability reports and websites of companies across the globe (e.g., Biloslavo and Trnavcevic 2009; Chaudhri and Wang 2007; Gill et al. 2008; Herzig and Godemann 2010; Stiller and Daub 2007), employing content analysis or rhetorical analysis (e.g., Bowers 2010; Ihlen 2009; Livesey 2002). Chapter 7 discusses how supply chain relationships are discussed within sustainability reports. *Action Plan:* Identify the issues covered in the sustainability report of an organization you admire or which is an industry leader.

Tyson Foods and Sustainability Report Content Deciding what to include in a report can be challenging. Kevin Igli, Senior Vice President and Chief Environmental Health and Safety Officer at Tyson Foods, Inc., told me:

Sustainability reports are tricky. There is a balance you have to find about transparency, confidentiality, and what you are really comfortable sharing about your company in a very open forum. As companies go through this process it takes an army of people to have a team to draw together all the different resources and pieces. We call them wheel owners. We draw a big pie chart and divide it up around the company in terms of the different aspects that go into our sustainability reports. One of the things that is unique to Tyson's sustainability approach, is that it is not a green report. We do talk about the planet. We do talk about environmental stewardship. We talk about a lot of other things. Sustainability at Tyson is very broad. We touch on animal welfare, we touch on human nutrition, we touch on hiring practices, and we touch on our team members' Bill of Rights. So we take sustainability as a very broad topic. We don't just look at it as greening the planet or being good environmental stewards. That is a very important thing, but that is only one part of what we view sustainability to be. We are pretty clear on that. When you go through our report we talk about the broad approach and why we do that.

Benefits of Sustainability Reporting

Sustainability reports provide organizations with an outlet for telling their story and give corporate leaders a public forum for stating their commitment to good corporate citizenship (Ferns et al. 2008). The organization has complete control over the content. While paid advertisements and press releases can be rebutted in the same medium (e.g., editorials, news stories, op-eds), sustainability reports are relatively protected from critical analysis. Voluntary social reporting also creates precedents and thresholds that push other firms to adopt more systematic practices and to share more information (Livesey 2002). Report similarity in the same industry and in a similar value chain (due to mimetic isomorphism) challenges the idea that the reports can provide a competitive advantage due to differentiation (Ferns et al. 2008), an idea which often emerges as a business case for sustainability argument. However, similar reports do reduce the likelihood an organization will be uniquely targeted by activists. Sustainability reports that match stakeholder needs, if read, can result in positive consumer opinions, enhanced stakeholder trust, higher employee satisfaction, increased community support, access to markets in new countries, and improved corporate brand management.

Sustainability reports which discuss climate change potentially can influence public attention to and understanding of the issue. Looking at how 30 of the world's largest corporations addressed climate change in their nonfinancial reports, Ihlen (2009) identified the rhetorical strategies used to persuade a range of stakeholders (e.g., competitors, suppliers, investors, governments) that the corporation was dealing with the issue appropriately. Four topics appeared central to the reports: (1) acknowledgment that the environmental situation is grave, (2) statements indicating that the corporation is in line with the scientific consensus and the international political process on curbing emissions, (3) assurances that the corporation plans to take measures to reduce company emissions, and (4) contentions that the climate challenge poses an opportunity for business. However, nothing in the reports suggested that these corporations were engaged in the radical rethinking of systemic problems we need if humanity is to seriously address climate change.

Moving to Online Sustainability Reports

We frequently learn about an organization's commitment to sustainability through electronic sources such as Internet search engines and websites (Herzig and Godemann 2010). In 2012, 80 % of the US population had access to the Internet. Given the plethora of electronic sources, the number of watchdog organizations (e.g., SourceWatch), and the speed with which a stakeholder can investigate a firm, comprehensive and truthful online reporting is vital if an organization is to successfully manage its corporate reputation (Ferns et al. 2008). Walker and Wan (2012) evaluated the sustainability statements appearing on the websites of 103 chemical, energy, mining, and forestry companies listed on the *Financial Post's* top 500 Canadian companies for 2008. They selected Canadian companies because Canada, unlike the USA, signed the Kyoto Protocol and has publically committed to emission reductions. The topics discussed on these websites included energy conservation, management of greenhouse gases, stakeholder engagement, environmental audits, recycling, technology development, environmental management systems, waste management, employee training, environmentally friendly products, innovation, carbon capture and recovery, and life-cycle analysis. As you would expect, industry sector influenced the topics appearing on the websites.

Tyson Foods' Printed and Online Reports Many organizations are augmenting their written reports with electronic website-based versions or moving their reporting completely online. Tyson Foods, Inc., published its first sustainability report in 2005. Their first two reports were high-gloss printed documents. They switched the next two reports to the Internet. All four reports are archived on their public website and are available in English and Spanish. Kevin Igli, Senior Vice President and Chief Environmental Health and Safety Officer at Tyson Foods, Inc., described the evolution of sustainability reporting at Tyson Foods, Inc., saying:

> Our most recent report is online and is a completely different format. It is tabbed and searchable so you can go in and pick and choose what you want to read, when you want to read it. That has made a difference in terms of people going and looking at what they want to look at. Because not everyone wants to read everything that is in there. So what will happen is they will go in and say, 'well I want to read about animal welfare' and they will start reading it and say, 'did you see this stuff on products?' All of a sudden they get interested and they are reading and reading. So it is a lot better format because you can come and go as you need. You can download a copy onto your IPad or whatever device you use, onto your phone. So electronic media and electronic format, we have learned a lot about what that means and how you make it easier for people to learn. ... It is all part of learning how to communicate. Tyson has a Facebook, we have Twitter, so we are engaging in all kinds of different directions and it has really been helpful. ... We want people to engage more and learn more.

Electronic reporting reduces information dissemination costs, potentially increases the reach of the message, and allows for more versatility in changing content. The ability to provide ongoing, additional, or updated information allows companies to respond to greater demands for information and convey a more complete vision of their sustainability performance. Audio, video, and interactive tools are available to both raise awareness and help convey complex ideas. Specific information can be directed toward targeted stakeholders (e.g., investors, donors, suppliers) (Herzig and Godemann 2010). Internet communication also can be structured to increase comprehension. The Internet's hypertextuality allows for the use of search functions, sitemaps and navigation alternatives (e.g., linked indexes), and glossaries, which can increase access to and the comprehensibility of the information. Customized reporting approaches allow for user-specific creation and interactive choice. A dialogue-based online relationship can include a range of mutual, asynchronous ways to solicit feedback (e.g., mail-to-1 guest books, and online surveys) and influence long-term and continuous (e.g., discussion forums and bulletin boards). A mutual, synchronous dialogue, with a high degree of spontaneity, is also possible (e.g., chats, audio- and videoconferencing). Online discussions can help organizations understand their stakeholders' attitudes and information requirements. Although, the Internet provides tools organizations can use to increase stakeholder dialogue, few appear to be using these options.

I have always admired Ray Anderson for the actions he took to put his carpet company, Interface, on the path toward sustainability and for his willingness to speak very publically about his decision and subsequent actions. Interface's Mission Zero goal is to eliminate all of its negative environmental impacts by 2020 through the redesign of processes and products, the pioneering of new technologies, and the reduction or elimination of waste and harmful emissions while increasing the use of renewable materials and energy sources. Sustainability is a main category appearing on their homepage. In 2013, clicking the Sustainability tab allowed viewers to read the Interface story, learn about their progress and challenges, and provide questions, suggestions, and advice. Several things about this website appeal to me. The use of the framing metaphor of a journey and the graphic of the mountain communicate effectively about the ongoing and difficult nature of the process. Understandable metrics are provided which show what they have accomplished in several important areas (i.e., energy, climate, waste, facilities, and transportation). Finally, their public outreach in the form of their speakers bureau shows their continuing willingness to help other businesses become more sustainable.

Not all websites are created equal. Poor websites reduce an organization's ability to successfully communicate with interested stakeholders. Practitioners might find this an interesting checklist to use when planning their organization's website. *Action Plan:* Use the following questions to evaluate the website of your organization or an industry leader.

- How easy is it to access sustainability-related information on the main webpage? How many clicks does it take to get to the sustainability page?
- What are the main topics on the sustainability page? Within each topic, how much information is provided?
- Is there a clearly stated sustainability vision? Is it attributed to the CEO? Are clear ways for achieving this vision listed?
- Are any statements provided by key stakeholders in addition to the CEO? Which ones? What are they saying?
- Are clear goals provided? Is a timeline for achieving the sustainability-related goals provided? Does the company indicate if and how they are meeting each goal? Do they provide evidence to support their claim the goals are being met?
- Is the information clearly targeted to specific stakeholders (e.g., investors, employees, customers, etc.)? Which stakeholders are being specifically addressed?

- Are issues related to the triple bottom line (people, planet, profits) addressed? What are they reporting in each area?
- Do they have a sustainability report linked to the website? If so, for which years? In which language(s)?
- What other links exist on the sustainability page?
- Is the website visually appealing, easy to navigate, and easy to read? Are there clearly defined or labeled areas. Are downloadable pdfs provided?
- Is the writing clear and succinct? Are quotes provided and from whom? Are stories provided on the webpage?
- What is being highlighted in the pictures? Are there graphs and/or charts? What are the main points being conveyed by these pictures/graphs/charts?
- Can you provide feedback to the company? Are there questionnaires, forums, chat rooms, or any other forms of interactivity? Can visitors sign in or get on a mailing list for updated information?
- If you want to contact the company, is a general email address provided? A specific email address provided? Are any specific people identified as sustainability-related contacts?

3.3.2.3   Certifications

Many organizations mention certifications in their sustainability reports, on their websites, and in other materials they publish for internal and external stakeholders. My interviewees representing cities (e.g., Portland, Fayetteville), state governments (e.g., South Dakota), college campuses (e.g., University of Colorado, Boulder; University of Arkansas), and nonprofits (e.g., Ecotrust, Heifer International®) talked with me about their LEED-certified buildings. The Arbor Day Foundation certifies cities and towns. The certification or endorsement an organization seeks depends on its sector (e.g., government, industry) and its output (e.g., services, consumer products). For example, quickly skimming the Tyson Foods 2012 Sustainability Report, I saw they had Global Food Safety Initiative certification and 73 of their US facilities had British Retail Consortium Global Standard for Food Safety certification. In this section, I identify some sustainability standards and certifications and talk about factors influencing the decision whether or not to certify.

What Are Sustainability Standards and Certifications?

Sustainability standards are voluntary, usually third party-assessed, norms and standards relating to environmental, social, ethical, and food safety issues associated with an organization's performance or specific products. The range of new sustainability-related standards significantly increased over the last decade.

The trend started in the late 1980s and early 1990s with the introduction of ecolabels and standards for organic food and other products. Some of the common ones involving consumable products are the Fairtrade label, the Rainforest Alliance, and Organic. Builders would be most familiar with the USGBC's LEED standards and FSC certification. Previously, I talked about FSC certification, and identified how AASHE awards universities with STARS.

Other organizations award important certifications. The International Organization for Standardization (ISO) provides requirements, specifications, guidelines, or characteristics that can be used to ensure that materials, products, processes, and services are safe, reliable, and of good quality. The ISO publishes and sells over 19,500 international standards but does not certify companies. Management system standards include ISO 9001, ISO 14001, and ISO 31000. The ISO 14000 family of standards (14001:2004, ISO 14004:2004) focus on environmental management and provide tools for organizations seeking to identify and control their environmental impact, as well as improve their environmental performance. In terms of the social dimension of sustainability, the ISO 26000 standards provide guidance for how businesses and organizations can operate in a socially responsible way. Standards for operating in a socially responsible way are also offered by Social Accountability International (SIA). SIA publishes the SA8000 standard, an auditable social certification standard for judging decent workplaces. Structures and procedures that companies must adopt are described and continually monitored. The group AccountAbility publishes their A1000 Stakeholder Engagement Standard (AA1000SES), an open-source framework which supports the group's AA1000APS Principle of Inclusivity. Blackburn (2007) describes a wide range of other certifying organizations.

Numerous groups stand ready at the international level to certify that organizations are operating in environmentally and socially sustainable ways. Community-level sustainability initiatives also exist. I found local certification programs occurring in two communities I traveled through, Missoula, MT, and Lincoln, NE. These programs epitomize sustainability-related communication challenges and opportunities at the local level. Missoula, MT, is in a state characterized by small businesses. The community started investigating the idea of sustainability in 2000 and by 2002 a group had formed which was interested in promoting sustainable business. Today their Sustainable Business Council (SBC) has about 180 paying business members. The SBC works to advance sustainable commerce by supporting and connecting businesses that strive to become more sustainable and educating consumers about sustainable business options.

Susan Anderson, who was their administrative coordinator when I was in Missoula and a founding board member, told me:

We have tried really hard, coming from a community that had some polarization with the environmental movement, to make certain they [business owners] understand that we are not out to put them out of business. We want to help them do business better. We want to help them prepare for the future that is coming. I think the key behind that is when you look at markets right now the messages are not correct. There are signals that are not in the market that really need to be in terms of the impact of environmental damage or the impact of social damage on the long term viability of the business. . ... A friend of mine, who wrote the book, *Green Business Practices for Dummies*, often says to people to get their attention: 'Is your business ready for a carbon-constrained environment?' Right now we're actually seeing a drop in our fuel prices. . ... But they are not going to stay that way. So people who use the savings they get right now to get ready for the future are the ones that I think are going to survive better, healthier in the long term.

Susan described their Strive Toward Sustainability program which includes an education, self-certification and a graduated recognition program, a membership appreciation event, and annual sustainability awards. The Strive Toward Sustainability program begins with a half-day workshop. Then companies complete a self-assessment, select areas for improvement, and receive information on how to improve. The program has a set of seals that can be put in the windows of local businesses and appear on the SBC's website. Their program was replicated in several other communities, including Lincoln, NE. Carrie Hakenkamp, Executive Director of WasteCap Nebraska, described how the self-assessment covered nine areas: management, energy, facilities, transportation, waste, community, customers, employees, and purchasing. There are standards and questions addressing each standard. If companies meet the minimum standard in an area, they receive a badge in recognition. If they meet the standards in enough badge areas, then they earn a gold seal. Carrie explained:

When you read those questions you learn about things, the gaps that you have, and get some ideas on other projects you could do. . ... What we have found is that so many companies have already done something. They have a recycling program. They have switched out their lights. But most people in the company don't even realize it. So now you have an internal marketing program. You have something that demonstrates your values as a company to your employees, your customers and people on the outside. It is just a really great first big step. The first step is the hardest, but once you start down that path then [they think] here is something else we can do, and here is something else. And it gets easier to implement projects because more people understand what your values are.

WasteCap Nebraska ultimately discontinued using the Strive for Sustainability program and began working with B Lab, a nonprofit organization that created the B Impact Assessment and the B Corp Certification. Certified B Corps are companies that use the power of business as a force for good. B Corps are different from ordinary businesses because they meet a higher standard of social and environmental performance, transparency, and accountability. There are more than 1100 Certified B Corps in 120 industries in 40 countries, including Patagonia, Ben and Jerry's, Method, Dansko, and Etsy. More than 20,000 companies have taken the free, online, and confidential B Impact Assessment. Companies can compare themselves to other companies that have taken it. To be certified, a company must score at least 80 out of 200 points and be audited. WasteCap Nebraska combines the B Impact Assessment with their "Good Company" green team training, which helps green teams organize and learn a system for selecting, implementing, and managing projects and measuring and reporting achievements. It includes training and information about writing a sustainability plan and report and about audience analysis and communication.

The Decision to Certify or Not?

Organizations have various reasons why they seek certification including in response to industry isomorphism, to symbolically communicate their legitimacy, to utilize a framework to guide internal processes, to reap reputation and efficiency related benefits, in response to a contractual or regulatory requirement, to meet customer preferences, to aid in risk management, and to help motivate staff by setting clear goals. Cities use frameworks to guide decisions which will hopefully enhance the livable of their communities. In this section, I pay particular attention to financial reasons, employee productivity, and stakeholder perceptions.

Businesses often adopt ISO1400 and similar international management certifications when they are dependent on international trade or need bank financing since financial institutions are concerned with long-term financial and environmental risks which might threaten their investments. Others seek certification believing employee productivity may increase. Surveying 5,220 French firms, Delmas and Pekovic (2013) investigated the relationship between ISO 14001 certification and labor productivity. Firms that adopted environmental standards enjoyed one standard deviation higher labor productivity than firms that did not adopt such standards. The authors concluded the adoption of an environmental standard signals that an organization is committed to improved environmental performance which is likely to lead to a more positive reputation. If employees feel proud of their organization, they often perform better at work, engage in more cooperative and citizenship-type behaviors, and become ambassadors for their employer. Also, environmental standards require the adoption of management practices such as new environmental policies, internal assessment (e.g., benchmarking, accounting procedures), environmental performance goals, internal and external environmental audits, and even employee incentive programs based on an organization's environmental performance. Companies are required to identify, minimize, and/or seek to manage environmental risks, liabilities, and impacts. Efficiency may improve as new systems gather additional information used to monitor environmental performance. Communication, cross-functional teams, and employee training are essential elements of ISO 14001 implementation. This can result in more effective and knowledgeable employees, increase human capital, facilitate knowledge transfer and innovation, and thus increase productivity.

Certifications are symbols meant to denote an organization's legitimacy. If message receivers are aware of the meaning of the symbol, they will quickly process it and assume the organization is trustworthy in multiple areas. Parguel et al. (2011) used attribution theory (see McDermott 2009) as the basis of their study of the role of independent sustainability ratings on 177 French consumers' responses to companies' CSR communication. Attribution theory explains that when people communicate, their decisions are based on the attributions they make for their own and others' behaviors. The authors distinguished two types of causal attributions for CSR communication: attributions to the dispositions of the actor (intrinsic motives) and attributions to environmental factors (extrinsic motives). When faced with CSR communication, consumers must decide whether it conveys a genuine environmental consciousness (intrinsic motive) or is an attempt to take advantage of sustainable development trends (extrinsic motive). If an organization makes claims about its CSR engagement but provides no other information, consumers assume extrinsic motives. They may believe the organization is greenwashing if they don't green highlight. Sustainability ratings from independent agencies provide information which suggests intrinsic motivation.

In building their argument, Parguel et al. (2011) refer to the three elements of Kelley's covariation model: consensus, distinctiveness, and consistency. Do others in similar circumstances adopt the same behavior (consensus), does this behavior occur only within a particular situation (distinctness), and is this behavior repeated across time (consistency)? Consumers are more likely to attribute extrinsic motives if most of an organization's competitors engage in the same behavior at the same time (social consensus), if the company only engages in one CSR activity (distinctiveness), or when CSR engagements are infrequent company practices (no consistency). A positive global sustainability rating, provided by an independent and credible third party, suggests the company enacts sustainability in various ways (i.e., its engagements are nondistinctive) and on a frequent basis (i.e., its engagements are consistent). Consumers are more likely to attribute the CSR message to the organization's intrinsic motives resulting in an increase in positive brand evaluation. However, poor sustainability ratings should drive CSR engagement attributions to more extrinsic motives. The researchers conclude that sustainability ratings can deter greenwashing and encourage virtuous firms to continue their CSR practices. Attribution theory also has implications for how employees and supply chain members respond to CSR-related messages.

For some organizations, the decision to seek external certification is not easy. The management team of a plastics recycling organization I consulted with disagreed as to whether or not to certify an innovative new product they had developed. Opponents expressed concerns about recouping the costs associated with the certification process, whether or not most clients would even care and whether clients who did care would commit to doing enough future business to make the certification process financially worthwhile. Given the confusing array of certifications available, others were unsure which certification was appropriate and would be most persuasive to potential customers. As part of the decision-making process, we identified the most common certifications appearing on the websites of their competitors and potential customers. Some organizations developed their own certifications, a clear case of greenwashing since they lacked the legitimation associated with external evaluation. *Action Plan:* Investigate which certifications appear most useful in your particular industry. Consider talking with your management about seeking certification.

3.3.2.4 Architecture and Visual Rhetoric

The last type of public communication I discuss involves architecture. Spaces can function as a nonhuman agent that speaks on behalf of organizational actors

(Fairhurst and Putnam 2014) to a large number of people. With architecture, the message is transmitted over a long time period. In this section, I discuss architecture as a form of visual rhetoric used to influence perceptions of legitimacy. Visual rhetoric involves the study of visual imagery and is a newer, but flourishing, area within the discipline of rhetoric (Foss 2005; Hill 2009). Rhetorical theorist Kenneth Burke encouraged the analysis of symbols in all their forms, including architectural styles. Human experiences that are spatially oriented can only be communicated through visual imagery. Visual rhetoric is defined "as a product individuals create as they use visual symbols for the purpose of communicating" (Foss 2005, p. 143), although it also applies to an analytic perspective. Three elements must be present for a visual image to qualify as visual rhetoric: (a) it must be symbolic and use arbitrary symbols to communicate, (b) it must involve human interaction either in the process of creation or interpretation, and (c) it must be presented for the purpose of communicating with an audience. Audience members, as lay readers of the visual message, bring their own experiences and knowledge to the interpretation process. Meaning is attributed to an image. The function of a visual image is on the action it communicates.

Noting that public buildings exist as visual rhetoric, Berman (1999) introduced the idea of strategic architectural communication as a form of public relations. Architecture involves structural and symbolic elements. Buildings are often designed in light of their potential for communicating values. Their structure and aesthetics are designed to persuade (e.g., the building is sound, the organization is credible, the organization cares about sustainability). Business strategy often employs constructing buildings which function as visual persuasion (Parker and Hildebrandt 1996). Architects link top managements' desire for a building that represents their company's mission, goal, and even its power and strength to the structure they design. There is a conscious rhetorical strategy behind the design, the situation or audience to which it responds, and the ways it seeks to realize its intentions.

LEED-Certified Corporate Offices

Three of my interviews took place in LEED-certified buildings. In each case, my interviewee made sure I realized the building was certified and externally directed company messages (e.g., websites, corporate publications) identified the building as LEED certified. In addition, I visited two communities (i.e., Boardman, WA; Greensburg, KS) which used buildings as a form of strategic communication about sustainable communities to deliver a standardized message to a wide group of stakeholders.

My interview with Steve Denne, the COO of Heifer International®, occurred in Little Rock in their LEED Platinum-certified office overlooking the Arkansas River and the adjacent Clinton Presidential Library. The first LEED Platinum building in the South Central USA was built on one of the largest brownfield locations in Arkansas, a 20-acre site which had been a railroad switching yard polluted with creosote and diesel fuel. Before construction, 75,000 tons of contaminated soil were removed. The 94,000-square-foot narrow, curving, four-story building was completed for $19 million or $189 per square foot—important for an organization supported by individual donors. Since Heifer teaches sustainability-related values around the world, they wanted to demonstrate their values through the building design. Delving into the organization and its history, the building design team found a guiding metaphor in a statement attributed to Heifer founder Dan West: "In all my travels around the world, the important decisions were made where people sat in a circle, facing each other as equals" (Bond 2007). That sentiment was reflected through a set of concentric circles that create a sense of unity among the site's elements. Rippling outward from their center at a public entrance commons, the circles also illustrate the cycle of giving that Heifer calls "passing on the gift."

In Lincoln, NE, Assurity Life Insurance began recycling in the early 1990s, has a very active Green Team, and recently constructed a LEED Gold-certified home office in an urban renewal area. Assurity Life employs approximately 420 employees and provides group and individual life insurance, disability income protection, critical-illness policies, and annuity products with approximately 420,000 customer accounts nationwide. Their home office is 175,000 square foot. Like the Heifer International® building, sustainability was stressed during design and construction. It utilizes energy- and water-efficient technologies and employs external landscaping to manage storm water runoff and minimize watering. Internally, the building's open office design facilitates communication and innovation. More than 80 % of the workspaces have daylight sufficient for working; 93 % have a view of the landscape. Employee health was considered during the design phase to ensure good indoor air quality and by locating the building near public transport and the city trail system.

When the building opened in 2012, Tom Henning, Assurity Life Insurance Chairman, President, and CEO, was quoted as saying, "The Assurity Center represents our company's commitment to being a leader in corporate sustainability. Creating a high-performance, environmentally friendly facility supports our corporate goal of ethical behavior." I interviewed Bill Schmeeckle, their Vice President/ Chief Investment Officer, and asked him what prompted the company to build a LEED-certified office building. He replied:

> We were blessed by having a president and CEO who took sustainability very seriously. And we wanted to. When you go through the process of designing and constructing a building, it is quite a process. At the end of the day we wanted the facility to represent [the company] as forward thinking, forward looking sustainable thinkers. How do we want to look at the end of the day when we move into the building? That drove a lot of the conversation and a lot of the sustainable features that we put in the facility. There are things that you do from a sustainable point not because they are cost efficient but because you feel like they are the right things to do. There are some things that are frankly just too cost-inefficient that you can't do them. It's a little bit of a balancing act whenever you are trying to implement sustainability features in a building.

Later in the interview I asked Bill, "How does your organization define sustainability and has that definition changed over time? Do you have an official definition of sustainability that all employees might be aware of?" He responded:

> I don't know that we have an official sustainability mission statement. What I would say is the fact that we achieved that LEED Gold Certification was our president's way of defining sustainability for the company. That is a nationally recognized system and you could look at it and say 'those people are sustainability conscious'. So I think that seeking out that Gold LEED certification was probably the best way we went about defining sustainability.

Foss (2005) writes that three elements must be present for something to qualify as visual rhetoric. In Bill's description, we learn the home office was designed to communicate to external and internal stakeholders. The Heifer International® and Assurity Life Insurance examples reflect the link between values, architecture, and communication. LEED certification is simply an arbitrary set of letters which communicate a quality, as do a building's design elements. The planning and design processes associated with both buildings involved intentional human interaction aimed at shaping interpretations. While both buildings include eco-efficiencies, sit lightly on reclaimed land, and have health benefits, both also act as media for persuasion. The legitimacy of both organizations is increased if their key stakeholders (e.g., employees, donors, clients, collaborating institutions, communities) see them as forward-looking, forward-thinking sustainable organizations.

Architecture and Community Rhetoric

Visual rhetoric in the form of public buildings can be used intentionally to create a sense of community (Freschi 2007). During my trip, I observed three interesting examples of this form of sustainability communication. Two involved single buildings (the SAGE Center and Ecotrust) and the other involved the town of Greenburg, KS. For many, sustainability is about empowering communities to adapt and innovate in the face of environmental pressures. Oakley Brooks, Senior Media Manager of Ecotrust, explained:

> There is a fair amount of work going on in both the private sector and in environmental organizations to put place back in the discussion of how people live their lives and how they do business. There are some digital apps coming out [e.g., NextDoor, Placely]. ... So you are talking about every day minutia but you are pinning it to the map to bring out the local stories. ... It's like creating a relationship so that people on the ground tell the story in this ongoing way, in this self-organized way..., It's about trying to bring things back to a very specific geography to understand how the social and economic and ecological systems interact at one place.

That was what I witnessed at the three places I describe next.

The SAGE Center, Broadman, OR Boardman is a small town of about 3,400 located along the Columbia River. It is home to the largest inland river barge terminal in the USA. The Port of Morrow, covering more than 7,000 acres, is the region's economic development leader and promotes:

economic expansion and creation of family wage jobs through maintaining a positive business environment; developing water sources; providing and expanding utility services; expanding regional transportation hub roles and capabilities; and fully developing industrial, commercial, and community development potential while supporting the region's quality of life. (http://www.portofmorrow.com/)

The port opened the SAGE (Sustainable Agriculture and Energy) Center in June 2013 to provide visitors a glimpse into the region's history, agriculture, industries, energy production, sustainability practices, and transportation infrastructure. It was designed to be a channel for communicating informative and persuasive messages. Area farms and dairies, food-processing facilities, advanced energy projects, and various operations coming through the port are promoted. Inside, a large sign proclaims "Green Is Good for Business." Writer Terry Richard (2013) described the center's exhibits as feeling like being in an infomercial. Inside the center, a world map lights up to show where the region's products go. A simulated balloon ride allows visitors to float above the region. Visitors take turns driving a tractor simulator and learn about water-saving irrigation practices. One powerful exhibit shows local youth interviewing employees working at the various plants and farms about their jobs. Wandering through the interactive displays, you learn that the region grows ½ of Oregon's apples and 10 % of the peas grown in the USA, wastewater from the food-processing plants irrigates local farms, and low-carbon and cellulosic ethanol is produced nearby. The SAGE Center is a powerful example of visual rhetoric. It is clearly designed to be persuasive. It communicates to businesses that the port is a legitimate partner for sustainability-related enterprise.

Ecotrust, Portland, OR Ecotrust's office is in the heart of Portland, OR, in the Jean Vollum Natural Capital Center, a warehouse built in 1895. In 2001, it became the nation's first LEED Gold Historic Renovation. Visitors can pick up a field guide that encourages them to "Hike the Building!" to learn of its history and about issues related to transportation, bioswales, green building, community, energy, materials, eco-roofs, neighborhood, and FSC wood. In the field guide's community section, you read:

> The process of identifying a community of tenants to inhabit the restored building was critical to the seeding of this human ecosystem. The result is that today you'll find within these brick walls a roster of non-profit, for-profit, and public institutions that embody a powerful new vision for a sustainable society.

Oakley Brooks, Senior Media Manager, explained:

> Ecotrust thinks of itself, not just as an environmental organization but as a social enterprise or economic development organization/conservation organization. So whenever we do things we think of the social and economic impacts in addition to the environmental impacts. This building was designed to be a hub for local enterprise for environmentally responsible enterprises. We had that economic and social element in mind when it was renovated. We wanted to bring like-minded businesses and organizations here and cross pollinate with ideas and provide a place for those kinds of folks and groups to gather. . . . It is a show piece of how we weave the different social, economic, and environmental bottom lines into what we do.

In the building's lobby, there is a coffee shop for building residents and the public to gather. The interior space of the building is designed to promote community and facilitate communication. Reaching outward, the space around the building was promoted as a gathering spot drawing in people who lived in its proximity as well as those who shared the organization's values. Behind the building, a patio and urban garden provide a space which is periodically opened to the general public as part of Ecotrust's summer concert series. A concert was scheduled for the evening I met Oakley. *Best Practice:* Provide signage in LEED buildings telling your story; make sure to educate employees regarding its features; train them to be tour guides. This way your building's story can be an inspiration to others.

Greensburg, KS At 9:50 p.m. on May 4, 2007, the Greensburg community of 1,400 people experienced an EF5 tornado which destroyed the town's infrastructure and over 90 % of its homes and buildings. The twister was 1.7 miles wide as it swept through town with 205-mph winds. Having 20-minute warning from the National Weather Service, only 11 people died. Within days of the storm, area residents Daniel Wallach and Catherine Hart, along with Kansas Governor Kathleen Sebelius, began discussing how Greensburg might be rebuilt in a sustainable manner. At a gathering attended by about 500 displaced residents, Major Lonnie McCollum announced that the town would be rebuilt in an environmentally friendly manner. Sustainability was woven through the fabric of a community rebuilt as a living laboratory. Greensburg is a model green community and a town for the future. Visitors can visit 30-government, for-profit, and nonprofit organizations on the GreenTour, and numerous existing and new homes have been built to more sustainable standards. Since the vision was to power Greensburg with 100 % renewable energy, 100 % of the time, the Greensburg Wind Farm was built about three miles out of town. GreenTown was designed as an independent nonprofit business to provide education and support to residents, business owners, and municipal leaders as they sought to build back sustainably. This nonprofit reaches out to other destroyed communities (e.g., Joplin, MO). Greensburg is a symbol of how an entire town can rebuild in an environmentally sustainable way to withstand future tornados which will most certainly tear across the Kansas plains. Such a public statement conveys legitimacy to large-scale projects conceived in other towns seeking to adapt to climate change due to global warming. All who drive through Greenburg are exposed to sustainability-focused messages.

3.4 Stakeholders

For nearly two decades, the management literature has discussed the stakeholder concept to help managers think about how to manage relationships with those groups or individuals which can effect or be effected by the achievement of an organization's objectives. Organizations must strategically communicate with their stakeholders if they are to establish social legitimacy, as well as operate effectively. Since the 1990s, stakeholder theory has been applied to sustainability-related studies investigating CSR, natural resource management, public health, and sustainable development. It has been applied in traditional business settings, policy making, NGOs, and community-based or activist organizations (Dempsey 2009).

Most research discussing stakeholders refers to R. Edward Freeman's (1984) book *Strategic Management: A Stakeholder Approach*. Freeman, a professor of business administration, proposed a model of stakeholder management that focused on identifying and responding to multiple stakeholders. Subsequent researchers identified key stakeholder groups including, but not limited to, stockholders, employees, government officials, supply chain members, investors, and communities. Other important stakeholders when communicating about the environment may include citizens and community groups, environmental groups, scientists, corporations and lobbyists, anti-environmental and climate change critics, news media and environmental journalists, and public officials (Cox 2013).

Climate disruptions are expected to force organizations to expand their definitions of stakeholders to include emission trading groups and scientific expert groups, among others (Brown and Flynn 2006). Two significant environmental meta-trends (i.e., decreased freshwater access and global climate change) are threatening to disrupt organizational operations due to reduced resource supply (e.g., less clean water to manufacture Coke, inadequate water for citizens living in Denver, CO) and the potential for a displaced workforce (e.g., migration due to rising waters along the coasts). Climate change is expected to result in increased healthcare costs, increased insurance costs (both business and property), disrupted access to and higher costs for supply chain materials, and changed tax structures as the public sector strives to manage intensified weather and climate conditions while providing infrastructure and social-service support.

Is the natural environmental a stakeholder? In order for an entity to be a stakeholder, the following attributes must be present: power to influence the organization, a relationship perceived to be legitimate, and/or the urgency of the claim. "The salience of a particular stakeholder to the firm's management is low if only one of these attributes is present, moderate if two attributes are present, and high if all three attributes are present" (Driscoll and Starik 2004, p. 879). Definitive stakeholders possess all three which makes them a management priority. Driscoll and Starik make a compelling argument that we need to give stakeholder status to the natural environment since it is the source of resources. Manufacturing organizations exchange more with the natural environment than with any other stakeholder group. However, traditionally, the environment is not included in theoretical discussions as stakeholders because we tend to privilege the concept of human agency. Although the natural environment's claims are often seen as legitimate and urgent, the natural environment is dependent upon the actions of other stakeholders for protection. Many people only act to protect the environment after dominant stakeholders support it or if managerial values are pro-environmental.

Friedman and Miles' (2002) stakeholder analysis framework offers us another way to classify power and dependence within stakeholder relationships. They propose four types: opportunistic, unacknowledged, mutually beneficial, or concessionary. No clear contractual obligation exists between parties in an implicit relationship (i.e., it is opportunistic and unacknowledged). Most organizations exist in an opportunistic or unacknowledged relationship with their natural environment. Mutually beneficial relationships can exist when both parties share compatible objectives and parties call on and work in terms of each other. In such relationships, each party has something to lose if the relationship deteriorates although some have more to lose than others. This is closer to the mindset of the green organizations identified by Biloslavo and Trnavcevic (2009) or as promoted by the radical and imaginative *Discourses* identified by Dryzek (2005). Like mutually beneficial relationships, concessionary relationships are considered necessary by both parties. However, the goals among parties are incompatible; these relations "occur when material interests... are necessarily related to each other, but their operations will lead to the relationship itself being threatened" (Friedman and Miles 2002, p. 6). Compromise is a key component of concessionary relationships. In terms of the *Discourse* of practicality, compromise rarely benefits the natural environment (see Sect. 2.3.4.3)—often occurring between human stakeholders seeking to protect their own interests. In terms of our understanding of the natural environment—stakeholder power relationship, managers might consider pervasiveness (the degree the stakeholder impact is spread over distance and time) and dependence (Driscoll and Starik 2004). Viewed with those lenses, nature holds the balance of power. In terms of urgency, management needs to consider consequential, if not catastrophic, ecological problems and the likelihood of issues occurring like shortages of fuel, fish, and forest stocks. "Cognizant organizations will already factor likely-occurring environmental phenomena into their respective plans before such shortages occur" (p. 63). Both Coca-Cola and Tyson Foods are aware of their need to protect water purity and plan for water scarcity. Driscoll and Starik suggest including an additional criteria for stakeholder status into the equation: proximity. *Action Plan:* Work with your executive team to identify key stakeholders, including the natural environment, important to your organization's operations in light of the environmental challenges most likely to impact your institutional field and geographic location.

3.4.1 Adapting Messages to Stakeholders

The power relationships between stakeholders influence how messages are formed. I asked NRDC Senior Scientist Allen Hershkowitz to discuss the challenges he faces in sharing his messages regarding sustainability with outside groups like the sports industry. I would classify these relationships are opportunistic or perhaps mutually beneficial. Allen said:

> When you deal with the private sector... one of the initial challenges is to get their attention based on the factual integrity of the issue. Then you have to acknowledge their agenda and their agenda is basically increasing the revenue of their operation. So we have to be very

sensitive to cost configurations and other business aspects (e.g., supply chain reliability, product quality). There are many criteria that go into encouraging a business to shift to environmentally preferable products. Vending issues, existing sponsor and vendor relationships, a lot of complicated issues that effect the viability of any proposal to move an organization toward sustainability. We have to do that in the context of competitive business and what the generation needs and existing vendor relationships and try to encourage engagement and visibility and branding protections. There are a lot of challenges in advancing our issue. ... So getting each organization to understand the risks that exist, to their public health, to their image, and doing it in a way that is not threatening and that is collegial. All of these things present challenges to communicating about environmental stewardship.

Prior to Cheney's (1991) seminal book on organizational rhetoric, most research into external corporate discourse failed to account for the complexity of communication whereby multiple official and unofficial speakers send multiple orchestrated and unorchestrated persuasive messages to multiple audiences, nor did it account for how institutional norms operating within an organizational field constrain a corporate actor's voice. Building on Cheney's work and on institutional theory, Allen and Caillouet (1994) investigated how one embattled reuse–recycle organization utilized different impression management strategies when speaking with different stakeholders. Today researchers investigating image building, corporate issue management, corporate advocacy, public relations, and crisis communication acknowledge the need to strategically adapt messages to various stakeholders. For example, Peloza et al. (2012, p. 86) reviewed research suggesting "employees are more likely than customers to require justification of sustainability initiatives, and be more positively influenced by a higher degree of fit between the initiative and the core business of the firm."

One theory provides a useful framework for thinking about how to tailor messages which allow for coordinated action. Von Kutzschenbach and Bronn (2006) utilized the co-orientation model of communication in order to provide guidance for improving forest owners' communication with their end consumers. They write, "Sustainability communications require a systematic approach in which all the communication activities are directed toward achieving increased understanding between the organization and its relevant stakeholders" (p. 304). The co-orientation model began as an attempt to explain dyadic communication investigating an individual's perception on an issue and his or her perceptions of what significant others thought about the issue. It expanded into the organizational arena in the form of organizational co-orientation theory (see Taylor 2009). Three basic levels of communication can occur between parties: congruency, agreement, and accuracy. Communicators can be in true consensus (i.e., both agree and know how each other feels), dissensus (i.e., both disagree but know how each other feels), false conflict (i.e., they think they disagree but they do not), or false consensus (i.e., they think they agree but they do not). Effective communication with stakeholders requires developing internal and external strategies to improve short-term accuracy and to increase long-term levels of agreement. Transparency, information exchange, and credibility are important. Creating shared definitions and increased accuracy improves the relationship and can influence party willingness to engage in dialogue. Feedback can be used to integrate additional relevant stakeholders into the discussion. Communicating, negotiating and contracting, relationship management, and motivating stakeholders are all important strategies. This theory is useful in suggesting the need to be systematic in how organizations think about their stakeholder-directed communication.

Lessons from the Field: South Dakota and Colorado Those seeking to communicate about sustainability must tailor their messages to address different understandings and should be prepared to explain the various dimensions of the concept so that those with different understandings can relate to it (Linnenluecke et al. 2009). Although he generally stresses efficiencies and economics when talking with stakeholders, Mike Mueller, Sustainability Coordinator for the South Dakota Bureau of Administration, is aware of the need to adapt his message based on the audience. He told me:

> Our current administration is very focused on efficiencies, so whenever we can make strides, it is always warmly received when we can say we are saving energy, we are saving money. . . . It is, of course, important to understand the audience you are trying to communicate with. I think that the effort of sustainability is broad enough that you can tailor a message no matter how resistant the person may be to certain parts of sustainability. If you know that they are interested in a certain result, then that's the way to message. I have also found that data specific always reaches a broader audience. It is easy to look at someone and say, 'The reason that we don't have irrigation of that grass anymore, this is prairie grass, is because we are using six hundred thousand gallons less water on that lawn.' It looks just as green, although it is green for a more compressed time period. When you start talking numbers to people, they get it, tell them how much we spent on irrigation on this campus to keep it green, it is an attention grabber.

Message modification also emerged in another of my interviews. Founded in 1876, the University of Colorado, Boulder, was the first US university to rank gold through the AASHE STARS program, the first to establish a recycling program, and the home of a student-powered Environmental Center since 1970. It has been rated as the greenest campus in the USA. Moe Tabrizi, former Assistant Director of Engineering and Campus Sustainability Director at the University of Colorado, Boulder, generally sought to send out a broad but consistent message when he spoke. He told me:

> We create a messaging that appeals to the broad range of students, staff and faculty. Part of the message has to do with the impact on the environment and global warming, part has to do with the financial impact (i.e., energy, resource and paper waste), [part involves meeting campus-wide goals], and part has to do with operational efficiency. When you throw a big net, it is easy to bring a lot of people under the tent. So I think that has been our messaging strategy. And it's fairly effective. It has appeal to the staff, appeal to the students, and appeal to the faculty.

Moe talked about messaging when preparing to design a LEED-certified building saying:

> The argument is the cost of ownership over time is less. It is a better, healthier building to be in. It is a more productive place to work in terms of lighting and views, having a cleaner environment in terms of indoor air quality. Also, it is in support of campus carbon neutrality

and our long term goal of energy conservation. So when you put that kind of discussion in front of people, generally people are very supportive. But if you just present that from a narrow perspective you may not have as much support in the broad audience.

When talking to design and construction folks, Moe said his usual example involves the 500,000-square-feet campus Engineering Center built in the mid-1960s for about $15 million:

> So I ask the folks, 'What if we had spent one percent or two percent of the construction costs at that time and improved the building's efficiency—energy, water, insulation? What do you think that positive impact would be?' People can't answer that question. Then I tell them that we have owned that building for 50 years and it is going to continue to be a part of the campus for 10, 15, 20 years, and the cost of utilities for that building is roughly $2 million a year. That gets people's attention. . ... You have to have a meaningful example, a meaningful conversation based on your audience, based on who is who in the zoo.

3.4.2 Communication and Stakeholder Engagement

In light of global warming's threat to humanity (Frandsen and Johansen 2011) and their desire to mitigate risks and capitalize on opportunities, businesses, policy makers, and the public will need to engage in increased collaboration (see Sect. 7.2). Stakeholder engagement refers to those practices that an organization takes to involve stakeholders in a positive manner in organizational activities (Greenwood 2007). Stakeholder engagement practices may involve public relations, customer service, supplier relations, management accounting, and human resource management. Public forums, NGO partnerships, community newsletters, and media relationships are important communication avenues. Communication can be used to gain consent, achieve control, enlist cooperation, enact accountability, increase participation, enhance trust, or serve as a mechanism of corporate governance. Although communication is a morally neutral practice, moral dimensions may emerge depending on power dynamics within the stakeholder–organization relationship and a situation's urgency.

Who communicates about sustainability when seeking to engage stakeholders? Although top management establishes the sustainability-related vision for their organization, an organization's sustainability personnel are tasked with disseminating the message. Ultimately, multiple employees will discuss sustainability initiatives with those stakeholders they interact with. But that chorus comes later after sustainability permeates its way into an organization's culture, processes, and tasks. Often, few staff are completely devoted to communicating about sustainability initiatives and their background in strategic communication is limited.

Communication Assistance in Portland, OR, and Boulder, CO Increasingly, city sustainability communicators are members of the Urban Sustainability Director's Network or the International Society of Sustainability Professionals. When it comes to messaging, they may need the assistance of skilled communicators. Susan Anderson, Director of Portland's Bureau of Planning and Sustainability, described how she hired someone from a marketing firm "and this is not something big cities usually do. The focus on understanding messaging, understanding communication is huge in this group. So we have spent a good bit of time finding out what works and what does not work in terms of changing behavior." David Driskell, Boulder's Executive Director of Community Planning and Sustainability, works with marketing firms when communicating with citizens about sustainability:

> It's totally ramped up our game as a local government. Obviously the private sector is doing much more sophisticated things than we do. But I think it's really changed our expectations on what we do to communicate with our public. We are using a lot more online tools and social media. It has been pretty exciting. I have just come to appreciate our communications team. We would not be here without them.

Public relations and communication professionals can make good corporate sustainability communicators skilled at facilitating stakeholder engagement (Signitzer and Prexl 2008), once they understand about sustainability. They know about differentiated target group analysis and segmentation, have experience building relationships and engaging in boundary spanning, have education and experience which allows them to better communicate complex issues in a differentiated way to various stakeholders, have expertise communicating with internal stakeholders, have a consensus-orientation reaction to conflicts, and have experience in making a public case. But before communication personnel can effectively discuss sustainability-related issues, they need to understand them (Bortree 2011). This will influence their attitude toward environmental issues and may improve their motivation to develop campaigns and other communication around environmental issues. Bortree gathered data from 320 members of the Public Relations Society of America. The public relations practitioners rated their own knowledge of environmental issues as above the midpoint. They said they knew the most about recycling, waste reduction, and energy efficiency and the least about offsetting energy usage and green packaging. They learned about environmental issues through reading and watching the news, visiting websites, and talking with colleagues. In response to an open-ended question, they identified turning to other information sources including blogs, social media sites, local nonprofit alliances, meetings, magazines, newsletters, and other sources of employee communication. Organizations hoping to engage stakeholders around sustainability initiatives should use the team approach and pair up sustainability and communication professionals. Or they could seek someone who is trained in both areas as universities increasingly develop courses in sustainability communication.

3.5 Concluding Thoughts

Narrative theory provides us with insight into how communication channels such as architecture, sustainability reports, websites, and public meetings can be persuasive by means other than rational, scientific appeals (Browning 2009). Rhetorical scholar Walter Fisher (1987) discussed the homo narrans metaphor arguing that people are essentially storytellers. Corporate communicators find storytelling to be a useful way to think about the messages they create for external and internal stakeholders as they seek legitimacy (Norlyk et al. 2013). We use stories to provide good reasons to guide our decision making. What counts as a good story is influenced by factors such as history and culture. Listeners judge whether or not stories are coherent (narrative probability) and ring true (narrative fidelity).

Authors have sought to identify basic archetypal themes appearing in stories. For example, Booker (2006) identified seven themes—three of which are especially useful when thinking about organizational actions related to sustainability and/or in the face of climate change: overcoming the monster, rebirth, and quest. The first is the classic underdog story; the second is a story of renewal; and the third tells of a mission from point A to point B. Sustainability stories can fit into all three themes. Oakley Brooks, Senior Media Manager of Ecotrust, said:

> A lot of organizations are putting more and more effort into story-telling. . .. There is an opportunity to connect stories . . . to the work that makes change. You know, the change work organizations typically do and by doing that . . . connecting to the public in new ways. The public is searching for big, hopeful, narratives now and that is what we are trying to provide . . .. Once you are done with the story, what happens next? How do you connect all the people that really got into that story to do the work?

Organizational stories are powerful and memorable. They stimulate our senses and arouse our emotions, and because they tap into well-established story types, they activate schemas in our brains (Browning 2009). *Action Plan:* Look at some sustainability-related organizational documents and identify the stories they are telling. Listen to the sustainability-related stories circulating in your workplace. How do you judge their coherence and truthfulness? Are they inspiring you to think/ act in a more sustainable way? How might you use stories differently as a form of strategic communication?

References

Allen, M. W., & Caillouet, R. H. (1994). Legitimation endeavors: Impression management strategies used by an organization in crisis. *Communication Monographs, 61*, 44–62.

Axelrod, R. A. (2000). Brave new words: The financial value of environmental communications. *Environmental Quality Management, 9*, 1–10.

Berman, S. (1999). Public buildings as public relations: Ideas about the theory & practice of strategic architectural communication. *Public Relations Quarterly, 44*, 18–22.

Biloslavo, R., & Trnavcevic, A. (2009). Web sites as tools of communication of a 'green' company. *Management Decision, 47*, 1158–1173.

Blackburn, W. R. (2007). *The sustainability handbook: The complete management guide to achieving social, economic and environmental responsibility*. London: Earthscan.

Bond, P. (2007). *Case study: Circle of life*. GreenSource Magazine. http://construction.com/GreenSource/projects/case_studies/2007/0701_COL/0701_mag_COL.asp. Accessed 30 Jan 2013.

Booker, C. (2006). *Seven basic plots: Why we tell stories*. London: Bloomsbury Academic.

Bortree, D. S. (2011). The state of environmental communication: A survey of PRSA members. *Public Relations Journal, 4*, 1–17.

Bowers, T. (2010). From image to economic value: A genre analysis of sustainability reporting. *Corporate Communications: An International Journal, 15*, 249–262.

Brown, B., & Flynn, M. (2006). The meta-trend stakeholder profile: The changing profile of stakeholders in a climate- and water-stressed world. *Greener Management International, 54*, 37–43.

Browning, L. (2009). Narrative and narratology. In S. W. Littlejohn & K. A. Foss (Eds.), *Encyclopedia of communication theory* (Vol. 2, pp. 673–677). Los Angeles, CA: Sage.

Bullis, C., & Ie, F. (2007). Corporate environmentalism. In S. May, G. Cheney, & I. Roper (Eds.), *The debate over corporate social responsibility* (pp. 321–335). New York: Oxford University.

Campbell, K. S., Follender, S. I., & Shane, G. (1998). Preferred strategies for responding to hostile questions in environmental public meetings. *Management Communication Quarterly, 11*, 401–421.

Carrington, T., & Johed, G. (2007). The construction of top management as a good steward: A study of Swedish annual general meetings. *Accounting, Auditing & Accountability Journal, 20*, 702–728.

Caster, T., & Cooren, F. (2006). Organizations as hybrid forms of life: The implications of the selection of agency in problem formulation. *Management Communication Quarterly, 19*, 570–600.

Chaudhri, V., & Wang, J. (2007). Communicating corporate social responsibility on the Internet: A case study of the top 100 information technology companies in India. *Management Communication Quarterly, 21*, 232–247.

Cheney, G. (1991). *Rhetoric in an organizational society: Managing multiple identities.* Columbia: University of South Carolina Press.

Cheney, G., & Christensen, L. T. (2001). Organizational identity: Linkages between internal and external communication. In F. M. Jablin & L. L. Putnam (Eds.), *The new handbook of organizational communication* (pp. 231–269). Thousand Oaks, CA: Sage.

Cooren, F. (2009). Actor-network theory. In S. W. Littlejohn & K. A. Foss (Eds.), *Encyclopedia of communication theory* (Vol. 1, pp. 16–18). Los Angeles, CA: Sage.

Cooren, F., Thompson, F., Canestraro, D., & Bodor, T. (2006). From agency to structure: Analysis of an episode in a facilitation process. *Human Relations, 59*, 533–565.

Cox, R. (2013). *Environmental communication and the public sphere* (3rd ed.). Washington, DC: Sage.

Dart, R., & Hill, S. D. (2010). Green matters? An exploration of environmental performance in the nonprofit sector. *Nonprofit Management & Leadership, 20*, 295–314.

Delmas, M. A., & Pekovic, S. (2013). Environmental standards and labor productivity: Understanding the mechanisms that sustain sustainability. *Journal of Organizational Behavior, 34*, 230–252.

Dempsey, S. E. (2009). Stakeholder theory. In S. W. Littlejohn & K. A. Foss (Eds.), *Encyclopedia of communication theory* (Vol. 2, pp. 929–930). Los Angeles, CA: Sage.

Domenec, F. (2012). The "greening" of the annual letters published by Exxon, Chevron and BP between 2003 and 2009. *Journal of Communication Management, 16*, 296–311.

Dowling, J., & Pfeffer, J. (1975). Organizational legitimacy: Social values and organizational behavior. *Pacific Sociological Review, 18*, 122–134.

Driscoll, C., & Starik, M. (2004). The primordial stakeholder: Advancing the conceptual consideration of stakeholder status for the natural environment. *Journal of Business Ethics, 49*, 55–73.

Dryzek, J. S. (2005). *The politics of the earth: Environmental discourses* (2nd ed.). Oxford: Oxford University Press.

Elkington, J. (1999). *Cannibals with forks: The triple bottom line of 21st century business.* Gabriola Island, BC: New Society.

Emtairah, T., & Mont, O. (2008). Gaining legitimacy in contemporary world: Environmental and social activities of organisations. *International Journal of Sustainable Society, 1*, 134–148.

Fairhurst, G. T., & Putnam, L. L. (2014). Organizational discourse analysis. In L. L. Putnam & D. K. Mumby (Eds.), *The SAGE handbook of organizational communication* (pp. 271–296). Thousand Oaks, CA: Sage.

Ferns, B., Emelianova, O., & Sethi, S. P. (2008). In his own words: The effectiveness of CEO as spokesperson on CSR-sustainability issues: Analysis of data from the Sethi CSR Monitor. *Corporate Reputation Review, 11*, 116–129.

Fisher, W. R. (1987). *Human communication as narration: Toward a philosophy of reason, value, and action*. Columbia: University of South Carolina Press.

Foss, S. K. (2005). Theory of visual rhetoric. In K. Smith, S. Moriarty, G. Barbatsis, & K. Kenney (Eds.), *Handbook of visual communication: Theory, methods, and media* (pp. 141–152). Mahwah, NJ: Lawrence Erlbaum.

Frandsen, F., & Johansen, W. (2011). Rhetoric, climate change, and corporate identity management. *Management Communication Quarterly, 25*, 511–530.

Freeman, R. E. (1984). *Strategic management: A stakeholder approach*. New York: HarperCollins.

Freschi, F. (2007). Postapartheid publics and the politics of ornament: Nationalism, identity, and the rhetoric of community in the decorative program of the New Constitutional Court, Johannesburg. *Africa Today, 4*(52), 27–49.

Friedman, A. L., & Miles, S. (2002). Developing stakeholder theory. *Journal of Management Studies, 39*, 1–21.

Gatti, M. (2011). The language of competence in corporate histories for company websites. *Journal of Business Communication, 48*, 482–502.

Gill, D. L., Dickinson, S. J., & Scharl, A. (2008). Communicating sustainability: A web content analysis of North American, Asian and European firms. *Journal of Communication Management, 12*, 243–262.

Global Reporting Initiative. (2013). https://www.globalreporting.org/. Accessed 20 Dec 2013.

Greenwood, M. (2007). Stakeholder engagement: Beyond the myth of corporate responsibility. *Journal of Business Ethics, 74*, 315–327.

Guthrie, J., & Farneti, F. (2008). GRI sustainability reporting by Australian public sector organizations. *Public Money & Management, 28*, 361–366.

Hannan, M. T., & Freeman, J. (1977). The population ecology of organizations. *American Journal of Sociology, 82*, 929–964.

Henly, A., Hershkowitz, A., & Hoover, D. (2012). *Game changer: How the sports industry is saving the environment*. NRDC Report R:12-08-A. http://www.nrdc.org/greenbusiness/guides/sports/game-changer.asp. Accessed 20 Dec 2013.

Herzig, C., & Godemann, J. (2010). Internet-supported sustainability reporting: Developments in Germany. *Management Research Review, 33*, 1064–1082.

Hill, C. A. (2009). Visual communication theories. In S. W. Littlejohn & K. A. Foss (Eds.), *Encyclopedia of communication theory* (Vol. 2, pp. 1002–1005). Los Angeles, CA: Sage.

Hunter, T., & Bansal, P. (2007). How standard is standardized MNC global environmental communication. *Journal of Business Ethics, 71*, 135–147.

Ihlen, O. (2009). Business and climate change: The climate response of the world's 30 largest corporations. *Environmental Communication, 3*, 244–262.

Lammers, J. C. (2009). Institutional theories of organizational communication. In S. W. Littlejohn & K. A. Foss (Eds.), *Encyclopedia of communication theory* (Vol. 2, pp. 520–523). Los Angeles, CA: Sage.

Lammers, J. C., & Garcia, M. A. (2014). Institutional theory. In L. L. Putnam & D. K. Mumby (Eds.), *The SAGE handbook of organizational communication* (3rd ed., pp. 195–216). Thousand Oaks, CA: Sage.

Linnenluecke, M. K., Russell, S. V., & Griffiths, A. (2009). Subcultures and sustainability practices: The impact of understanding corporate sustainability. *Business Strategy and the Environment, 18*, 432–452.

Littlejohn, S. W. (2009). Speech act theory. In S. W. Littlejohn & K. A. Foss (Eds.), *Encyclopedia of communication theory* (Vol. 2, pp. 918–920). Los Angeles, CA: Sage.

Livesey, S. M. (2002). The discourse of the middle ground: Citizen Shell commits to sustainable development. *Management Communication Quarterly, 15*, 313–349.

Maharaj, R., & Herremans, I. M. (2008). Shell Canada: Over a decade of sustainable development reporting experience. *Corporate Governance, 8*, 235–247.

Makower, J. (2009). *Inside Walmart's sustainability consortium.* http://www.greenbiz.com/blog/2009/08/17/inside-walmarts-sustainability-consortium. Accessed 26 Nov 2013.

McDermott, V. M. (2009). Attribution theory. In S. W. Littlejohn & K. A. Foss (Eds.), *Encyclopedia of communication theory* (Vol. 1, pp. 60–63). Los Angeles, CA: Sage.

Meyer, J. W., & Rowan, B. (1977). Institutional organizations: Formal structure as myth and ceremony. *American Journal of Sociology, 83*, 340–63.

Mitchell, M., Curtis, A., & Davidson, P. (2012). Can triple bottom line reporting become a cy-cle for "double loop" learning and radical change? *Accounting, Auditing & Accountability Journal, 25*, 1048–1068.

Norlyk, B., Lundholt, M. W., & Hansen, P. K. (2013). *Corporate storytelling.* http://www.lhn.uni-hamburg.de/article/corporate-storytelling. Accessed 3 June 2014.

Parguel, B., Benoit-Moreau, F., & Larceneux, F. (2011). How sustainability ratings might deter 'greenwashing': A closer look at ethical corporate communication. *Journal of Business Ethics, 102*, 15–28.

Parker, R. D., & Hildebrandt, H. W. (1996). Business communication and architecture: Is there a parallel? *Management Communication Quarterly, 10*, 227–242.

Peloza, J., Loock, M., Cerruti, J., & Muyot, M. (2012). Sustainability: How stakeholder perceptions differ from corporate reality. *California Management Review, 55*, 74–97.

Pfeffer, J., & Salancik, G. R. (1978). *The external control of organizations: A resource dependence perspective.* New York: Harper and Row.

Richard, T. (2013). *Sage Center at Boardman soars with hot air balloon 'ride' over Morrow County.* http://www.oregonlive.com/travel/index.ssf/2013/10/sage_center_at_boardman_soars.html. Accessed 27 Nov 2013.

Salimath, M. S., & Jones, R., III. (2011). Population ecology theory: Implications for sustainability. *Management Decision, 49*, 874–910.

Sethi, S. P. (1979). A conceptual framework for environmental analysis of social issues and evaluation of business response patterns. *Academy of Management Review, 4*, 63–74.

Signitzer, B., & Prexl, A. (2008). Corporate sustainability communications: Aspects of theory and professionalization. *Journal of Public Relations Research, 20*, 1–19.

Stiller, Y., & Daub, C.-H. (2007). Paving the way for sustainability communication: Evidence from a Swiss Study. *Business Strategy and the Environment, 16*, 474–486.

Suchman, M. C. (1995). Managing legitimacy: Strategic and institutional approaches. *Academy of Management Review, 20*(3), 571–610.

Taylor, J. (2009). Organizational co-orientation theory. In S. W. Littlejohn & K. A. Foss (Eds.), *Encyclopedia of communication theory* (Vol. 2, pp. 709–713). Los Angeles, CA: Sage.

The Sustainability Consortium. (2013). http://www.sustainabilityconsortium.org/. Accessed 26 Nov 2013.

von Kutzschenbach, M., & Bronn, C. (2006). Communicating sustainable development initiatives: Applying co-orientation to forest management certification. *Journal of Communication Management, 10*, 304–322.

Walker, K., & Wan, F. (2012). The harm of symbolic actions and greenwashing: Corporate actions and communications on environmental performance and their financial implications. *Journal of Business Ethics, 109*, 227–242.

Walls, J. L., & Hoffman, A. J. (2013). Exceptional boards: Environmental experience and positive deviance from institutional norms. *Journal of Organizational Behavior, 34*, 253–271.

Zhang, Y., & O'Halloran, K. L. (2013). 'Toward a global knowledge enterprise': University websites as portals to the ongoing marketization of higher education. *Critical Discourse Studies, 10*(4), 468–485.

Zoller, H. M., & Tener, M. (2010). Corporate proactivity as a discursive fiction: Managing environmental health activism and regulation. *Management Communication Quarterly, 24*, 391–418.

Chapter 4
Understanding Pro-Environmental Behavior: Models and Messages

Abstract Certainly societal *Discourses* shape our individual worldviews, and our behaviors are influenced by our organizations' quest for legitimacy during stakeholder interactions. However, when individuals communicate about sustainability-related initiatives, part of that exchange is influenced by each person's environmental values, attitudes, and beliefs. What influences an individual's pro-environmental values and behaviors? How can communication facilitate individual-level behavioral change? This chapter identifies factors influencing an individual's pro-environmental values and behaviors; discusses the tentative link between values, attitudes, and behaviors; and identifies how communication can be used to influence individual-level behavioral change. Literature is reviewed which identifies and discusses pro-environmental values and beliefs and defines pro-environmental behaviors. Various persuasion and social influence theories are reviewed to help practitioners better understand how to stimulate pro-environmental behaviors. Key models of pro-environmental behavior are identified. Information is drawn from social marketing, health-related models, stages of change models, energy use reduction models, and communication campaign literatures. At the end of each block of theories, ways these theories can guide practice are highlighted. The chapter ends by focusing on concrete message strategies and the importance of interpersonal communication. Interview data is drawn from Sam's Club; Bayern Brewing; the South Dakota Bureau of Administration; the State Farm Insurance processing facility in Lincoln, NE; the Neil Kelly Company; the City of Fayetteville; the University of Colorado, Boulder; and the University of Colorado, Denver.

Lewis and Clark as Individuals Meriwether Lewis was a boyhood neighbor of Thomas Jefferson. When Jefferson was elected president, he offered Lewis a position as his secretary aide. He wrote, "Your knolege of the Western country, of the army, and of its interests and relations has rendered it desireable for public as well as private purposes that you should be engaged in that office." Although Lewis was introverted and moody, he was also philosophical, had a speculative mind, and dealt well with abstract ideas. Lewis offered the role of co-commander to his friend and former commanding officer on the Northwest Campaign, William Clark. Clark was an excellent cartographer and a practical man of action. He was extroverted,

© Springer International Publishing Switzerland 2016
M. Allen, *Strategic Communication for Sustainable Organizations*, CSR, Sustainability, Ethics & Governance, DOI 10.1007/978-3-319-18005-2_4

even-tempered, and gregarious, qualities which helped him manage the other Expedition members. Both men were relatively young, experienced woodsmen, seasoned army officers, cool in a crisis, and quick to make decisions. What motivated these two individuals to take on this uncertain journey? Perhaps it was because each was asked, desired adventure, or sought the promised rewards. After the Expedition returned, Lewis became Governor of the Louisiana Territory until he died 3 years later at 35. Clark was promoted to brigadier general and appointed to the Superintendency of Indian Affairs. After over 210 years, we still talk about these two *individuals* who embarked on their perilous journey West.

4.1 Individual Values Regarding Sustainability

Sustainable development can be represented as a statement of values or moral principles unique to each individual which influence how that individual views the world. Byrch et al. (2007) explored the meaning of sustainable development held by 21 New Zealand thought leaders who were promoting either sustainability, business, or sustainable business. Those who promoted sustainable development emphasized limits to the Earth's resources, humanity's dependency on the environment, providing now for future generations, and the environmental domain. These ideas are similar to the NEP (see Sect. 2.2.1). Individuals who promoted business generally emphasized the economic domain, focused on how things should be, stressed that a healthy growing economy precedes environmental and social improvement, and expressed faith that economic growth could achieve sustainable development. Those who promoted sustainable business showed a combined emphasis on the environmental domain and the achievement of sustainable development, a focus on problems and solutions, and a mix of less radical solutions to achieve sustainable development. Common across the groups was the belief that humanity should benefit from sustainable development and the lack of an expressed concern for social equity and the world's poor.

Byrch et al. (2007) based their work on the cognitive framework or theory of knowledge described by Kenneth Boulding who argues that a person's behavior is largely governed by his or her subjective knowledge structure or image. An image is similar to schema, a more common term in the communication, psychology, and cognitive science literatures. The schema concept first appeared in Frederic Bartlett's learning theory and was later popularized by Jean Piaget in his theory of the intellectual development stages children go through. Schemata are mental structures we use to organize categories of information, create relationships among the categories, and process incoming information. Messages are received as information and filtered through a person's schemata. What an individual knows about sustainable development and what he or she says about it is the product of his or her culture and personal experiences filtered through a personal worldview which includes values, attitudes, and beliefs and is held in his or her schemata. So let's look more closely into values, attitudes, and beliefs.

4.1.1   Differentiating Values, Attitudes, and Beliefs

Multiple definitions of values exist in the broader literature. Dietz et al. (2005) provide a definition of values which involves our sense of what something is worth, our opinions about that worth, and the moral principles and standards relevant to our social group. Values are concepts or beliefs about desirable end states or behaviors that go beyond specific situations to influence how we behave and evaluate behavior. Values are different from attitudes and beliefs. Attitudes are positive or negative evaluations of something very specific. The most commonly used measure of pro-environmental attitudes has been the NEP scale (Bissing-Olson et al. 2013). Beliefs are our understandings about the state of the world or the facts as we see them. For example, I may think that it is very important to protect forests (value), but strongly oppose purchasing forest credits as a way to offset my own organization's CO_2 emissions (attitude), due to my skepticism regarding whether or not climate change is really occurring (belief).

A range of environmental values exist. A central topic in environmental ethics is whether the environment has intrinsic value or simply has value because it serves as a means to human ends (e.g., financial benefits, resource availability, or instrumental utility). Is the environment a stakeholder (see Sect. 3.4)? In economics, market prices and contingent valuation are ways to place value on the environment (i.e., ask people what they would pay to leave a forest ecosystem intact). Researchers have suggested at least three value bases for environmental concern: self-interest (i.e., it influences me and those I care about), humanistic or social altruism (i.e., it influences my community; it influences humanity), and biospheric altruism or biocentrism (i.e., it influences other species or ecosystems). Only biospheric altruism acknowledges the intrinsic value of the environment and has been shown to contribute significantly to the formation of environmental beliefs (Stern et al. 1995). Situational cues and the context can influence the relative importance of each form of altruism at any given time, although basically an individual's value set is relatively stable across time barring a significant change in circumstances (Dietz et al. 2005).

4.1.2   Research into Environmental Values

For almost 50 years, researchers representing a range of disciplines including ethics, social psychology, economics, political science, and sociology have discussed factors they believe influence environmental values (e.g., religious values, altruistic values, traditionalism, or openness to change) (Dietz et al. 2005). Gender (women generally are more pro-environmental), education and/or income (more resources allow individuals to express their pro-environmental values), and religious orientation have been linked with pro-environmental values. Other research links environmental values to early childhood experiences. For example, Chawla (1999) investigated how early experiences influence an individual's predisposition to learn about the environment, feel concern for it, and act to conserve it. She found the most frequently mentioned influences were childhood experiences in nature, experiences of pro-environmental destruction, pro-environmental values held by close family members, membership in pro-environmental organizations, role models (friends or teachers), and education. Allen et al.'s (2013) research reinforced the role of family on young people's pro-environmental attitudes. They found a parent's environmental self-efficacy was a strong predictor of the child's environmental self-efficacy as measured by the statement "My actions can influence the quality of the environment." A parent's environmental prioritization also significantly predicted the child's environmental prioritization. Prioritization was measured by the statement, "Global warming should be a high priority for our next president." A strong relationship emerged between self-efficacy and prioritization for both the parents and the children. "Self-efficacy refers to beliefs in one's capabilities to mobilize the motivation, cognitive resources, and courses of action needed to meet given situational demands" (Wood and Bandura 1989, p. 408).

Unfortunately research investigating environmental values is generally either based on self-reported behaviors (e.g., do you commonly recycle?), behavioral intentions (e.g., would you recycle if bins were provided?), or attitudinal measures (e.g., I believe climate change is a major threat). Dietz et al. (2005) provide a nice review of past research and emerging trends when investigating environmental values. Until recently, few studies investigated the relationship between values and actual behavior. Behavioral intentions (verbal commitment and environmental behavioral intention) are the most studied proximal antecedents of pro-environmental behavior. When you read studies testing various models and theories, notice that often researchers only measure the process through behavioral intention. It can be extremely difficult to gather actual behavioral data.

4.2   Stimulating Pro-Environmental Behaviors: Persuasion and Social Influence Theories

More current research explores the link between values and pro-environmental behaviors than seeks to identify the antecedents of environmental values. Pro-environmental behavior is behavior that consciously seeks to minimize the negative impact of one's actions on the natural and built world (e.g., minimize resource and energy consumption, reduce waste production) (Kollmuss and Agyeman 2010). Citizens engage in four distinct types of environmentally significant behavior (Stern 2000): environmental activism, nonactivist behaviors, private-sphere environmentalism, and other environmentally significant behaviors. Environmental activism often involves participation in social movements. Nonactivist behaviors in the public sphere can include active environmental citizenship (e.g., contributing to environmental organizations) as well as support for or acceptance of public policies (e.g., approval of environmental regulations). The outcome of public policy change can be significant since policies change the behaviors of many people and organizations simultaneously. Private-sphere environmentalism includes the purchase, use, and disposal of personal and household products having an environmental impact. The environmental impact of such behaviors is significant only in the aggregate. In terms of other environmentally significant behaviors, individuals can influence the actions of the organizations to which they belong. Such behaviors "can have great environmental impact because organizational actions are the largest direct source of many environmental problems" (p. 410).

Although Stern's typology was focused on citizens, some organizations seek to mobilize their customers (e.g., Aspen Ski Company), others in their business community (e.g., Kelly Company), and members of their industry (e.g., Portland Trail Blazers) to become active. In terms of public policy, we read about how more than 500 businesses signed a Climate Declaration in 2013 urging US policy makers to capture economic opportunities associated with addressing climate change. Finally, the company that cleans Assurity Life's LEED-certified company headquarters can only use green organic, cleaning solutions, and all paper products (e.g., paper towels, tissue) are Green Seal certified, meaning that they are 100 % recycled and have no chlorine, bleach, harmful chemicals, or pigment dye. So the literature focusing on citizen's behavior has implications for our discussion of sustainability initiatives within and between organizations.

4.2.1 Persuasion Theories and Research

Although many people in the USA report they hold pro-environmental values, many others do not. As a result, researchers have focused on identifying how to stimulate pro-environmental behaviors. Persuasion plays a role in creating, reinforcing, or modifying beliefs, attitudes, or behaviors. It is the underlying motivation for much of our communication. We develop our identity and what we perceive to be important values through our communication with others whose views we respect. Also, communities develop shared values. In this section, I review several of the major persuasion and social influence theories used to explain a range of behaviors (e.g., purchasing decisions, pro-environmental behaviors). Later, models specific to influencing pro-environmental behaviors in the general public are provided.

Early persuasion research included a model called the message-learning approach which investigated four sets of factors likely to influence persuasion: the source, the message content, the channel, and the receiver (Seiter 2009). A CEO's credibility influences how stakeholders perceive his or her company's sustainability efforts. Researchers have found speaker credibility, likeability, attractiveness, and similarity to the audience influence persuasion. In terms of the message content, more persuasive messages present all sides of the argument while refuting opposing sides, present stronger arguments either first or last, and use only moderate fear appeals. Inoculation theory identified how to use messages to block behavioral change and/or refute opposing arguments. The message-learning model proposed persuasion occurred in a series of steps, with each step being less likely to occur. First, messages must gain attention, be understood, be accepted, be retained, and then be acted on. Incentives at each stage enhance an appeal's persuasiveness. Soon after developing the message-learning model, Hovland worked with Sherif to develop social judgment theory. You read about social judgment theory in the earlier discussion of how sustainability is a contested term.

Another set of persuasion theories focused on consistency (e.g., cognitive dissonance theory, attribution theory, the self-perception theory of attitude change, balance theory, congruity theory) (Seiter 2009). People want to be consistent. When our attitudes and behaviors are not aligned, we experience psychological stress. We seek to reduce the stress using a variety of techniques including changing our attitude, changing our behavior, or engaging in rationalization. Cognitive dissonance theory, developed by Leon Festinger, predominated persuasion research through the 1970s. It was the theoretical base for research investigating water conservation (e.g., Dickerson et al. 2006) and environmentally responsible behavior (e.g., Thogersen 2004). Earlier you read about attribution theory in relationship to Parguel et al.'s (2011) study of the influence of independent sustainability ratings on French consumers' responses to companies' CSR communication. Like Festinger, Daryl Bem challenged the assumption that attitudes cause behavior in his self-perception theory of attitude change. He investigated how people attribute internal causes to their own behavior in the absence of external causes. This theory helps us understand the effectiveness of the foot-in-the-door tactic where a persuader makes a small request which, if agreed to, often leads to compliance with a larger request. Compliance with the small request allows people to see themselves as helpful, agreeable, or altruistic. Often in campaigns designed to influence pro-environmental behavior, someone will be asked to sign a pledge in hopes that small action will open the person up to making a larger behavioral change when asked. Research suggests that if a person engages in a new behavior, he or she will come to believe that the new behavior is consistent with his or her attitudes and values.

Dual-process theories of persuasion (e.g., the elaboration likelihood model, the heuristic-systematic model) were developed to better understand how people process persuasive messages. You read about the elaboration likelihood model in Peloza et al.'s (2012) investigation into how stakeholders decode and interpret sustainability messages. Chaiken's heuristic-systematic model is similar but talks about how we simultaneously process messages both systematically and using heuristics until we gather sufficient information on which to base our decision. Her heuristics involve communicator cues (e.g., trust in experts; listen to people I like), contextual cues (e.g., if other people think it's true, so do I), and message cues (e.g., more arguments are better arguments; arguments based on statistical support are better). Peloza et al. (2012) identified four heuristics they felt explained how stakeholders decode and interpret an organization's sustainability-related messages: sustainability initiative form, category biases, brand biases, and senior management image. These theories suggest we need to construct sustainability-related messages that provide sufficient information for people with an existing knowledge base who are able and motivated to process more complex messages. However, for people who are just learning about sustainability or who are not motivated to learn more, message heuristics are important.

Compliance-gaining research investigates persuasion in interpersonal contexts which is important to anyone seeking to intentionally promote sustainability initiatives within or between organizations. The focus of the compliance-gaining research is on behavioral conformity vs. attitude change (Gass 2009). Early researchers generated a confusing list of the strategies people said they would use to ensure compliance rather than investigating actual behavioral change. Cialdini integrated various studies and identified six basic principles: reciprocity (i.e., we often comply with those who have done us a favor), commitment and consistency (i.e., we want our behaviors to be consistent with our beliefs, attitudes, and values), scarcity (i.e., we value things we perceive to be in short supply), social proof (i.e., we compare ourselves to others and model others' behaviors), authority (i.e., we rely on source characteristics such as credibility), and liking (i.e., we respond to warm, ingratiatory behaviors and attractiveness). Other researchers investigated the verbal and nonverbal strategies used to gain compliance. In addition to the foot-in-the-door strategy (i.e., small initial request followed by a larger request), there is the door-in-the-face technique where the communicator makes a large initial request knowing it will be rejected but follows up with a second, more reasonable approach. People are more likely to comply with the second request for a number of reasons including our social tendency to make reciprocal concessions, our feeling of social responsibility to our friends, our assessment that the second request is more reasonable, and/or our feelings of guilt if we must say no a second time. Another verbal strategy is the disrupt-then-reframe strategy. Nonverbal strategies including a light touch, eye contact, smiling, mirroring, and mimicry are immediacy behaviors used to evoke compliance.

Compliance Gaining and the City of Fayetteville Many of my interviewees mentioned using one or more of the compliance-gaining strategies. I discussed compliance gaining with Peter Nierengarten, Director of Sustainability and Resilience for Fayetteville, AR, who told me:

> I am always a big fan of the carrot approach over the stick approach. It is usually better to try and work collaboratively, to find a benefit I can offer to whatever department head I am trying to convince. On occasion we can get the Mayor to say 'this is what we are going to do, this is how it is going to be' but we try not to regularly use that approach. In my experience with the city government, the triple-bottom-line approach tends to sell well. For example, when dealing with energy and retrofits on city buildings we show the improvement cost versus the total life cycle savings as a way of gaining buy-in for a project. We do not advocate for implementation of projects unless they meet a minimum return on investment.

Several of the compliance-gaining models focus on the face-related needs of the participants (e.g., politeness theory, goals-plan-action model). Face is the public performance of our identity; it is how we want others to see us. Important in our everyday life, it is especially important within an organizational context given individuals' goals for respect and/or power (Shimanoff 2009). Because sustainability-related initiatives may involve some conflicts depending on the worldview of the participants, awareness of Brown and Levinson's politeness theory is important. These two sociolinguists developed their theory after observing three different cultures. Positive face involves our desire to have others approve of and validate our identity, while negative face involves our wish not to be imposed on by others in ways that appear to disrespect our identity. We often engage in face-threatening acts (FTA) such as disagreeing with someone (threatens positive face) or enforcing obligations (threatens negative face). Other FTA involve contradicting, criticizing, interrupting, imposing, asking a favor, borrowing, giving advice, requesting information, and embarrassing someone. These are common behaviors. But depending on our relationship with the other person and the context, they can become problematic behaviors. Politeness theory focuses us on the communication strategies we can use to conduct our business in the least face-threatening way possible. We can use various politeness strategies with the most polite being not to do the FTA, followed by off-record strategies (e.g., hinting), negative redress strategies (e.g., apologizing, hedging, honorifics, use of past tense), and positive redress strategies (e.g., complements, slang, familiarity, giving reasons, reciprocal exchange, use of inclusion forms like the word *we*), with the least polite being bald-on-record strategies (e.g., making no effort to save face).

Face concerns also appear in the goals-plans-action model of interpersonal influence Dillard developed to explain how communicators develop and select their strategies (Gass 2009). People have primary goals (e.g., to increase pro-environmental behaviors within a work team) which are the catalysts for compliance-gaining efforts. But they also have secondary goals (e.g., saving face if their request is rejected). Dillard identified seven types of primary goals including to gain assistance, to give advice, to share an activity, to alter another's perspective, to change the relationship, to obtain permission, and to enforce rights and obligations. Our secondary goals include identity goals, conversation management goals, relational-resource goals, personal-resource goals, and affect management goals. Based on the primary and secondary goals important to the communicator, he or she will develop plans, strategies, and tactics for achieving the goals.

4.2.1.1 Best Practices Provided by the Persuasion Theories

The persuasion theories provide those interested in communicating about sustainability initiatives with a number of important insights. Because individuals want to appear consistent, it is important to frame messages so the requested behavior appears to match their values. The credibility and likeability of the speaker is important, as is the message content and sequencing. Be thoughtful about your message sequencing utilizing the goals-plans-action model. Think about your goals for the message sequence and whether or not a foot-in-the-door or similar technique might be useful. Messages should be crafted differently for those who are interested in learning more about sustainability and those who are unlikely to process complex messages. For the latter, attention to message heuristics is critical. In your inter-personal communication with others, recognize their face-related concerns and the utility of compliance-gaining strategies. Remember sometimes our behavior drives our attitude and value formation rather than the reverse, especially when we face novel situations.

4.3 Pro-Environmental Behavior

4.3.1 The Tentative Link Between Values, Attitudes, and Behavior

We commonly assume that values and attitudes influence behaviors and if we can change someone's values and attitudes, then we can expect their related behaviors to change. That certainly isn't the case. Persuasion theories have evolved over time to account for the weak link between values, attitudes, and behaviors. Decisions are influenced by much more than values, and many behaviors are not the result of thoughtful decisions but rather emotion, habit, and modeling. A large amount of persuasion research in the 1960s and 1970s focused on specifying the conditions under which the attitude–behavior link existed (Seiter 2009). The theory of rea-soned action (Fishbein and Ajzen 1975) and the theory of planned behavior (Ajzen 1991) were among the theories developed to identify the myriad of factors influenc-ing behavior. These theories and even more complex models of pro-environmental behavior follow.

The attitudes–behaviors–constraints theory of pro-environmental behavior (Guagnano et al. 1995) discusses how the link between attitudes and behaviors is often weakened due to constraints which often are not investigated. Most pro-environmental behavior models fail to account for individual, social, and institutional constraints and assume that people are rational and systematically use the information they receive (Blake 1999). It is not that people are irrational but rather we may feel that we lack the power needed to significantly address global or even local environmental issues.

Blake's (1999) description of three barriers to action—individuality, responsi-bility, and practicality—overlaps with barriers described by others (e.g., Kollmuss and Agyeman 2010; Stern 2000). Attitudinal barriers consist of attitudes, values, norms, beliefs, lack of motivation, and temperament. Kollmuss and Agyeman (2010) do an excellent job identifying some other individual-level barriers such as lack of internal incentives and negative or insufficient feedback about behavior. For those who do not have a strong environmental concern, environmental issues may be outweighed by other attitudes (e.g., political identification, desire for the good life). External or contextual forces may include nonsupportive interpersonal influences (e.g., persuasion), nonsupportive community or institutional expectations, government regulations and other legal factors, limited financial incentives and high costs, the action's difficulty, capabilities and constraints provided by technology and the built environment, and the lack of policies supporting the behavior. Personal capabilities are the third type of causal variable and include lack of knowledge, skills, ability, and resources. A feeling of limited responsibility, which is similar to an external locus of control, applies when people feel they cannot or should not have to engage in a pro-environmental behavior. Barriers related to practicality includes social and institutional constraints like lack of time, lack of money, and lack of information. A final type of causal variable involves habit or routine.

Habits, the City of Fayetteville, and the City and County of Denver Peter Nierengarten, Director of Sustainability and Resilience for Fayetteville, AR, illustrated the challenge habits can present to those seeking to promote sustainability within an organization. I asked him about the increase in energy usage appearing on the city website even after the city's energy efficiency efforts. He explained, "You implement some efficiency improvement and track it closely but sometimes an efficiency is undone by users. They flip back into old habits and waste [energy]." The challenge is to get city employees to develop new habits. Jerry Tinianow, Chief Sustainability Officer for the City and County of Denver, described how that is his focus saying:

> This approach that we are taking is so new and different. The idea of driving [sustainability] so deep in every city department that it becomes their habit, and then it becomes their instinct. I don't know if anyone has attempted to do that yet. So I don't really have anywhere else I can look to write out this script. I just have to believe in myself and my own approach and the backing that I have from the Mayor. …. The Mayor really wants to leave a legacy of sustainability and does not want this to disappear when the next mayor comes in. And I think his philosophy of trying to make it a habit and then an instinct is going to ensure that sustainability continues as a basic operating mode in Denver regardless of who is sitting in the mayor's office.

Practitioners concerned with promoting sustainability-related initiatives within and between organizations need to identify potential barriers which may prevent people from engaging in the actions being promoted. *Best Practice*: When possible, craft messages that show how something really isn't a barrier or which provide message recipients with a way to overcome a potential barrier.

4.3.2   Models of Pro-Environmental Behavior

Behavioral Change at Sam's Club Brian Sheehan, former Sustainability Manager at Sam's Club, shared his views on how to promote behavioral change among employees saying:

It's got to do those five things—it's got to be easy to understand and execute, it's got to be understood so that people can execute it. It's got to be desirable. It's got to be rewarding, and not necessarily financially but that does not hurt. And you have to be reminded about it, constantly or regularly. So unless it does all five of those things, we know that we don't get behavior change. So we are designing programs that hopefully achieve all five of those things.

Brian's ideas dovetail nicely with some of the elements appearing in the theoretical models we cover in this chapter.

Researchers have investigated a myriad of potential psychological, sociological, and communication antecedents to pro-environmental behavior. A number of excellent reviews of the various theories exist (e.g., Kollmuss and Agyeman 2010; Stern 2000). Some of the major theories and concepts are summarized here to help researchers and sustainability professionals better understand antecedents to pro-environmental behavior outside the workplace. For example, Kollmuss and Agyeman (2010) review theories for explaining the gap between environmental knowledge and awareness and the display of pro-environmental behavior: early linear progression models; altruism, empathy, and prosocial behavior models; and sociological models. In addition, several models commonly associated with health and energy efficiency campaigns are included along with information on social marketing. All of the models show validity in certain circumstances, but the question of what shapes pro-environmental behavior is such a complex one that it has yet to be visualized on one framework or diagram.

4.3.2.1   Linear Progression Models

The linear progression models suggest that if you provide individuals with information that changes their attitudes, this will change their behaviors. In order to act, we must be aware of environmental problems and the consequences of individual behaviors (Deci and Ryan 2000). Awareness is a precondition for pro-environmental action. The information we need in order to act must be specific. People must have the declarative and procedural knowledge necessary to act. In other words, they must know what to do and how to do it. Direct experiences have a stronger influence on people's behavior than indirect experiences (e.g., seminars, reading materials, classes).

Experiential Learning at the University of Colorado, Boulder, and Aspen/ Snowmass The importance of direct experiences, indirect learning, and rewards was illustrated by Moe Tabrizi, former Assistant Director of Engineering and Campus Sustainability Director for the University of Colorado, Boulder, who described his organization's Peak to Peak (P2P) program. His campus is one of the most sustainable in the USA, and P2P was designed to integrate sustainability into the larger campus learning environment. The P2P program began in 2011–2012 as a 2-day summer sustainability seminar for 40–50 faculty drawn from every academic program. It was so successful in the first year that the program was continued. Attendees are exposed to an in-depth discussion of the science of sustainability and to the best practices of sustainability on the campus. They tour some of the campus facilities and receive $500 in compensation from their provost for their time and their effort. Moe indicated the tours were an eye-opener for him:

> I assumed if you are a faculty on this campus, you knew every corner of the campus. That is absolutely a wrong assumption, because most of our faculty members are so focused on their students and their research, they just think of one path to [and from campus]. Just because we build the latest, greatest, greenest residence hall on one corner of campus, or we have a solar farm in another corner of the campus, there is no guarantee that they have seen it or that they have heard about it. So [the tour] is an opportunity [for them] to get out of the classroom and just see the campus.

Sustainability communicators need to provide direct experiential learning opportunities for their audience members and not just assume. Auden Schendler at Aspen/Snowmass reinforced the importance of both points. For 10 years Auden sent articles and did presentations internally but "one of the classic epiphanies for our CEO was going to a Fortune conference in California and meeting all these other CEOs who are worried about this issue and working on sustainability. They were not freaks and weirdoes. They were at Walmart and other big companies."

However, the impact of knowledge alone on behavioral change has not been supported by the research. In order for a message to be effective, people must not only attend to it but also cognitively process it in such a way as to lead to behavioral change. Over time, the linear models became more sophisticated. The theory of reasoned action (TRA) and the theory of planned behavior (TPB) (Fishbein and Ajzen 1975; Ajzen and Fishbein 1980) are influential social–psychological theories utilized by scholars in numerous disciplines. Indeed, TRA is considered by some to be the most influential attitude–behavior model in social psychology (Kollmuss and Agyeman 2010). It predicts behavior based on seven causal variables—behavioral intention, attitudes, subjective norms, belief strength, evaluation, normative belief, and motivation to comply (Greene 2009). TPB extends the TRA by adding an additional component, perceived behavioral control. Behavior is driven by intention which is influenced by a combination of attitudes (is this a good thing to do?), subjective norms (do others think I should do this?), and perceived behavioral control (can I do it?). Both models maintain that people are essentially rational, systematically using the information available to them so as to avoid punishments and seek rewards. These theories are of greater utility when behaviors have higher costs and individuals face strong constraints because that is when people devote attention and energy to processing messages.

Building on the TPB, Hines et al. (1986) published their model of responsible environmental behavior. Important antecedents to behavior include the message receiver's knowledge of issues, knowledge of action strategies, locus of control, pro-environmental attitudes, verbal commitment, and sense of responsibility. However, the relationships proposed in the model were only weakly supported, and additional situational factors influenced pro-environmental behaviors (Kollmuss and Agyeman 2010). These situational factors include economic constraints, social pressures and opportunities to choose different actions.

In order to replicate and extend the Hines et al. (1986) model, Bamberg and Moser (2007) published a meta-analysis drawing on 46 studies. Their key determinants of pro-environmental behavior were problem awareness, internal attributions, social norms, feelings of guilt, attitudes, moral norms, perceived behavioral control, and intention. Guilt was an interesting addition to their model. Guilt is an important prosocial emotion because it results in a felt obligation (moral norm) to act. It occurs when individuals recognize a perceived mismatch between their own behavior and social norms. Social norms directly influence moral norm development by indicating which behaviors respected others view as appropriate in a given context. The authors found that perceived behavioral control, attitudes, and moral norms explained 52 % of the variance in intention to act. In turn, intention directly explained 27 % of the variance in pro-environmental behavior. Feelings of guilt, social norms, internal attribution, and problem awareness predicted the moral norm construct (explaining 58 % of the variance). Social norms were directly associated with perceived behavioral control and attitude, and a direct association existed between guilt and attitude. Internal attribution was a significant predictor of social norms, moral norms, feelings of guilt, and attitude. This model is an important one because it largely validated the earlier model and summarized the last 20 years of this research stream.

Bamberg and Moser (2007) concluded that an individual's pro-environmental behavior is a mixture of self-interest (as illustrated by the TRA) and concern for other groups beyond the family including other species or whole ecosystems (as illustrated by Schwartz' norm-activation model which appears next). They write (p. 21):

> The intention to perform a pro-environmental behavioral option can be described as a weighted balance of information concerning the three questions 'How many positive/ negative personal consequences would result from choosing this pro-environmental option compared to other options?', 'How difficult would the performance of the pro-environmental option be compared to other options?', and 'Are there reasons indicating a moral obligation for performing the pro-environmental option?'

4.3.2.2 Altruism, Empathy, and Prosocial Behavior Models

A number of early theories and models focused specifically on altruism, empathy, and prosocial behaviors (Kollmuss and Agyeman 2010). Stern et al.'s (1995) early model was developed from the norm-activation model of Schwartz (1977). Schwartz argued that moral norms, or feelings of strong moral obligations that people experience, are direct determinants of prosocial behavior. Altruistic behavior should increase when a person becomes aware of other people's suffering and feels a responsibility to alleviate this suffering. Stern et al. expanded the altruistic orientation to include a social orientation, an egoistic orientation, and a biospheric orientation. Every person has all three orientations, but in different strengths. The early model was expanded into the value–belief–norm (VBN) model (Stern 2000) which essentially proposes that values relate to an individual's beliefs which then form intentions to act through norms. This theory combined three existing theories into a causal chain of five variables leading to behavior. Our values (i.e., self-interest, humanistic altruism, biospheric altruism) influence our worldview about the environment (general beliefs such as the NEP) which, in turn, influences our beliefs about the adverse consequences of environmental change on things we value. These beliefs then influence our perceived ability to reduce threats to the things we value, which influence our norms about taking action. These norms reflect our sense of personal obligation to take pro-environmental actions. The actions we can take may be activism, nonactivist public-sphere activities (e.g., voting), private-sphere behaviors (e.g., consumer choices), or behaviors in organizations. The intended pro-environmental behavior scale measures people's intention to sign a petition, to participate in an environmental protest, to disseminate information, and to write a public official (Cordano et al. 2003). Researchers find it to be a useful measure.

4.3.2.3 Several Sociological Models

Fietkau and Kessel combined sociological and psychological factors in their model of pro-environmental behavior (Kollmuss and Agyeman 2010). Their model includes five variables: attitudes and values, the possibility to act ecologically, behavioral incentives (e.g., social desirability, quality of life, monetary savings), perceived feedback about engaging in ecological behavior which can be intrinsic (e.g., the satisfaction of doing the right thing) or extrinsic (e.g., social, recycling is a socially desirable action; or economic, receiving money for collected bottles), and knowledge. Building on the inclusion of sociological factors, Kollmuss and Agyeman (2010) identify influences on pro-environmental behavior including demographic factors, internal factors, and external factors. Their model of pro-environmental behavior includes internal factors (e.g., personality traits, value system, and environmental consciousness), external factors (e.g., infrastructure), and old behavioral patterns. Some of the barriers influencing environmental consciousness include the emotional blocking of new knowledge, existing values that prevent learning, existing knowledge that contradicts environmental values, emotional blocking of environmental values/attitudes, and existing values that prevent environmental involvement. Additional challenges include lack of environmental consciousness, lack of internal incentives, negative or insufficient feedback about behavior, and lack of external possibilities and incentives. One of the primary strengths of this model is the attention to barriers appearing throughout the process. *Best Practice*: Identify and mitigate barriers to behavioral change.

Lubell et al. (2007) developed a model for explaining how people decide to act when they face a collective dilemma like global warming which involves massive populations, huge uncertainties, and relatively weak institutions. This is a severe collective-action problem since our individual actions have almost no influence on the problem, individuals cannot be certain others will engage in pro-environmental actions, and many of the recommended behaviors carry relatively high costs for the individuals who adopt them. Although rational citizens would probably choose to free ride on the efforts of others, many people support global warming policies and engage in sustainable behaviors. In seeking to explain why, the authors adapted the collective interest model of collective-action behavior. This model argues that people will engage in collective action when they assess the expected value of participation as being greater than the expected value of nonparticipation. Participation is shaped by people's belief in the value of the public good, their belief their participation will effect collective outcomes, and their assessment of the benefits and costs of participation. Key elements of the model include the perceived risk of global warming, personal efficacy in making a difference, group efficacy, engagement in discussion networks, environmental values, and demographic factors which can influence a citizen's ability to pay any costs associated with activism (i.e., education, income, age, gender, or ethnicity). Citizens who believe that global warming poses a very high risk to human welfare and the environment and who perceive their actions can make a difference will be more likely to support policies or take actions designed to reduce those risks. In deciding whether or not to act, citizens assess their community's level of social capital, the likelihood others in their community will reciprocate if they act, and the competence of policy elites. Political discussion networks provide citizens with resources such as positive reinforcement and access to information about preferred actions. Lubell et al. empirically tested the model in relation to three behaviors: policy support, environmental political participation, and environmental behavior related to global warming. They found different factors explained a significant percentage in the variance of the three behaviors. In general, those who believe the risk from global warming is high, believe their actions can make a difference, and hold pro-environmental values are more likely to support global warming policy and to take action. Higher education and income levels provide citizens with the civic skills and resources necessary to manage selective costs and recognize participation opportunities. Those interested in organizations in relation to community organizing, social movements, and collective actions should read Ganesh and Stohl's (2014) review.

4.3.2.4 The Importance of Motivation

Motivation is an important influence on behavior mentioned in numerous theories. Pelletier et al. (1998) operationalized three aspects of motivation in their development of the motivation toward the environment scale: intrinsic, extrinsic, and amotivation. Intrinsically motivated individuals participate because they find the activity pleasant or satisfying. Extrinsically motivated people do so hoping to experience positive consequences or avoid negative ones. Amotivated individuals do not see the behavior's consequences or fail to identify with reasons to continue the behavior. Their actions are mechanical and personally meaningless. Motivation ranges on a continuum depending on whether it is externally determined or self-determined (i.e., intrinsic). Self-determination theory (Deci and Ryan 2000) points out how self-determined motivation is better than externally determined motivation in terms of behavior maintenance, enhanced well-being, greater achievement, and deep information processing. People are motivated to internalize activities that will help them function successfully in their social world, even if the activities are not inherently interesting. Externally determined controlling events (e.g., rewards, punishments, imposed rules) increase temporary compliance but are less likely to lead to long-lasting commitment and investments in behaviors beyond those targeted by the controlling strategy. Although people might recycle for rewards (e.g., money), the behavior is more likely to persist if they have internalized the reason (e.g., they believe in the importance of a healthy world). Deci and Ryan propose that, in addition to intrinsic and extrinsic motivation, people have three main intrinsic needs: competence, autonomy, and psychological relatedness. Self-determined motivation is greater in social contexts which support our innate need for competence and autonomy and which provide us with sufficient information. Self-determination theory has been supported by research showing increased persistence over time in a new behavior, willingness to engage in more difficult behaviors, and the potential to engage in behavioral patterns that reflect a range of pro-environmental behaviors. The theory suggests that pro-environmental behaviors are facilitated when a good rationale for the activities is provided, the context points the way to more effectively meeting any challenges, and people can freely choose among different options.

This approach seems to guide the activities undertaken by the green team at the State Farm Insurance operating facility in Lincoln, NE. Mike Malone, team leader, described how one of the green team's earliest projects involved replacing Styrofoam cups in the break areas with individual coffee mugs and insulated water glasses. Mike said, "The cost of that was a [very small] fraction of what was being spent yearly on Styrofoam." The green team educated employees about the harms of Styrofoam, offered them a reasonable alternative, explained why they were making the change, explained the benefits of the change, and made it easy to use. "Then you get total buy-in," Mike explained.

4.3.2.5 Lessons Learned from Energy Reduction Studies

In a meta-analysis of 61 research studies and 57 feedback initiatives investigating advanced electricity metering initiatives and feedback programs on reduced residential energy consumption, Ehrhardt-Martinez et al. (2010) identified various motivational techniques which had shown some success. These include descriptive and normative feedback (Cialdini 2003; Schultz et al. 2007), goal setting, public commitments, and rewards (Ehrhardt-Martinez et al. 2010; Tiedemann 2010). A few years earlier, in order to identify how to design better interventions directed toward reducing residential energy use, Wilson and Dowlatabad (2007) reviewed models and theories coming from four diverse perspectives: conventional and behavioral economics, technology adoption theory and attitude-based decision making, social and environmental psychology, and sociology. They reviewed some of the models and theories we have already discussed (e.g., cognitive consistency, TPB, self-efficacy, stages of change, VBN theory) as well as one you will read about later (i.e., diffusion of innovation) (see Sect. 5.2). Their article provides a nice resource for students and scholars interested in learning more about these models and theories. They also provide their own model of the process.

4.3.2.6   A Workplace Model Focusing on Goals

Unsworth et al. (2013) developed a model building on the TPB and VBN theories which identified the psychological conditions under which an organization's sustainability-related interventions are most likely to succeed. They drew on theories of goal hierarchy, goal systems, multiple goals, self-concordance, and values. They argue that organizations promoting employee green behaviors face distinctive challenges because green behaviors and green goals are among many behaviors and goals employees continually manage and are often of low priority and inconsistently activated in a workplace. "Employees may be deciding between working on a report and walking to the recycling bin; while at work, they may be juggling their efficiency goals, their service and relationship goals, their family goals, their career ambition goals, and so forth" (p. 212). It is important to remember that people are more likely to engage in behaviors they see as self-concordant with their values. Leaders can increase employees' perceptions of the self-concordance of the pro-environmental behavior. Employees do not necessarily have to have altruistic or biospheric values. What is important is that the employee sees the proposed behavior as expressing as many personally relevant values or long-term goals as possible, even if they are egoistic values. If an intervention is to succeed, its goals should be efficacious and attractive, self-concordant, in limited conflict with other goals, able to spill over into related behaviors, and seen as achievable. Interventions can address the issue of goal conflict by using location-based cues to refocus people back on the pro-environmental goal. For example, on the Portland Trail Blazer Moda Center campus, trash cans have been replaced with "landfill-bound" and "recycling" receptacles.

4.3.2.7   Best Practices Provided by Pro-Environmental Behavior
Theories

These theories suggest individuals charged with promoting sustainability-related initiatives within and between organizations need to help people understand a problem exists and then consider arguments addressing how the proposed action is a good thing to do, how it will significantly address the problem, and how others think it is a good thing to do. Individuals need a strong rationale to act and they are concerned with the positive and negative consequence of an action for their own self-interest as well as for others they care about. People need to feel they have the responsibility and ability to take the proposed action. Emotions such as guilt and fear, within limits, may be stimulated by messages, if really necessary, and daily affect can stimulate pro-environmental behaviors. Recognize that message recipients make assessments: "How many positive/negative personal consequences would result from choosing this pro-environmental option compared to other options?"; "How difficult is performing the pro-environmental option compared to other options?" Behavioral incentives reappear but now we are also focused on the importance of feedback. Once again, we are directed to an individual's need for autonomy (an important face-related concern). Motivation is increased if people are emotionally involved, see the activity as pleasant or satisfying, or think it will help them function successfully in their social world. Old behavioral patterns are among the barriers that hamper individual change, and new behavioral patterns need to be formed and reformed. The collective-action model is useful for those who are interested in mobilizing action such as the ski industry efforts with POW or the Natural Resources Defense Council's work with groups such as the Green Sports Alliance. *Action Plan*: Building on the various theories and models reviewed in this section, create your own working model for what influences individuals' pro-environmental behaviors in your organization.

4.4 Pro-Environmental Behaviors and Communication

In this section, several additional models or theories are provided which deal with social marketing or have guided health and energy use communication campaigns. That is followed by information on message design and content. The section concludes with a discussion of the role of interpersonal communication.

4.4.1 Social Marketing

Kotler and Zaltman (1971, p. 5) coined the term social marketing and defined it as "the design, implementation, and control of programs calculated to influence the acceptability of social ideas and involving considerations of product planning, pricing, communication, distribution, and marketing research." Marketing concepts are integrated with other approaches to influence behaviors that benefit individuals and communities for the greater social good. This approach is rooted in social science and social policy as well as commercial and public sector marketing. Initially, the focus is on learning what a specific target audience thinks, wants, needs, and/or desires rather than moving directly to persuasive efforts. This focus is akin to the audience analysis public speakers conduct before constructing their messages. Data are gathered using focus groups and surveys. Then, communication materials (both informative and persuasive) are created which build upon what is known about the audience.

Since the 1980s, social marketing has been used to promote disaster preparedness and response, ecosystem and species conservation, environmental issues, global threats associated with antibiotic resistance, marine conservation and ocean sustainability, sustainable consumption, and other sustainability-related social needs (Lefebvre 2013). One branch of social marketing, community-based social marketing (CBSM), emerged to systematically foster more sustainable behavior. Developed by Canadian environmental psychologist Doug McKenzie-Mohr, CBSM focuses on helping communities reduce their impact on the environment. After focus groups and surveys uncover barriers, behavioral change is stimulated through the use of commitments, prompts, social norms, social diffusion, feedback, and incentives. This approach has been used to promote energy conservation (e.g., Schultz et al. 2007), enhance environmental regulation (e.g., Kennedy 2010), and stimulate recycling (e.g., Haldeman and Turner 2009). Individuals charged with communicating about sustainability initiatives to their internal and external stakeholders should consider this approach because it provides a way to systematically identify what each stakeholder group believes, knows, and desires in order to more effectively target messages aimed at changing their sustainability-related behaviors.

4.4.2 Health-Related Models

An early model designed to help researchers and practitioners understand and promote healthy behaviors provides several concepts potentially useful in promoting sustainability-related initiatives. The health belief model was developed in the 1950s by social psychologists at the US Public Health Service and updated in 1988. It remains one of the most widely used theories in health behavior research. The model identifies perceived seriousness, perceived susceptibility, perceived benefits, perceived barriers, perceived threat, self-efficacy, and cues to action as influences on an individual's likelihood of engaging in a particular health-promoting behavior. We may intend to engage in a new behavior but simply forget because it is not habitual. Therefore, behavioral cues are especially important. Cues can be internal (e.g., pain, emotional distress) or external (e.g., information from important others, signage). Self-efficacy was added to the model in 1988. Although this model was designed to influence a different set of individual-level behaviors, it reminds us to work to increase an individual's self-efficacy and to provide behavioral cues.

Behavioral Cues at the University of Colorado, Boulder Campus On-location cues are used extensively on the University of Colorado, Boulder, according to Moe Tabrizi, former Assistant Director of Engineering and Campus Sustainability Director. He told me:

> You can't walk around our campus without seeing a poster or sticker on the light switch.
> Whether they are intended for laboratories or the data center or just turning the lights off,

reporting water leaks and so forth, those are our attempts to impact or influence the behavior of our students, staff and faculty.

I asked him to comment on the success of such signage, and he provided some anecdotal evidence saying:

For example, in large classrooms where we did not have any posters or stickers or messaging to prompt people to turn off the lights or turn off the computers, a certain percentage of the lights were left on. In contrast, when you have those messaging, posters and stickers, 60–70 % of the lights were turned off after the class. So that is a good indication that message is being heard. When you place a poster in a laboratory research building and you let the researcher know that the fume hood is using three times as much energy as an average home you get a lot of feedback from the faculty member, who says, 'I have been teaching chemistry or researching in biochemistry and I never knew that. Thanks for telling me and I will make sure my students are closing the sash going forward.' So that's pretty good reinforcement [that the messaging campaign is working].

As I prepared to leave his office, Moe gave me copies of stickers they had placed across campus and a pledge card they had used. I describe this communication material in detail in hopes sustainability coordinators will find the information useful. Light switch stickers were blue with yellow letters, measured 2.5″ by 1″, and included a graphic of the globe and the words "Please turn off the lights when you leave," "Turn off climate change," or "Report energy and water waste (phone number) (hotline reporting website address)." A similar sticker read "See a water leak or energy waste? Report it!" and "You can turn off climate change" and provided the phone number and hotline website information. These messages capture attention through their use of color the first time someone sees them and function as an unconscious heuristic thereafter. The use of the word *please* is a politeness strategy designed to minimize the negative face threat of the command. People were told their initial action could make a difference and their assistance was solicited as monitors of potential waste providing them with a secondary action they could take. Signage for the bathrooms announced the new water-saving toilets and provided a graphic showing people how to use them. How-to information influences self-efficacy. The prefolded yellow 8.5″ by 5.5″ pledge card was part of a 2003 campaign. The top half of the detachable card asked people to make a pledge to reduce energy use, provided them with four sample actions and told what impact each made in terms of conservation percent (e.g., "Screen savers do not save energy. Enable the sleep mode in your PC monitor and two others (saves 2 %)"), and directed them to a website for more suggestions. A Ghandi quote told the reader, "You must be the change you wish to see in the world." The bottom half told them their signature would result in the Vice Chancellor of Administration committing $5 to energy conservation/renewable energy projects on campus. Under the message "I commit to reducing my CU-Boulder campus energy usage by 10 % by taking actions such as those listed above," there was a place for people to sign and indicate if they were students, faculty, or staff. The bottom half of the card was preaddressed on one side and could be dropped in campus mail. Looking at the pledge card as a message, people were asked to pledge, told how to participate, provided with actions they might take, given information regarding the action's significance, and asked to take an initial commitment-related action by signing the card. The top half of the card could be kept as a reminder for alternative actions they might take.

The transtheoretical model of behavior change was developed by Prochaska and colleagues beginning in 1977 and is another dominant model for those interested in health-related behavioral change. Individuals are at different stages in their readiness to engage in a recommended behavior. The model assesses an individual's readiness to act and provides information on how to guide the individual through ten stages of change. The stages of change include precontemplation, contemplation, preparation, action, maintenance, and termination. At each stage, the theory includes strategies communicators can use with the target audience. For example, during the precontemplation stage, people should be encouraged to become more mindful about their decision making and reminded of the multiple benefits of changing their behaviors. Activities designed to help people move through the stages include consciousness raising, self- and environmental reevaluation, self- and environmental liberation, contingency management, and helping relationships (Silk 2009). Within the context of pro-environmental behaviors, this stage model reminds us to assess where our target individuals are in the process of adopting new behaviors and to think strategically about the pro-environmental messages they are receiving. Interventions are more effective if they match an individual's stage of change. As people move toward action, they rely more on commitments, conditioning, environmental controls, and support. So practitioners interested in changing individual behaviors need to secure commitments and create opportunities for people who are adopting new behaviors to receive social support from peers. Staats et al. (2004) use that approach in their design of an intervention package which combined information, feedback, and social support to improve pro-environmental household behavior. Their intervention included group discussions, block captains, information received from friends, and commitments such as pledges. In Sect. 6.5.3.2 you will read about how Walmart makes social support available to associates who are interested in creating their own personal sustainability projects.

4.4.3 Communication Campaign Interventions

Communication campaign interventions fit into a number of categories. Some appeal to values and attempt to change broad worldviews and beliefs; others provide education to change attitudes and provide information; some utilize an incentive structure by providing rewards or penalties; and finally some rely on community management including the establishment of shared rules and expectations. Intervention studies have investigated a range of antecedents and consequences of pro-environmental behavior including commitment, goal setting combined with feedback, information, modeling, rewards, and tailored messages (see Unsworth et al., 2013, for relevant sample studies). For example, Stern (2000) provides a set of principles for interventions seeking to change environmentally destructive behavior:

- Identify target behaviors that are environmentally significant in terms of impact.
- Identify the responsible actions and actors.
- Set realistic expectations about outcomes.
- Attempt to understand the situation from the actor's perspective. Gather feedback from the targets about causal variables.
- Identify perceived and actual barriers to change and try to remove them.
- Use multiple intervention types to address factors limiting behavioral change. For example, provide information, incentives, and/or reminders.
- When limiting factors are psychological, get the individual's attention and make limited cognitive demands (e.g., use simple messages, provide heuristics).
- Stay within the bounds of an individual's tolerance for intervention.
- Apply principles of community management (credibility, commitment, face-to-face communication).
- Use participatory decision making if possible.
- Continually monitor responses and adjust accordingly.

4.4.3.1   Best Practices from Health and Communication Campaign Literatures

In addition to Stern's (2000) recommendations, sustainability communicators are reminded to conduct research using focus groups and/or surveys and then design messages for specific audiences. The health belief model reminds us to include information about an issue's seriousness and the susceptibility of the individual and what he or she cares about, as appropriate, in our messages. It also posits that a cue, or trigger, is necessary to prompt engagement in new behaviors. The stages of change model describes how people vary in terms of their readiness to engage in a recommended behavior. Different strategies need to be utilized at each stage. Goal setting, social support from peers, and soliciting commitments can be powerful tools for reinforcing behavioral change.

4.4.4   Message Design and Content

Discussing how we currently lack frames designed to tackle climate change-related challenges, Lakoff (2010) suggests that in the short term, while frames hopefully are being built to reframe challenges on a deeper level (something akin to the Space Race or citizen mobilization during World War II), it is important to talk about values, not just facts and figures, use simple nontechnical language, and appeal to emotions. Other things that might be stressed are empathy (which has a physical basis in the human mirror neuron system). Empathy links us to other beings and the natural world. The argument could be to take personal responsibility for taking care of yourself (e.g., maintaining your health) and taking care of others (e.g., protecting their health). Lakoff also suggests arguing for the ethic of excellence which calls on us to improve the environment or at least preserve it, starting with our actions (e.g., conserve energy). He provides his readers with some additional short-term suggestions while noting that it is really building up effective long-term messaging that matters:

1. Talk at the level of values and frame issues in moral terms. Distinguish values from policies. Several of my interviewees mentioned the need to balance the head and the heart when communicating about sustainability. For example, Steve Denne, the COO of Heifer International®, talked about how Heifer International® "thinks about the connection between head and heart in all people. We realize that connecting at the emotional level is important, but it is not sufficient." Communicating with stakeholders about both the head and the heart is challenging. He described a partnership they had with a for-profit organization where both organizations were able to measure progress to their key goals. Heifer International® could measure changes in farmer income levels, farmer nutrition, and farmer environmental practices. Their partner could measure product quality and quantity. The first three tied to Heifer values; the last two were especially relevant to their partner.
2. Go on the offense. Don't accept the other side's frames. Don't negate them or repeat them. That just activates their frames in the listener's brain.
3. Provide a structure for what you are saying. Find general themes or narratives that incorporate the points you want to make. Tell stories that exemplify your values and arouse emotions. If you give numbers and facts, reframe them so their overall significance can be understood.
4. Context matters. Be a credible messenger, have good visual aids, and be aware of your body language.
5. Address everyday concerns. Use nontechnical words people understand. Susan Anderson, Director of Portland's Bureau of Planning and Sustainability, told me, "Even in Portland not everybody cares about the environment. But people care about kids, and grandkids and the future. That is what we talk about. It is the future, prosperity, health, and family—those words. Things that people actually care about."

Balancing Passion with Facts and/or Process at the City of Boulder David Driskell, Boulder's Executive Director of Community Planning and Sustainability, described how city planners are trained to be objective and process oriented. They design a process, facilitate input, conduct analysis, and make recommendations. However, the people in the Boulder Office of Environmental Affairs are visionaries. "They feel their job is to go out and convince everybody in the community that this is the way to go. And that approach does not work too well in a community like Boulder. We have done a lot of teaming subject matter experts with folks who are process-oriented." He explained how Boulder's disposable bag fee came into existence. A staff member from the Office of Environmental Affairs identified a problem and proposed a ban on plastic bags and a 25 cent fee on paper bags. As evidence, the staff member provided information on what another community had done. The staff member had been trained to be a scientist, to understand the environmental impact of a decision but not necessarily the policy perspective and the economic and social aspects of the issue. David said he told her, before we get to the solution, let's define the problem. What is the problem we are trying to solve? Is it plastics in the environment? Is it the environmental impact of disposable bags? Is it a reuse culture that we are trying to create? What are the metrics for success? Where is the analysis of this? What is going to be the cost impact on consumers? What is going to be the impact on grocery stores? How is this going to be implemented? Have we done outreach with the stores that are going to be impacted? He explained:

> It was not in that person's skillset to think about the process for developing that, so we teamed her with a person in comprehensive planning who had a lot of experience doing policy projects in the city. . .and that staff person who was doing the project just blossomed, she is actually now seen as an expert in the state of Colorado. Denver is looking at modeling an ordinance on what she did.

This example illustrates how individuals proposing sustainability-related initiatives need passion, but they also must possess sufficient data to support their claims and to back their warrants as discussed in Steven Toulmin's model of argument (1958). Toulmin said good, realistic arguments typically consist of six parts: data (the facts or evidence used to prove the argument), claim (the statement being argued), warrants (the logical statements bridging the claim and the data), qualifiers (statements limiting the conditions under which the argument is true), rebuttals (counterarguments), and backing (statements that support the warrants).

Message design and content issues also appear in several theoretical models. For example, Silk (2009) discusses how Bandura's social cognitive theory and Witte's extended parallel process model can inform message content. Social cognitive theory contends that people learn from observation; reinforcement or punishment impacts behavior; and learning is more likely if we identify with a role model and possess self-efficacy. The theory supports message design strategies that include message sources with whom the audience can identify, demonstrations of recommended actions, and the use of reinforcement or punishment as motivators. Witte's model discusses the use of fear appeals. Threats can influence perceived severity, perceived susceptibility, response efficacy, and self-efficacy. If threat is high and self-efficacy is low, people will avoid the recommended behavior or reject the message. Messages which convey threats must also include recommended actions that people can realistically take to address the threat. For example, following the airing of Al Gore's *An Inconvenient Truth*, the Climate Reality Project Gore started continued to recommend specific actions citizens can take on their website.

4.4.4.1 Messages Emphasizing Normative Beliefs, Altruism, Gain vs. Loss, and Intrinsic vs. Extrinsic Appeals

Normative Beliefs

Normative beliefs appear in most of the models and theories previously identified. When you invoke norms, you indicate which behaviors are commonly approved or disapproved of by a specific group or culture. People use their perceptions of peer norms as a standard to which they compare their own behaviors. Social-norms marketing campaigns have emerged as an alternative to more traditional approaches (e.g., information campaigns, moral exhortation, fear-inducing messages) (Schultz et al. 2007). Coming from a background in psychology and marketing, Cialdini (2003) discussed how persuasive messages need to invoke two kinds of norms shown to motivate human action: injunctive norms and descriptive norms. Injunctive norms help the message receiver identify which behaviors people typically approve or disapprove of. Descriptive norms influence perceptions of which behaviors are typically performed. People tend to do what is popular and socially approved. For example, descriptive norms are evoked by the seals appearing on the windows of organizations participating in the Strive toward Sustainability programs in Missoula, MT, and Lincoln, NE (see Sect. 3.3.2.3). The more businesses showing these seals, the more sustainable operations become seen as the norm within that community. However, in situations where many people are engaging in socially censured conduct (i.e., littering), Cialdini recommends that audiences be reminded of the injunctive norm (i.e., littering is bad) and not the descriptive norms. When the prevalent behavior is environmentally beneficial, messages should include descriptive norms along with injunctive norms (assuming most people approve of the proposed action). Although providing descriptive normative information may decrease an undesirable behavior among individuals who perform that behavior at a rate above the norm, the same message may actually increase the undesirable behavior among individuals who perform that behavior at a rate below the norm. People who do better than average may regress down to the mean (Schultz et al. 2007). People who are doing better than average should receive injunctive normative messages conveying social approval while those who are doing worse than average should receive messages of disapproval.

Altruism and Gain vs. Loss

A number of the theories and models we reviewed earlier focused on altruism, empathy, and prosocial behaviors. Some messages focus on the benefits or consequences experienced by the person performing the behavior; others include the effect of the behavior on significant others (e.g., friends, family, or the community at large). Loroz (2007) investigated the role reference point (self or self and other) and message frame (positive or negative) had on resulting attitudes and behavioral intentions. Framing research also investigates whether or not messages which discuss benefits gained (positive frame: "Think about what we will gain") or consequences suffered from failure to act (negative frame: "Think about what we will lose") might influence the persuasiveness of a message in a particular decision context. People more actively cognitively process messages that talk about the negative consequences they will face if action isn't taken unless the negative consequences exceed the individual's fear threshold as discussed in Witte's extended parallel process model. On the other hand, it is likely that people will respond more to messages which stress benefits experienced by both self and others (e.g., future generations, neighbors). In two small studies focusing on messages seeking to change health and recycling behaviors, the researcher found behavioral intentions were higher after reading the negative-self and positive-self and other messages than when reading negative-self and other or positive-self messages. Loroz suggested that if a communicator wants to design a message that centers on how pro-environmental behaviors can influence future generations, then he or she should stress the benefits experienced rather than the dire consequences of failing to act.

Intrinsic vs. Extrinsic Appeals

Some message strategies used to motivate people can lead to pro-environmental behavior, but long-term maintenance of these behaviors is problematic. Often people's behaviors return to baseline once the external motivation is removed. Pelletier and Sharp (2008) argue that self-determined motivation can be increased by framing messages as a function of the intrinsic (i.e., health, personal growth) vs. extrinsic (i.e., financial incentives, fame) gains or losses as well as by accounting for the underlying processes of behavioral change. They provide an example of how using a car or using public transportation can be framed in four different ways: (a) intrinsic gains (e.g., public transportation reduces carbon gas emissions (CGE) and improves your health), (b) extrinsic gains (e.g., public transportation reduces CGE and saves you money), (c) intrinsic risks (e.g., using your car increases CGE and worsens your health), and (d) extrinsic loss (e.g., using your car increases CGE and costs you money). Framing a goal as a function of extrinsic motivations should result in lower levels of self-determined motivation, less engagement in the activity, and less persistence in the new behavior. A focus on intrinsic motives should facilitate the development of autonomous motivation and behavioral maintenance. Psychologists have identified three stages to the process: a detection phase, a decision phase, and an implementation phase. How a message is framed (intrinsic vs. extrinsic) during the detection phase will influence subsequent decisions made during the decision and implementation stage. For example, an emphasis on financial costs during the detection phase will lead to goals and solutions with financial implications in the decision phase and then the maintenance of financial incentives to initiate behavior during the implementation phase. During the detection phase people are more sensitive to messages that help them gather information to determine whether or not there is a problem. The messages should frame the problem as important and provide people with a rationale to act. People should be more open to messages which emphasize the costs—what is to be lost by failure to adopt the proposed behavior. But once someone is aware of the risk, additional risk information will have limited influence on their behavior. Indeed, people develop a defensive avoidance response to similar risk-focused messages. Once they are aware of a risk, people are more open to information about specific behavioral options and how they can effectively address the problem. If people see a problem as important, they are more sensitive to messages that help them decide if they should take action and, if so, what action to take. Messages need to provide information that helps people make decisions about the feasibility, desirability, and effectiveness of a behavior. At this point, gain-framed messages should resonate because they stress the benefits of adopting a specific behavior. The health literature suggests gain-framed messages influence the development of personal goals that are reflected in an individual's intentions to act. Finally, in the implementation phase, people are more open to messages that provide them with information about how to implement, maintain, and integrate the behavior into their lifestyle. This should include information about where, when, and how a behavior might be implemented. People also need to set personal goals and commit to specific ways to achieve these goals. Doing so helps create a bridge between intention and action. If people find the new behavior pleasant, their commitment strengthens, assuming they remember to engage in it as the correct time—hence the utility of behavioral cues until a habit is formed.

Abstract vs. Concrete Action

In three studies, White et al. (2011) investigated when loss- versus gain-framed messages were most effective in influencing consumer recycling by examining the moderating role of whether a more concrete or abstract mind-set was activated by the message. They proposed the effect of message framing on conservation intentions, and behaviors will be moderated by whether a person considers recycling in terms of concrete actions (e.g., How will I go about recycling?) or more abstract purposes (e.g., Why will I recycle?). The researchers argued a loss-framed message would be most effective when paired with a mind-set that engages lower-level concrete thinking because when facing potential loss people seek immediate and concrete action strategies. On the other hand, a gain-framed message would be most effective when matched with a mind-set that engages high-level, abstract thinking. Gain frames activate more abstract, distal, and higher-level thinking. They found evidence for their matching hypothesis where a pairing of loss- and gain-framed messages that activates more concrete (abstract) mind-sets leads to enhanced processing fluency, increased efficacy, and more positive recycling intentions.

4.4.4.2   Additional Persuasive Arguments

Clark (1984) provided a set of persuasive message strategies for public speakers to use when seeking to move people toward action. They include arguments that there is a problem growing in magnitude, the time to act is limited, and failure to act now harms the people and things we love. All of these arguments and more apply to the current situation facing humanity. As I sat in the auditorium at the Walmart home office in Bentonville, AR, listening to speakers at the 2013 Sustainability Summit, I heard Leslie Dach, Walmart's Executive Vice President of Corporate Affairs and Government Relations, utilize many of Clark's persuasive strategies as he discussed Walmart's aspirational goal to utilize 100 % renewable energy. He said renewable energy's "time is now," "we refuse to wait," "we can utilize our size and scale," "future generations depend on us," "we are doing what is right," "using cutting edge technology, we can drive down costs," "we have the opportunity and the responsibility," and "every kilowatt we don't have to use we don't have to pay for." Dach, who oversaw the company's sustainability efforts between 2006 and 2013, identified three reasons why Walmart should move toward the renewable energy aspirational goal: (1) renewable energy and energy efficiency increases productivity, decreases costs, and is a way the company can control rising energy costs; (2) renewable energy and energy efficiency addresses climate change (80 % of Walmart's operational greenhouse gases come from its stores); and (3) renewable energy and energy efficiency is good for communities reducing air and water pollution and allows community members to feel better, influences the health of future children, and helps people feel better about Walmart. Dach's arguments support the business case for sustainability. Blackburn (2007) provides support for seven business case arguments involving increased reputation and brand strength; more competitive, effective, and desirable products and services; access to new markets; increased employee productivity; lessened operational burden and interference; lower supply chain costs; lower cost of capital; and less legal liability.

Over the years I have developed a list of arguments my students might consider using when they seek to persuade others to be pro-environmental. You might find them helpful as well:

- Argue preservation is important.
- Argue that our international or business competitors or neighbors or most people or *cool* people are doing it.
- Argue for the sublime (feelings of awe and exultation some people experience when in God-created nature).
- Argue that it is the right thing to do.
- Use visual rhetoric picking a good condensation symbol or drop a mind bomb (i.e., simple images that change how viewers see a situation).
- Argue that it is in the public health interest; use risk-based messages.
- Argue the proposed action is just common sense or that it just makes good business sense.

- Argue that the problem is large, the situation is deteriorating rapidly, actions are doable (and easy or fun), and the actions will make a difference.
- Utilize messages which induce shock, shame, or guilt.
- Argue the target action makes life more beautiful and abundant.
- Argue that the proposed action is the result of collaborative action.
- Indicate the action is based on community advisory committee decisions.
- Share a story promoting altruistic tendencies or showing how a solution was effective.
- Become a voice for vulnerable populations being effected by environmental challenges (e.g., drought, limited water, polluted community).
- Argue the action is consistent with their own values.
- Appeal to the irreparable (i.e., we must act now or this will be forever lost).
- Use terms including fair balance, wise, and effective action.
- Focus on how the action will have multiple benefits far into the future.
- Talk about the real-world day-to-day effects of either the problem or the solution.
- Assure your audience of your shared values and good intentions.

4.4.5 The Role of Interpersonal Communication

Although many of these theories of pro-environmental behavior look at social norms, few really delve into interpersonal communication with the exception of the health-related models which discuss the utility of social support. Environmental risk is a social construction influenced by how people talk about perceived threats as much as by what they personally experience in their daily interaction with the physical world (Cantrill 2010). People are generally unaware of potential actions they can take. This lack of awareness limits what most people can do when confronting environmental change.

We turn to others whom we trust for advice (e.g., families, friends, and coworkers). Within these social interactions, our perceptions of risk can be amplified or diminished by our perceived trust in what the others are saying, by our belief that we or those we care about may be harmed, or by our lack of proximity to the threat. For example, I have given tours of our eco-efficient home to showcase environmental sustainability to community members and students (my own and local architecture students). I describe efforts we took to reduce our water consumption and how we considered utilizing rainwater recovery technology. When I provide the tour, I hope people perceive me to be a trusted and credible source, but few see water scarcity as a real threat to them, unlike it is to people living in other parts of the USA and across the globe. In any social network, the arguments that get talked about the most are the ones that people turn to when making up their minds, regardless of the soundness of the argument (see Sect. 7.1.3.4). So, despite periodic media coverage during times of drought, few homeowners actively talk about or conserve water in my part of Arkansas because that is not part of the normal community *Discourse.*

However, a change in awareness and action is possible because even environmentally apathetic publics interact with others in their social and professional circles who are knowledgeable and concerned about a particular environmental subject (e.g., water conservation given the scarcity of clean water globally) or an issue relevant to local people (e.g., a local drought). This is a powerful force amplifying environmental risks and opportunities—simply get people talking. Gatekeepers are members of a person's primary group who influence others' beliefs, attitudes and behaviors. We need research to see how much people actually are talking about the environment. Kassing et al. (2010) developed the environmental communication scale. This 20-item measure assesses environmental communication along three dimensions: practicing, dismissing, and confirming. The practicing and dismissing dimensions assess the extent to which people engage in or avoid conversations and media reports about environmental issues. Practicing sample questions are "I enjoy listening to discussions about the environment" and "Listening to discussions about environmental issues energizes me." Dismissing sample questions include "I ignore people who talk about the environment" and "I skip over news stories about the environment." The confirming dimension taps people's attitudes regarding the importance and necessity of engaging in environmental communication. Sample questions include "Discussing the environment is important" and "It is necessary to discuss environmental issues." Their scale can be modified to assess intraorganizational or interorganizational communication prior to and then immediately after an intervention designed to promote an organization's sustainability-related initiative. *Best Practice*: It is important to get people in your organization or community talking about the environment. Seek to create a positive pro-environmental "buzz."

4.5   Concluding Thoughts

This chapter focused on factors influencing pro-environmental attitudes and behaviors, primarily among citizens rather than employees. Such background knowledge is useful for sustainability communicators, as is information on how to use communication to promote pro-environmental behaviors. Although some people come into their organizational roles self-identifying as environmentalists, many others do not.

In 2014, researchers gathered online survey data from more than 48,000 consumers in 20 countries. They found 64 % of their respondents in China self-identified as environmentalists, more than twice as many as in Europe and the USA (Nicolaou 2014). That same year, the Pew Research Center surveyed 1,821 US adults finding that 32 % of the Millennials (ages 18–33), 42 % of Gen X (ages 34–49) and Boomers (ages 50–68), and 44 % of the Silent Generation (69–86) self-identified as environmentalists. Organizational leaders seeking to promote sustainability initiatives cannot assume that their younger employees will automatically embrace pro-environmental initiatives just because the broader societal *Discourses* are increasingly addressing sustainability. Knowledge about sustainability must be integrated into existing organizational processes and organizational cultures and climates must be designed to support sustainability initiatives. Employees must be empowered and a sustainability-focused "buzz" created.

References

Ajzen, I. (1991). The theory of planned behavior. *Organizational Behavior and Human Decision Processes, 50*, 179–211.

Ajzen, I., & Fishbein, M. (1980). *Understanding attitudes and predicting social behavior*. Englewood Cliffs, NJ: Prentice-Hall.

Allen, M. W., Wicks, R., & Schulte, S. (2013). Online environmental engagement among youth: Influence of parents, attitudes and demographics. *Mass Communication and Society, 16*(5), 661–686.

Bamberg, S., & Moser, G. (2007). Twenty years after Hines, Hungerford, and Tomera: A new meta-analysis of psycho-social determinants of pro-environmental behavior. *Journal of Environmental Psychology, 27*, 14–25.

Bissing-Olson, M. J., Iyer, A., Fielding, K. S., & Zacher, H. (2013). Relationships between daily affect and pro-environmental behavior at work: The moderating role of pro-environmental attitude. *Journal of Organizational Behavior, 34*, 156–175.

Blackburn, W. R. (2007). *The sustainability handbook: The complete management guide to achieving social, economic and environmental responsibility*. London: Earthscan.

Blake, J. (1999). Overcoming the 'value–action gap' in environmental policy: Tensions between national policy and local experience. *Local Environment, 4*, 257–278.

Byrch, C., Kearins, K., Milne, M., & Morgan, R. (2007). Sustainable "what"? A cognitive approach to understanding sustainable development. *Qualitative Research in Accounting & Management, 4*, 26–52.

Cantrill, J. G. (2010). Measurement and meaning in environmental communication studies: A response to Kassing, Johnson, Kloeber, and Wentzel. *Environmental Communication, 4*, 22–36.

Chawla, L. (1999). Life paths into effective environmental action. *Journal of Environmental Education, 31*, 15–26.

Cialdini, R. B. (2003). Crafting normative messages to protect the environment. *Current Directions in Psychological Science, 12*, 105–109.

Clark, R. A. (1984). *Persuasive messages*. New York: Harper & Row.

Cordano, M., Welcomer, S. A., & Scherer, R. F. (2003). An analysis of the predictive validity of the new ecological paradigm scale. *Journal of Environmental Education, 34*, 22–28.

Deci, E. L., & Ryan, R. M. (2000). The "what" and "why" of goal pursuits: Human needs and self-determination of behaviour. *Psychology Inquiry, 11*, 227–268.

Dickerson, C. A., Thibodeau, R., Aronson, E., & Miller, D. (2006). Using cognitive dissonance to encourage water conservation. *Journal of Applied Social Psychology, 22*, 841–854.

Dietz, T., Fitzgerald, A., & Shwom, R. (2005). Environmental values. *Annual Review of Environment and Resources, 30*, 335–372.

Ehrhardt-Martinez, K., Donnelly, K. A., & Laitner, J. A. (2010). *Advanced metering initiatives and residential feedback programs: A meta-review for household electricity-saving opportunities. Report Number E105*. Washington, DC: American Council for an Energy-Efficient Economy.

Fishbein, M., & Ajzen, I. (1975). *Belief, attitude, intention, and behavior: An introduction to theory and research*. Reading, MA: Addison-Wesley.

Ganesh, S., & Stohl, C. (2014). Community organizing, social movements, and collective action. In L. L. Putnam & D. K. Mumby (Eds.), *The Sage handbook of organizational discourse* (pp. 743–766). Thousand Oaks, CA: Sage.

Gass, R. H. (2009). Compliance gaining strategies. In S. W. Littlejohn & K. A. Foss (Eds.), *Encyclopedia of communication theory* (Vol. 1, pp. 155–160). Los Angeles, CA: Sage.

Greene, K. (2009). Reasoned action theory. In S. W. Littlejohn & K. A. Foss (Eds.), *Encyclopedia of communication theory* (Vol. 2, pp. 826–828). Los Angeles, CA: Sage.

Guagnano, G. A., Stern, P. C., & Dietz, T. (1995). Influences on attitude-behavior relationships: A natural experiment with curbside recycling. *Environmental Behavior, 27*, 699–718.

Haldeman, T., & Turner, J. (2009). Implementing a community-based social marketing program to increase recycling. *Social Marketing Quarterly, 15*, 114–127.

Hines, J. M., Hungerford, H. R., & Tomera, A. N. (1986). Analysis and synthesis of research on responsible environmental behaviour: A metaanalysis. *Journal of Environmental Education, 18*, 1–8.

Kassing, J. W., Johnson, H. S., Kloeber, D. N., & Wentzel, B. R. (2010). Development and validation of the environmental communication scale. *Environmental Communication, 4*, 121–141.

Kennedy, A. (2010). Using community-based social marketing techniques to enhance environmental regulation. *Sustainability, 2*(4), 1138–1160.

Kollmuss, A., & Agyeman, J. (2010). Mind the gap: Why do people act environmentally and what are the barriers to proenvironmental behavior? *Environmental Education Research, 8*, 239–260.

Kotler, P., & Zaltman, G. (1971). Social marketing: An approach to planned social change. *Journal of Marketing, 35*, 3–12.

Lakoff, G. (2010). Why it matters how we frame the environment. *Environmental Communication: A Journal of Nature and Culture, 4*, 70–81.

Lefebvre, R. C. (2013). *Social marketing and social change: Strategies and tools for improving health, well-being and the environment*. San Francisco: Jossey-Bass.

Loroz, P. S. (2007). The interaction of message frames and reference points in prosocial persuasive appeals. *Psychology & Marketing, 24*, 1001–1023.

Lubell, M., Zahran, S., & Vedlitz, A. (2007). Collective action and citizen responses to global warming. *Political Behavior, 27*, 391–413.

Nicolaou, A. (2014). *In China, 64 percent say they are environmentalists – Report*. http://news.yahoo.com/china-64-percent-environmentalists-report-010358076--sector.html.   Accessed 6 June 2014.

Parguel, B., Benoit-Moreau, F., & Larceneux, F. (2011). How sustainability ratings might deter 'greenwashing': A closer look at ethical corporate communication. *Journal of Business Ethics, 102*, 15–28.

Pelletier, L. G., & Sharp, E. (2008). Persuasive communication and proenvironmental behaviors: How message tailoring and message framing can improve the integration of behaviors through self-determined motivation. *Canadian Psychology, 49*, 210–217.

Pelletier, L. G., Tuson, K. M., Green-Demers, I., Noels, K., & Beaton, A. M. (1998). Why are you doing things for the environment? The motivation toward the environment scale (MTES). *Journal of Applied Social Psychology, 28*(5), 437–468.

Peloza, J., Loock, M., Cerruti, J., & Muyot, M. (2012). Sustainability: How stakeholder perceptions differ from corporate reality. *California Management Review, 55*, 74–97.

Schultz, P. W., Nolan, J. M., Cialdini, R. B., Goldstein, N. J., & Griskevicius, V. (2007). The constructive, destructive, and reconstructive power of social norms. *Psychological Science, 18*, 429–434.

Schwartz, S. H. (1977). Normative influences on altruism. In L. Berkowitz (Ed.), *Advances in experimental social psychology* (Vol. 10, pp. 221–279). New York: Academic.

Seiter, J. S. (2009). Social judgment theory. In S. W. Littlejohn & K. A. Foss (Eds.), *Encyclopedia of communication theory* (Vol. 2, pp. 905–908). Los Angeles: Sage.

Shimanoff, S. B. (2009). Facework theories. In S. W. Littlejohn & K. A. Foss (Eds.), *Encyclopedia of communication theory* (Vol. 1, pp. 374–377). Los Angeles: Sage.

Silk, K. J. (2009). Campaign communication theories. In S. W. Littlejohn & K. A. Foss (Eds.), *Encyclopedia of communication theory* (Vol. 1, pp. 87–91). Los Angeles, CA: Sage.

Staats, H., Harland, P., & Wilke, H. A. (2004). Effecting durable change: A team approach to improve environmental behavior in the household. *Environment and Behavior, 36,* 341–367.

Stern, P. C. (2000). Toward a coherent theory of environmentally significant behavior. *Journal of Social Issues, 56*(3), 407–424.

Stern, P. C., Dietz, T., & Kalof, L. (1995). Values, beliefs, and proenvironmental action: Attitude formation toward emergent attitude objects. *Journal of Applied Social Psychology, 25,* 1611–1636.

Thogersen, J. (2004). A cognitive dissonance interpretation of consistencies and inconsistencies in environmentally responsible behavior. *Journal of Environmental Psychology, 24,* 93–103.

Tiedemann, K. H. (2010). Targeting residential energy use behavior. In K. Ehrhardt-Martineq & J. A. Laitner (Eds.), *People-centered initiatives for increasing energy savings* (pp. 1–18). Washington, DC: American Council for an Energy-Efficient Economy.

Toulmin, S. (1958). *The uses of argument.* Cambridge: Cambridge University Press.

Unsworth, K. L., Dmitrieva, A., & Adriasola, E. (2013). Changing behavior: Increasing the effectiveness of workplace interventions in creating pro-environmental behaviour change. *Journal of Organizational Behavior, 34,* 211–229.

White, K., MacDonnell, R., & Dahl, D. W. (2011). It's the mind-set that matters: The role of construal level and message framing in influencing consumer efficacy and conservation behaviors. *Journal of Marketing Research, 48,* 472–485.

Wilson, C., & Dowlatabad, H. (2007). Models of decision making and residential energy use. *Annual Review of Environmental Resources, 32,* 169–203.

Wood, R., & Bandura, A. (1989). Impact of conceptions of ability on self-regulatory mechanisms and complex decision making. *Journal of Personality and Social psychology, 56*(3), 407–415.

Chapter 5
Transformational Organizational Change, Reinforcing Structures, and Formal Communication

Abstract Given the changes forecast to result from global warming, scholars' and practitioners' interest in sustainability and organizational change is increasing. Although sustainability-related changes can be piecemeal and incremental, my interest is on transformational change. Transformational organizational changes begin when key individuals become aware of new processes, technologies, opportunities, constraints, and expectations. Once awareness occurs, the challenge becomes transforming information into useable knowledge and diffusing it throughout the system. Factors influencing the adoption of an innovation are reviewed. The characteristics of change adopters and stages of change are identified. Important communication roles during times of change (e.g., board members, top executives, change agents, sustainability champions), the process of communicating about change, guidance for change communicators, and formal structural and communication efforts to facilitate change efforts are discussed. Formal ways to embed a focus on sustainability within an organization include changing an organization's structure (e.g., creating new roles, creating new inter- and intraorganizational coordinating structures) and designing pathways (e.g., mission and vision statements, goals and plans, formal communication channels). In addition to transformation change, theories or theoretical concepts highlighted include diffusion of innovation theory, the absorptive capacity concept, sensemaking theory, structuration theory, systems theory, transformational leadership, the communication approach to leadership, models of communication and change, and ethos. Interview data spotlights the City of Boulder, the City of Portland, the City and County of Denver, Sam's Club, Assurity Life Insurance, the HEAL project, the University of Colorado, Portland, the Neal Kelly Company, Tyson Foods, the Portland Trail Blazers, and Aspen Skiing Company.

Lewis and Clark, Information and Change Lewis and Clark began their 1804 expedition during a time of great change in the USA. The Library of Congress had just been established (1800), West Point Military Academy opened (1802), Ohio became our 17th state (1803), and New Jersey abolished slavery (1804). With the purchase of the 530,000,000 acre Louisiana Territory, the USA doubled in size. A major goal of the Expedition was to bring back information on the inhabitants, plants, animals, minerals, geography, and weather of the West. Between 1804 and

© Springer International Publishing Switzerland 2016
M. Allen, *Strategic Communication for Sustainable Organizations*, CSR, Sustainability, Ethics & Governance, DOI 10.1007/978-3-319-18005-2_5

1806, Lewis gathered specimens including boxes of seeds, dried plants, soils, and minerals; live animals and birds; Native vocabularies, pots, bows and arrows, baskets, quilled and beaded clothing, and painted buffalo robes. Some of these artifacts "were the first and last evidence of cultures that would soon perish due to disease and cultural devastation" (Hunter 2009). Over time, physical specimens and artifacts were lost because of the lack of awareness of their historical importance and an infrastructure to keep them safe. This example illustrate how systems (e.g., the USA) change, and how information gathered to meet goals can be lost if it is not integrated into a system when it can be changed into knowledge.

5.1 Transformational Change

"Transformational change is corporation-wide and is characterized by radical shifts in business strategy, reorganization of systems and structures, and changes in the distribution of power across the whole organization" (Robinson and Griffiths 2005, p. 205). If an organization makes a serious commitment to address climate change, this will mean changing their organizational capabilities, culture, structure, and processes (Okereke et al. 2012)—in other words they must engage in transformational change. They must develop new capabilities to assess climate change-related opportunities and risks and to evaluate response options. "The capabilities challenge of climate change is particularly formidable because in addition to the high level of uncertainty, the phenomenon embodies complex technical and multifaceted dimensions ranging from physical science through management to ethics and philosophy" (p. 13). Effective strategy must be formed from often competing voices as internal tensions result from differences in levels of knowledge, risk exposure, and training.

Transformational change can occur in response to specific as well as holistic sustainability initiatives. For example, Delmas and Pekovic (2013) found organizations that sought ISO 14001 certification often also developed new environmental policies, engaged in internal assessment (e.g., benchmarking, accounting procedures), set environmental performance goals, conducted internal and external environmental audits, created cross-functional teams, engaged in more formal and informal communication, and developed employee incentive and training programs. Organizations develop a process of proactive holistic organizational management actions. Blackburn (2007) and Strandberg Consulting (n.d.) provide resources for organizations interested in creating a holistic system.

Transformational Change and the Portland Trail Blazers Justin Zeulner, former Senior Director of Sustainability and Public Affairs, shared a glimpse into how the Trail Blazers began their transformational change process. Initially, their focus was on recycling, but early on Justin's executive management wanted to find out what the Trail Blazers' carbon footprint was and how they compared to other corporations in Portland, as well as nationally. "We had an executive team, and a president saying, 'This is important to us, we would like to understand how we can minimize our environmental impacts and how to have a positive relationship to our community and environment at the same time'." The Trail Blazers hired consultants who led a retreat and helped them measure their carbon footprint and assess their existing policies and procedures. Early on at the 2-day retreat, the consultants asked everyone to close their eyes, imagine a sustainable future, and then describe what that [sustainable future] looked like. "Everyone had their own version. But everyone was involved from the very beginning," Justin explained. Management commissioned a LEED assessment of the Moda Center campus. A consulting organization conducted a Scope 3 analysis. Looking at the analysis of the Trail Blazers' carbon footprint showed they generated 20,000 metric tons of carbon. The Scope 3 analysis also showed 70 % of their transportation-related carbon footprint was generated by people traveling to and from the Trail Blazers' campus. Their consultants assessed existing procedures, processes, and policies. Justin explained, "It was very important to see what kind of policies we had in place. Are we a sustainable organization? Are we environmentally friendly? And if we weren't, then what would that look like?" New policies, programs, and procedures were created and quickly implemented. An internal sustainability team of 35 individuals representing every department and all organizational levels worked together to develop a set of sustainability goals, known as a sustainability charter. Their *Sustainability Charter* established a vision of the Trail Blazers being the leader of sustainability in the sports and entertainment community. Actions were designed to "minimize all of our impacts [energy, water, waste, transportation, purchasing] and to try to benefit our community so that we are a climate-positive organization. That was the driving vision," Justin explained.

As of 2013 the Trail Blazers were communicating their measurable successes on their website. For example, in terms of the transportation portion of their carbon footprint, the Trail Blazers' action strategies included subsidizing transit passes for staff, utilizing bikes and electric vehicles for on-site operations, improving the bike infrastructure for employees and fans, installing electric vehicle charging stations, providing reserved VIP parking for electric and hybrid vehicles, participating in the local Bicycle Transportation Alliance's Bike Commute Challenge, working with local transportation officials to encourage the use of public transit, providing funds to support fareless travel within the surrounding Lloyd District Business Improvement District, and supporting the development of the Eastside Portland Streetcar extension. Reflecting back on their progress toward sustainability, Justin said:

> To me it's a tremendous example of how to succeed with this. It was not perfect by any means. I would say there were some ways to navigate this a little bit differently that would have been more optimal... But I don't know if it is possible. Every organization, whether you are nonprofit, for profit or public sector, everybody has this sort of struggle communicating.

However, to Justin, having the support and encouragement of the team's owner and from top management coupled with the sustainability team's participation in goal and strategy development made for a winning combination.

5.2 Diffusion of Innovations

Transformational change begins when organizations become aware of new processes, technologies, opportunities, constraints, and expectations. Institutional theory helps us understand how new ideas and practices spread throughout a social system. You read about how Walmart supported the development of The Sustainability Consortium to drive change on a global scale (see Sect. 3.2.3).

Diffusion of innovation theory is a key theory which sheds additional insight into the process. Sociologists beginning with the French sociologist and legal scholar Gabriel Tarde originated the concepts basic to this theory (Singhal 2009). Anthropologists, rural sociologists, and agricultural officials expanded the theory focusing on things like the diffusion of the horse among the Plain Indians and farming and family planning practices in Third World countries. Everett Rogers, a professor of communication studies, popularized and expanded the theory into the social sciences with his book *Diffusion of Innovation*. Published initially in 1962, the book is now in its fifth edition (Rogers 2003). To date, the theory has appeared in over 4,000 articles published by scholars representing a wide range of disciplines. Researchers have applied it to the diffusion of sustainability-related initiatives, the spread of organic farming, sustainable prevention innovations, and renewable energy technologies (e.g., Craig and Allen 2013; Smerecnik and Andersen 2011).

Some individuals and organizations adopt innovations earlier than others. Innovativeness is "the degree to which an individual or other unit of adoption is relatively earlier in adopting new ideas than the other members of a system" (Rogers 2003, p. 22). Change agents from outside a social system bring awareness of the innovation to the social system—first through gatekeepers and then through opinion leaders. Rogers proposed an S curve of adoption whereby within any social system there are innovators (2.5 %), early adopters (13.5 %), early majority (34 %), late majority (34 %), and laggards (16 %). Innovators take risks, are younger, have more resources, and interact frequently with other innovators. Early adopters also are opinion leaders who are often younger, have more resources, and are integrated into communication networks. But they are more judicious in their choice of which innovations to adopt. Peter Nierengarten, Director of Sustainability and Resilience for the City of Fayetteville, AR, discussed the S curve. He said:

> The challenge is reaching beyond the 50 % [the innovators through the early majority] to the more conservative demographic. Those are the ones that you have to be particularly interested in when designing your communication messages. That is why we went with this livable community messaging idea. Regardless of what you define livable as, who could be against friendly, livable, hospitable communities?

Awareness of the S curve made its way into how Peter thinks about communicating with stakeholder groups in Fayetteville.

Organizations which adopt sustainability initiatives differ from those which do not. Looking at how for-profit organizations embrace sustainability, Hannaes et al. (2011) also appear to be influenced by the diffusion of innovation theory. They identified two distinct types of organizations: embracers and cautious adopters (laggards). They concluded that the practices of the embracers may provide a snapshot of the future of management. Companies which embrace sustainability see the payoff in terms of intangible advantages, process improvements, the ability to innovate, and the opportunity to grow. Cautious adopters see it in terms of risk management and efficiency gains. Embracers tend to recognize that sustainability strategies have the potential to deliver new customers and markets, innovate existing business models, increase market share and profit margins, and provide a competitive edge. Realizing it is difficult to quantify the outcome of sustainability activities, embracers remain enthusiastic and take a leap of faith. They show six characteristics: they move early, even if information is incomplete; they balance broad long-term visions with projects offering concrete, near-term wins; they drive sustainability top-down and bottom-up; they aggressively de silo sustainability integrating it throughout company operations; they measure everything and, if necessary, create measures; and they value intangible benefits. Initially, embracers focus on waste reduction and efficiencies (e.g., water, energy, materials). These low-hanging fruits are used to make the initial business case for implementing sustainability strategies. For example, Clorox began by measuring its carbon footprint and laying the groundwork on projects so that executives would later adopt greenhouse gas (GHG) reduction, solid waste reduction, and water reduction goals.

The diffusion of an innovation is a five-step process beginning with knowledge followed by persuasion, decision, implementation, and confirmation (Rogers 2003). During the knowledge stage, people become aware of the innovation but know little about it. Many people rarely seek more information. During the persuasion stage, some people become interested and actively seek out additional information. In the decision stage, people weigh the advantages and disadvantages (personal and for the system) of adoption and decide to either adopt or reject the innovation. During the implementation stage, people judge the innovation's usefulness and may continue to seek additional information. Finally, during the confirmation stage, people decide whether or not to continue using the innovation. An innovation can be rejected at any stage.

In the next section, I incorporate diffusion of innovation theory with additional research discussing awareness and knowledge. Then, the focus shifts to how elements of the innovation and the social system can influence an innovation's adoption.

5.2.1 Information and Useable Knowledge

Ultimately, if it is to be adopted within a larger social system, people must be aware that the innovation exists. Awareness is part of the knowledge stage of innovation dissemination (Rogers 2003). People become aware through mediated and interpersonal communication channels. Justin Zeulner, former Senior Director of Sustainability and Public Affairs, became aware of what sustainability entailed after coworkers started asking why the Trail Blazers were not more actively recycling, his senior management asked him to learn more about sustainability, and he began attending sustainability summits. Knowledge, interest in learning more, and the perceived importance of adopting the innovation become important during the persuasion and decision stages of the process. Justin explained that as Trail Blazers staff started getting more energized about recycling they began coming to him and saying,

> Hey, what are we doing with that? I've got to throw that away but I don't want to throw it away. What do I do with that? So I said, 'I don't know, why don't I find somebody that takes used blinds and builds something else out of them' and it slowly evolved into 'why are we using these cleaning chemicals? Why don't we use greener? Why don't we think about the lighting?' It was so exciting to keep doing all of that and pretty soon I started realizing it was connecting personally to the values that I was raised with. And I started just loving the connection and I started educating myself more and more, going to summits and conferences, meeting great people like Allen Hershkowitz at the Natural Resources Defense Council, and getting a chance to hear about what is really going on globally.

5.2.1.1 Where Employees Learn About Sustainability-Related Issues

In their study of how environmental champions convince and enable organizational members to turn environmental issues into successful programs and innovations, Andersson and Bateman (2000) found champions frequently scan multiple information sources and this active scanning increases the likelihood of a successful championing episode. Attending industry and environmental conferences, reading periodicals, and working with consultants were particularly important information sources. Others gathered resources from national and state environmental groups and from competitors. In one Fortune 100 organization, Craig and Allen (2013) asked those employees who rated themselves as more knowledgeable about sustainability and who perceived their company to be more involved in sustainability initiatives where they became aware of information about sustainability. Professional/industry associations, faith-based institutions, and customers were three important information sources.

I asked my interviewees to identify information sources they turned to when seeking to learn more about sustainability. The Urban Sustainability Directors Network was mentioned by my interviewees from the cities of Fayetteville, AR, and Portland, OR. Small businesses receive information from groups such as WasteCap Nebraska and the Missoula Sustainability Council (see Sect. 3.3.2.3). Paul Hawken's *The Ecology of Commerce* (1994) set Ray Anderson, the CEO of Interface, on a new course. *Action Plan*: Identify the information sources you use to learn more about sustainability; seek to expand these sources.

An organization's external environment is brimming with sustainability-related information, ambiguity, and uncertainty. Many of our information seeking actions occur to help us create orderliness from a chaotic information environment. As individuals we construct, rearrange, single out, and ignore features of our external environment. Through talk we then modify and transmit this order as we work together to formulate goals, plans, and strategies. The same process occurs when we attempt to manage our intraorganizational information environment. In his sensemaking theory, Karl Weick (1969) identified how organizational members construct meaning, search for patterns, deal with surprises, and interact as they seek a common understanding which allows them to take action, especially in the face of high risk and complex situations (Dervin and Naumer 2009). Three important concepts in sensemaking theory are enactment (we focus on parts of our environment), selection (we decide how to act in the face of ambiguity), and retention (if an action works we retain it). Sensemaking theory provides the theoretical basis for several studies mentioned in this chapter (e.g., Benn et al. 2013; van der Heijden et al. 2012).

5.2.1.2 Those Who Process External Information to Create Knowledge

A key precursor to successfully absorbing knowledge involves employees' existing related knowledge. "Prior related knowledge confers an ability to recognize the value of new information, assimilate it, and apply it" (Cohen and Levinthal 1990, p. 128) and "relevant knowledge and skill is what gives rise to creativity, permitting the sorts of associations and linkages that may have never been considered before" (p. 130). It permits individuals to understand underlying assumptions and interconnections. Malcolm Gladwell makes a similar and compelling argument in his book *Blink* (2007). He argues that people (e.g., physicians, musicians) can quickly gauge what is really important and make swift decisions based on minimal data relying on their intuitive judgment, something he calls thin-slicing. However, this intuitive judgment is developed by experience, training, and knowledge. The broader employees' educational background and specific competencies are, the wider the pool of knowledge available to the group/organization and the more individuals can contribute to networks of those with similar competencies. Relational network knowledge also is important. "Critical knowledge does not simply include substantive, technical knowledge; it also includes awareness of where useful complementary expertise resides" (Cohen and Levinthal 1990, p. 133).

Simply giving individual employees access to new information and interpersonal connections isn't sufficient to influence the adoption of an innovation. David Driskell, Boulder's Executive Director of Community Planning and Sustainability, described how Susan Anderson and Michael Armstrong from Portland's Bureau of Planning and Sustainability visited Boulder to share information about the *Portland Plan*. Their presentation resonated with administrators, city staff, and others because of its focus on equity and creating connected and socially thriving communities. Having the speakers on location, meeting with groups of people, and sharing what each city was doing over a 2-day period "created some really deep learning... That was hugely beneficial," David explained. He said that when only one person attends a learning opportunity,

It is hard to disseminate that learning into the organization. . . even though we try to whenever anyone goes to any state or national training. We have them do a brown bag lunch when they come back and share their learning. But still only a handful of people are able to go to that and it just doesn't feel like we get the traction.

Key Point: It takes effort to ensure new ideas spread throughout an organization.

5.2.1.3   Absorptive Capacity Is Critical

Absorptive capacity refers to an organization's ability to identify and value new information, combine it with existing knowledge, and use the combined knowledge to drive innovation (Cohen and Levinthal 1990). An organization that is actively seeking external information will be more capable of valuing and acquiring useful knowledge. Absorptive capacity occurs as part of a four-step process—acquisition, assimilation, transformation, and exploitation (Zahra and George 2002). Acquisition involves an organization's capability to identify and acquire externally generated knowledge; assimilation refers to an organization's routines and processes that allow it to analyze, interpret, and understand the new information; transformation involves an organization's capability to develop and refine the routines for combining new knowledge with existing knowledge; and exploitation refers to the development or existence of routines that leverage the existing knowledge and integrate it so it can be useful to the organization. Focusing here on the acquisition phase, the absorptive capacity literature describes how more exposure to the external environment provides opportunities for better access to information which may stimulate innovation. An organization's ability to monitor and scan the external environment for potentially useful information and then to allow organizational members to synthesize this information into concepts and/or ideas is important. Proactive organizations create pathways for importing new information. Without such pathways, knowledge will not be absorbed. Discussing knowledge and knowing from an organizational communication perspective, Kuhn (2014) shifts us from focusing on knowledge as the cognitive domain of an individual to the practice by which knowledge contributes to organizational effectiveness through networks of communication relationships.

5.2.2   Factors Influencing Innovation Adoption

Diffusion of innovation theory (Rogers 2003) identifies four main elements which influence the spread of an innovation: the characteristics of the innovation, the characteristics of the social system (i.e., its norms on diffusion, the perceived consequences of innovation), the communication channels used to disseminate and evaluate information (e.g., opinion leaders and change agents), and the process occurring over time (e.g., the types of innovation decisions). This next section focuses on the innovation and social system characteristics before moving on to discuss key communication channels.

5.2.2.1 Innovation Characteristics

Important characteristics of an innovation include the relative advantages it brings to goal achievement; whether it is compatible with the existing values, past experiences, and needs of potential adopters; whether it is simple to use; whether users can try the innovation before adopting it; and whether others will see the innovation. Visible innovations increase both positive and negative communication within personal networks.

In their investigation of how the characteristics of sustainability innovations influenced their adoption, Smerecnik and Andersen (2011) focused on seven sustainability innovations being made by North American hotel and ski industries: sustainability management, environmental communication, managing resort pollution, resource conservation, water recycling, energy conservation, and guestroom sustainability. Most of these concepts are self-explanatory. However, sustainability management refers to the creation of an environmental committee, a written environmental policy, an environmental impact assessment report, a GHG emissions or carbon footprint assessment, a program to reduce environmental impacts, the use of external consultants, sending officials to sustainability conferences, and/or the adoption of a sustainability certification program. Environmental communication involved environmental training of staff, environmental education of guests, environmental statements in public messages or resort descriptions, routine meetings to discuss environmentally related issues, community environmental support, involvement or advocacy, and dialogue with others in the industry about environmental sustainability. The authors found an innovation's simplicity was the best predictor of its adoption. They suggested emphasizing simplicity and ease of adoption whenever possible. Usually, relative advantage, which involves business case for sustainability arguments, is the strongest predictor of adoption. However, they found relative advantage only correlated with the sustainability management and environmental communication initiatives. They discussed why compatibility and trialability should influence an innovation's adoption, but found neither did in their study. *Best Practice*: Sustainability communicators should design messages addressing a proposed innovation's relative advantage, simplicity, compatibility, observability, and trialability, as appropriate.

Compatibility and Observability at Sam's Club Brian Sheehan, former Sustainability Manager at Sam's Club, talked about compatibility and visibility in terms of measurement and feedback messages. As Walmart sought to ensure its buyers seriously considered sustainability they integrated it into the tools buyers used every day. On the operations side, when market managers tour a Sam's Club, they receive a report showing how well that Club is executing around various sustainability initiatives. Brian explained,

> We've taken what we think are key performance indicators for Clubs, if you are doing this thing well you are probably executing most of your sustainability program well, and put them right on the front page [of the report] along with how well sales and associate engagement are going.

Employees can see how they are doing on the key sustainability performance indicators (KPI) (i.e., observability). If they are not doing so well, that becomes an improvement opportunity. Brian said, "It is important to make it [sustainability] clearly visible and integrated into all of the standard business practices." For example, the front of the report provides a recycling score for super sandwich bales (these bales capture most of the plastics coming out of a Sam's Club). Clubs receive scores that indicate the number of tons recycled and how that Club ranks relative to its region, market, and division. *Best Practice*: KPIs are useful because they are measures or mileage markers that indicate whether procedures are actually working to help a company meet its goals.

5.2.2.2   Social System Characteristics

Organizations differ in how actively they seek external information and in their openness to innovation. An organization's strategic posture is comprised of its tendency toward innovation, its response orientation in comparison to peer institutions (proactiveness), and its propensity toward risk (Covin and Slevin 1989). Proactive organizations often develop the capabilities needed to acquire and assimilate new knowledge. Increased exposure to new knowledge builds future capabilities. When risk-taking is encouraged, organizations are more apt to innovate. For example, Google engineers spend 80 % of their time working on the core business and 20 % working on a company-related project which they find personally interesting (Danet et al. 2013). Promising new ideas are quickly launched and minimal soul searching occurs if the new ideas fail. Failure is tolerated as a form of innovation waste. Fast failures allow Google to move on to the next idea. Sometimes Google appears not to engage in the reflection needed to learn from mistakes and abandons the fast fail approach when it really matters. But system characteristics at Google do support innovation.

Structural barriers and barriers to organizational learning can hamper the full integration of sustainability within an organization. It is difficult to identify ways to integrate sustainability into preexisting organizational systems and structures and to implement and embed sustainability practices to the point that employees have embraced sustainability (Benn et al. 2013). Therefore, sustainability professionals should be aware of what influences an organization's absorptive capacity. The absorptive capacity literature talks about various capabilities which can help organizations turn information about an innovation into useful knowledge which can then be used to improve existing operations or develop new alternatives. Cohen and Levinthal (1990) call these important capabilities combinative capabilities and write "an organization's absorptive capacity is not resident in any single individual but depends on the links across a mosaic of individual capabilities" (p. 133), something Kuhn (2014) discusses in terms of networks. Combinative capabilities can take three forms: coordination, socialization, and systems capabilities. The first two will be discussed later (see Sects. 5.4.1 and 6.2.2.2). Systems capabilities involve the integration of knowledge through policies and procedures and "provide a memory for handling routine situations" (Van den Bosch et al. 1999, p. 556). They can minimize the need for further communication and coordination among sub-units. However, in organizations where there is a high degree of rules, procedures, required approvals, and red tape, employees are less motivated or free to engage in spontaneous knowledge scanning, information absorption, and innovation. On the other hand, goals, policies, and procedures can help promote sustainability across an organization.

5.2.2.3   Systems Thinking

In the nineteenth century biologist Ludwig von Bertalanffy developed general systems theory. In the late twentieth century, the theory was adopted by multiple disciplines. It has proven especially useful for natural sciences, communication (Littlejohn 2009), and organizational researchers. Systems theory focuses us on how a system is a set of integrated and interacting parts that together create a larger whole. Poole (2014) provides an excellent summary of the theory and its influence on organizational communication research and practice. All systems have four aspects—objects (i.e., parts or elements), attributes (i.e., characteristics of the object and the system), internal relationships (i.e., patterns of relationships), and an environment (i.e., influences that impact the system). Systems are distinguished by three qualities—wholeness and interdependence, hierarchy, and self-regulation and control.

This chapter's initial focus was based on the systems theory idea that organizations import resources from their external environments which help them adapt to conditions in their macro-environments (e.g., changing norms, new technologies, climate changes). It is important that people within an organization engage in systems thinking (Oncica-Sanislav and Candea 2010). Often we tend to focus on the parts rather than the whole and fail to see an organization as a dynamic long-term process existing within a macroenvironment.

Systems Theory, the City of Portland, and the City and County of Denver Knowledge of systems theory made an important impact of Susan Anderson's, Director of Portland's Bureau of Planning and Sustainability, thinking—especially the ideas of wholeness and interdependence. When I asked her how she personally defined sustainability, Susan said:

> I define it as two simple things. The first is everything is connected—the environment, economy, jobs, our personal health, the community's health. If you mess up one piece, somehow down the road the other pieces will get messed up. The second idea is that everything you do today effects tomorrow—how you got here, what clothes you have on, the foods you eat, the car you drove, the table we are sitting at, the building we are in. Everything. The decisions we make about our children, our families, our businesses,

they all have an impact on tomorrow. If we do things right, we may have a better tomorrow and if we do things wrong, we leave a big mess for our kids to clean up. Everything is connected and you cannot run the jobs program without having some environmental and health impacts. You can't run a health program without job impacts. People tend to think in silos because we are human and it is hard to think of more than one thing at a time. We have to think about the connections... To me, sustainability is about those connections.

Best Practice: In his description of learning organizations, Senge (1990) recommends that decision makers construct a systems map diagram that shows the key elements of systems and how they connect, and utilize information systems that measure the performance of the organization as a whole and of its various components.

Jerry Tinianow, Chief Sustainability Officer for the City and County of Denver, shared an example of the need for systems thinking saying, "Getting people to think systemically instead of in their own little area is another of the big challenges. But we have seen some early progress with that." He described the Denver Energy Challenge's success [in conducting home energy retrofits] in meeting their goals 3 months ahead of schedule, operating within their budget, and receiving awards from the federal government. He said:

But the fact of the matter is, if we were distributing necessary energy efficiency gains proportionately, we have to go from 2,000 homes in 3 years to 90,000 homes in 7 years. Well, that probably is not going to happen, especially since the federal grant that funded the Challenge has run out. So that is a real shock when you run those numbers because people are not used to thinking in those terms but what I saw [when I shared the numbers with them] was they immediately began to think systemically. Immediately. And one of the first things they said was maybe we should look at getting a disproportionate contribution from some of the other energy sectors where we produce bigger numbers. Maybe we really need to be focusing on commercial and industrial buildings. Which is exactly what they need to be thinking... Suddenly they are thinking across departmental lines, across disciplinary lines and so forth. I was very encouraged by that meeting, even though when you run the numbers by people initially, you know a lot of them are kind of crestfallen. A lot of people think that you are criticizing their prior work. Are you saying that there is no value in what we did? No, I am not saying that at all. What I am saying is, what you did was very valuable but what you were asked to do was not sufficiently ambitious. We have to think at a much bigger level. And to their credit, they are.

Systems theory is generally oriented toward the long-term view given than systems depend on resources in their external and internal environments so that they can adapt, survive, and potentially thrive. Corporate sustainability requires a broader vision and longer time horizon over which opportunities and risks are assessed and major action programs developed, as well as attention to a wider spectrum of stakeholders. Routinely people and organizations engage in convenient actions that produce improvements in the short term, even though these actions may lead to significant long-term costs. For example, making modest energy saving retrofits provide short- and near-term cost savings whereas also supporting and investing in wind and solar energy have the potential to result in long-term savings and a more dependable clean energy supply. *Action Plan*: Think about how to shift more of the focus of your organization onto long-term planning while simultaneously making significant short- and near-term changes.

5.2.2.4 Influential Communicators About Innovation

This section discusses the role of boards of directors, top executives, and other change agents and champions.

The Board of Directors

Board of directors can provide important strategic direction for how organizations respond to environmental issues and can provide access to external resources, including information and relational networks (Walls and Hoffman 2013). Walls and Hoffman indicated approximately 60 % of public companies have dedicated board committees charged with overseeing issues related to sustainability. However, survey data gathered from 2,587 respondents from commercial enterprises in 113 countries suggests that board engagement is much lower (Kiron et al. 2015). A report on sustainability and governance by the United Nations Environmental Programme Finance Initiative (UNEP 2014) called this a real leadership challenge. The report's authors analyzed 2011 Bloomberg corporate data on 60,000 businesses and found less than 2 % of companies that report environmental, social, and governance information had a director with responsibility for sustainability.

It is important that board members attend to sustainability issues. Board members have agency (i.e., the ability to act) and can interpret, construct, and enact their external institutional environment by paying selective attention to some issues, interpreting issues in nonroutine ways, and providing important strategic direction (Walls and Hoffman 2013). Companies that include sustainability as a top management agenda item are more than twice as likely to pursue strategic or transformative sustainability-focused collaborations with a broad array of partners (i.e., other companies, academic institutions, governments, NGOs, multilaterals) (UNEP 2014).

It is important to understand what barriers limit greater board engagement with sustainability. The greatest barriers seem to be that sustainability has an unclear financial impact, the lack of sustainability expertise among board members, the board's other priorities, a tendency to only focus on the short term, and the belief that boards should only focus on increasing shareholder value. Kiron et al. (2015) provide suggestions on how to address each barrier. Board members need to realize they do not have a legal fiduciary responsibility to focus only on shareholder value and that they can be held personally liable if they do not adhere to environmental regulation and sued if they do not recognize the implications of an organization's environmental actions.

Beyond simply looking at whether or not boards even consider sustainability-related issues, it is important to identify factors which influence their ability to recognize, frame, and interpret environmental issues and influence their organizations' actions (Walls and Hoffman 2013). Two aspects are particularly important: structural elements and intraorganizational factors. Structural elements involve interlocking directorship or network ties which are key channels for the collection of information and the dispersion of organizational practices. The information dispersed through these channels is often trusted and timely. Intraorganizational factors involve board members' skills and experience at retrieving, filtering, and interpreting information. Board members' specialized and innovative knowledge and experience allow organizations to innovate by breaking away from their institutional field's norms. The more experience board members have involving environmental issues, the more complex their personal knowledge structures and the more likely they are to deviate from the dominant institutional norms. If organizations want to innovate away from normative environmental practices, they must appoint directors with environmental experience. This is especially critical if an organization is centrally embedded in its institutional field, because centrality promotes conformity to institutional norms. Another way to increase board expertise is to create an external advisory board. For example, Kimberly-Clark created a seven-member external advisory board made up of experts in different aspects of sustainability. Sustainability needs to be integrated into the board duties and into established board committees (e.g., compensation, governance, audit, and nominating), or become the focus of a board subcommittee charged with identifying and addressing material sustainability challenges. *Action Plan*: Investigate the credentials of the people on your board of directors. Discuss how a different set of credentials and a different board configuration might benefit your organization.

Top Executives

Top executives increasingly recognize that environmental sustainability is important to the profitability of their companies. Many recognize that changing climate conditions, fresh-water scarcity, and high energy prices can have a negative impact on their company's long-term success. As the spokesperson for an organization, a top executive's conduct and level of commitment to ethical principles influences the organization's overall image (Ferns et al. 2008). Internally their leadership behaviors involve motivating employees, communicating ethical norms, and setting future direction.

Early on in the process of refocusing Interface, the world's largest manufacturer of commercial carpets and floor coverings, to become an innovator aspiring for climate neutrality, CEO Ray Anderson felt he must take an ethical stand and no longer be a plunder of the earth. He created a definition of sustainability for his company which added product and place to the three p's of people, planet, and profit. He worked with an advisory firm to create awareness of corporate responsibility, minimize environmental degradation, and introduce the Seven Fronts of Sustainability to his employees. He collaborated with sustainability experts to generate business changing ideas. For example, the science of biomimcry, which involves using nature's models, designs, and processes to solve human problems, focused Interface on designing individual carpet tiles so they resemble nature and can be replaced when worn. Interface then developed a 100 % recycled nylon carpet made from old carpets and discarded fishing nets. To gather fishing nets they worked with NGOs which were cleaning beaches in the Philippines, India, and Africa. Finally, they developed a new type of nylon, 63 % of which is made from castor oil. Castor oil comes from a growing crop which only needs water 1 day out of 25 (Danet et al. 2013). The process Ray began has led to continual innovations at Interface.

Leading Change at Bayern Brewing Organizational leaders make decisions which can shape the pro-environmental behaviors of others in their organization, their community, their consumers, and, potentially, their industry. Certainly Walmart's three aspirational goals show how this can be done on a large scale; however, small organizations also have a significant impact. While I was in Missoula, MT, I interviewed Thorsten Geuer, brew master, and Jared Spiker, sales and marketing manager, at Bayern Brewing. Bayern Brewing, the only German microbrewery in the Rockies for over 25 years, has 37 full-time and a number of part-time employees. This small company has sought to make a region-ally significant impact by changing the way they do business. Since water is critical when making beer, they teamed up with Montana Trout Unlimited, naming Danc-ing Trout the official beer of Montana Trout Unlimited. A percentage of all Dancing Trout merchandise and beer sales goes directly toward conserving and restoring Montana's cold water fisheries, riparian areas, and watersheds. Bayern contributes to the Bonneville Environmental Foundation water restoration program. Their improved water efficiency methods reduced the amount of water it takes to brew a gallon of beer down to around 5 gallons. Bayern recycles the water they use in brewing beer so it can be used by ranchers, by wildlife, or for recreation. Their spent grains go to a nearby family-owned ranch. But their biggest impact has to do with packaging. Thorsten told me:

> The biggest challenge that we have here in Montana is that Montana is an importing state. A lot of the raw materials (e.g., cardboard, six pack carriers, glass) are imported. We generate money in the state. We make 80 % of all our wholesale sales in Montana. A lot of those dollars go out of the state . . . That's when we started thinking about more than what you just mentioned [i.e., feeding supply products to animals, trying to have a zero water impact, buying water certificates]. So we started thinking, what else can we do? And the biggest thing we implemented over the last 4 or 5 years is to find a way we can get some of our packaging material back to the brewery and at least get one more use out of it.

For several years the company had bought back their 6-pack holders and/or gave customers who recycled bottles and 6-pack holders trade-in value for beer or merchandise in their tasting room. In 2010 they began recycling their own glass, making it the first brewery in Montana, and one of a few in the nation, that recycles (or reuses) all its packaging materials. Thorsten explained:

> We said we have so much glass coming back what if we purchased a bottle washer where the glass bottle, which is perfectly fine, will not be repurposed into road material or building material. It will be filled again with beer. So we purchased a bottle washer. We noticed that money is no longer going out of the community, out of the state. We keep the money in the state.

Bayern Brewing produces 50,000 bottles of beer a week, with 30–40 % of that going into recycled bottles. They saw no reason they could not utilize 100 % recycled bottles. "There is so much glass out there... There must be a way to get more glass back. That is what we are thinking about right now, how can we get more glass to come back to Bayern?" Considering that most beer bottles in the USA hold beer only once, imagine the impact if other breweries followed this brewery's lead. Their decision to purchase a bottle washer has implications for the Montana economy, has reduced energy use associated with transportation and glass recycling, and has stimulated recycling within their community. It also protects them from price fluctuations in packaging materials and reduces the freight changes they pay to import materials. This example illustrates how influential acts of human-material organizing emerge from leadership communication and how leadership communication has the potential for reflexivity, moral accountability, and change (Fairhurst and Connaughton 2014).

Leaders play various roles when it comes to an organization's ability to innovate and change (e.g., designers, stewards, and teachers) (Senge 1990) as well as promote sustainability (Egri and Herman 2000; Ones and Dilchert 2012; Robertson and Barling 2013). As designers, leaders and their top management teams create or endorse the purpose, vision, and core values by which their employees are to be guided (e.g., as a sustainable organization). They craft purpose stories which are 'the overarching explanation of why they do what they do, how their organization needs to evolve, and how that evolution is part of something larger' (Senge 1990, p. 346). As stewards they manage the vision to benefit others, listening to others' visions and changing their own as necessary. As teachers, leaders influence others' views. Leaders of innovative organizations focus mainly on providing purpose, developing systemic structures, and helping others develop systemic understandings.

Organizational communication research conceptualizes leadership as an act of transmission and negotiated meaning. It is an individually informed yet relational phenomenon (Fairhurst and Connaughton 2014) which enables collectives to mobilize in order to act. The management of meaning perspective of leadership (Smircich and Morgan 1982) is an important contribution of communication scholars (Fairhurst et al. 1997). It focuses us on how leaders frame and define reality for others. As leaders manage meaning they highlight particular aspects of the overall flow of experience occurring in a particular context. Those being led see the action being promoted as a sensible and viable action. Leaders must make actions personally meaningful to those who must implement them. Together leader and follower discuss implementation choices, possible futures, and next steps.

Julie Diegel, Director of Sustainability Programs at WasteCap Nebraska, talked about the leader's role with me saying,

> The head of the company has to share their values with upper management and upper management has to believe in them enough to share them with the people that they supervise. If they don't, there are going to be some gaps. But you need the leader(s) to take that stand and say it over and over again. [They] have to live it and say it and breathe it and repeat it again and again. That's how people learn and believe you are serious about

it. If you are the leader, you have to work harder than everyone so the people under you believe that your quality standards are high. That's how people understand what your values are and your commitment is. I don't know that there is one secret formula. It takes lots of communication and education and training. There are structural things you can do but I think that there is that human truth or passion that has to exist or it probably won't really fully take hold. You think about Ray Anderson. That is what he had. He was a real person with real honest values and people responded to that.

Ray Anderson understood how to transmit his shared vision because he was a transformational leader.

Transformational Leadership and Pro-environmental Behavior

Transformational leadership theory is important when discussing the creation and dissemination of a shared vision. The concept was introduced by Burns (1978) and extended by Bass (1985). Transformational leaders engage in inspirational motivation by creating an energizing vision, demonstrate confidence in themselves and their mission, set challenging goals, and enlist others to support their vision and mission. They stir emotions (e.g., trust, loyalty), evoke symbolic images and expectations, create a shared sense of identity, inspire desires, develop fresh approaches to long-standing problems, seek out risks where opportunities and rewards appear likely (Egri and Herman 2000) and encourage followers to generate new options.

Arguing employees' pro-environmental behaviors (e.g., recycling, conservation, waste reduction) contribute to the greening of organizations, positively effects climate, and prevents additional environmental degradation. Robertson and Barling (2013) developed and tested a model that linked environmentally specific transformational leadership and leaders' workplace pro-environmental behaviors to employees' pro-environmental passion and behavior. They studied 139 subordinate–leader dyads recruited in the USA and Canada. They defined environmentally specific transformational leadership as leadership behaviors focused on encouraging pro-environmental initiatives. Four behaviors are typically discussed in the transformational leadership literature: idealized influence, inspirational motivation, intellectual stimulation, and individualized consideration (Bass 1985). Idealized influence involves acting as a role model guided by a moral commitment to an environmentally sustainable planet and selecting actions benefiting the natural environment. Leaders high in inspirational motivation display passion and optimism which stimulate employee actions toward the collective good. Intellectually stimulating leaders encourage employees to think for themselves, question assumptions, and innovatively approach problems. Individualized consideration involves displaying compassion and empathy for employee well-being and providing help in employee development. Robertson and Barling found that when leaders believe respected others and family members care about the environment, this influences their transformational leadership and personal environmental behaviors. When leaders display environmentally specific transformational leadership behaviors and pro-environmental behaviors, this influences employees to feel more interested in and passionate about the environment which then leads to more pro-environmental employee behaviors. Environmentally specific transformational leaders can positively influence their employees' pro-environmental passion and behaviors when they (1) share their own environmental values, (2) convince employees they can achieve at an aspirational level, (3) help employees think about issues in new and innovative ways, and (4) have a relationship with employees.

Leadership at Assurity Life When traveling through Lincoln, NE, I interviewed Bill Scmeeckle, Assurity Vice President/Chief Investment Officer. Bill said, "We are fortunate that our President and CEO (Tom Henning) is very environmentally focused...[he is someone] who took sustainability very seriously and we wanted to." Bill said they built their LEED Gold certified office building partly due to their leader's vision. After construction, but prior to occupancy, Henning wrote a three-part blog outlining his sustainability-related philosophy and vision for the new building. He indicated he wanted to form a sustainability task force (i.e., green team) and asked for volunteers. That was when Tammy Rogers, Senior Information Technology Business Analyst, joined the green team. Tammy said, "He tasked us with promoting awareness about sustainability topics and helping people learn about things they did not know about." Henning's vision shaped the working life and environment of Assurity employees, inspiring many to be more interested in sustainability and providing some with an outlet where they think creatively about new issues and aspire to influence their organization's operations.

In a study of 73 leaders of nonprofit environmental and for-profit environmental product and service organizations, Egri and Herman (2000) investigated the leaders' personal values. The nonprofit environmentalist organizations focused on issues such as promoting alternative transportation, protecting and/or conserving natural resources, coordinating other environmental organizations, or education. The for-profit organizations manufactured products such as alternative fuels, nontoxic cleaners, and biological pesticides, or they offered services such as waste disposal or environmental facility design. Egri and Herman found the nonprofit leaders' personal values were more ecocentric (i.e., they saw intrinsic value in all living organisms and the natural environment), open to change, and self-transcendent than those of for-profit leaders. Both nonprofit and for-profit leaders held more ecocentric values than did the leaders of for-profit nonenvironmental organizations, although the for-profit environmental organizations leaders' values were more moderate than the nonprofit leaders. In both types of environmental organization, the leaders were master managers performing both transformative and transactional leadership behaviors. Egri and Herman concluded that nonprofits are highly receptive contexts for transformational leadership while the for-profit environmental organizations are only moderately receptive. An organization's size and age also influenced whether or not they supported transformational leadership. Egri and Herman developed a model of the values, personality characteristics, and leadership skills that typify environmental leaders. They investigated the influence of goals, production systems (e.g., low energy and resource use), and organizational systems (e.g., nonhierarchical structures, participative decision making, and decentralized authority) in the three organizational types. Environmental organizations generally had adaptation orientations; a boundary-spanning task structure; a simple, adhocracy, or network structure; and a clan mode of governance.

Leaders are inspired to adopt sustainability innovations for a variety of reasons, communicate their interest in a variety of ways, and have multiple lessons to share. What advice can these leaders share? *Best Practice*: Quinn and Norton (2004) gathered advice from leaders for others interested in embarking on the journey toward sustainability including:

- Make a focused commitment to long-term strategic thinking and action
- Look at the basic science to see where your business fits into the larger social and environmental context
- Incorporate sustainability principles into every facet of the business (e.g., corporate strategy, decision-making processes)
- Frame sustainability concepts positively (e.g., stay optimistic and hopeful)
- Make sure top management is fully committed to sustainability
- Communicate tirelessly to all employees
- Seek input and ideas for turning principles into action
- Be persistent
- Learn by doing
- Start with the low-hanging fruit
- Track your results
- Expect setbacks but celebrate successes
- Keep the momentum going
- Avoid the adversarial
- Forget about late adopters and resisters
- Have faith in the power of one person to make a difference and one company to significantly impact its industry and the world
- Remember to stress hope and sharing abundance rather than doing without
- Be nonproprietary and share information internally and externally

Interface is a good example of an organization which shares information internally and externally. They promote their speaker series where they share what they've learned along their journey to Mission Zero on the sustainability portion of the Interface website. Sustainability director Ramon Arratia described how Interface shows other organizations they can achieve zero impact (Danet et al. 2013). Arratia is quoted as saying:

> We work with consultants, giving our knowledge for free so they can use that to help their clients. Every year we take some consultants to Holland for 2 days, they can bring their clients if they want... Basically we are trying to give everything we have learned about sustainability so others can profit from it.

Sustainability Champions

Leaders can't change an organization's processes, procedures, products, and culture singlehandedly (Blackburn 2007). Within any social system (e.g., organization, culture), there are gatekeepers, opinion leaders, and change agents who can assist or block leaders who seek to create change. Ideally, "Gatekeepers, boundary spanners, and change agents are key to the ability of organizations to acquire, assimilate, transform, and exploit new knowledge" (Jones 2006, p. 368). The decision whether or not to adopt an innovation is significantly influenced by largely subjective, rather than scientific, discussions with peers and respected others. Opinion leader recommendations are critical. When a system's norms are changing, opinion leaders can become innovators.

A number of my interviewees referred to or were themselves opinion leaders or sustainability champions. According to Woodrow Nelson, Vice President of Marketing Communication, when the Arbor Day Foundation began its Tree City USA program, they partnered with the National Association of State Foresters and the U.S. Forest Service which had a few champions who saw the program's potential and were ready to help recruit cities. Working across organizations, these early champions created a large and dynamic movement. In 2015, there were 50 state coordinators employed by various governments or NGOs and the program exists in 3,400 towns, cities, and military bases which serve as home to more than 135 million Americans.

Championing the HEAL Program   One of my interviewees was a sustainability champion seeking to promote an initiative across the USA. Champions as individuals who, either through their formal roles and/or personal activism, attempt to introduce or create change in a product or process within their organization, or beyond (Andersson and Bateman 2000). Martha Jane Murray initiated and now administers the Home Energy Affordability Loan (HEAL) program through the Clinton Climate Initiative (CCI). Martha Jane, an architect, was one of five national USGBC core committee members who organized the national USGBC Challenge GreenBuild 2005 response during the Hurricane Katrina sustainable rebuilding effort. Martha Jane designed and piloted the HEAL program in the shoe manufacturing business, Neil Munroe Footwear; she and her husband co-own in Wynne, AR. Following an energy retrofit of their factory, they placed the first year's financial savings (about $40,000) into a revolving loan fund their employees could use to make energy retrofits in their own homes. The loans were paid back through payroll deductions. This model is being implemented across the USA. One of the early adopters was the L'Oreal manufacturing plant in North Little Rock, AR, since the program fits the parent organization's sustainability initiatives.

Sustainability champions recognize the significance of the issues. Martha Jane was quoted as saying (Paddack 2010),

> What continues to drive me is the belief that we can't do enough fast enough to avert the consequences of climate change for our children and future generations. My goal is to open people's minds to other possibilities for solving some of our most serious challenges: energy independence and national security, economic development and, most importantly,

connecting us to each other and the planet. I hope to demonstrate through the CCI that the employer-assisted energy-benefit program is an effective, scalable financing tool for residential energy-efficiency retrofits.

In their field study of 132 successful and unsuccessful championing episodes in US business organizations, Andersson and Bateman (2000) developed and refined a model of the championing process. The championing process involves (1) identifying/generating an issue or idea, (2) packaging it as attractive, and (3) selling it to organizational decision makers. Often champions are those who place the idea of sustainability on the business agenda and rally others to take action (Blackburn 2007). I asked Martha Jane how she promotes the HEAL program. She said:

One of the first things we do is try to get to the decision makers. So the first step is to educate the highest person available who can put us in touch with the decision maker. Our preference is to start with a C-level engagement (e.g., the COO, the CFO), the CEO would be optimal. With L'Oreal, we took the contract all the way up to the top of their corporate chain which was in France. Unless you get that level of buy-in it is very hard to move any programmatic changes forward.

Allen Hershkowitz, Senior Scientist with the NRDC, who is working with the entertainment and sports industry to drive a cultural change in how people view the environment, agreed. He told me, "At the end of the day when decisions about direction need to be made, they are typically made by the sports commissioner or the president of the company or the team owner. Those are the people responsible for the role of decision making."

How do champions frame their messages? Champions often initially package and sell sustainability-related initiatives by reflecting on idealized norms rather than cost–benefit frameworks (Bissing-Olson et al. 2012). Then, they rationalize actions within a cost–benefit framework, stress the situation's urgency, and/or argue the organization can have a significant impact on the local or global community (Andersson and Bateman 2000; Blackburn 2007). Small projects can be used to create learning which, if successful and resonant, can be disseminated across an organization. Those who seek to engage others should display and act from their own passion, share their own stories, support earlier adopters and latecomers, and stay positive (Werbach 2009). They should avoid preaching, using scary facts without offering good news, evoking authority, and not giving people a place to start. Substituting complex and technical jargon with understandable business terminology is important. Arguments need to be logical, provide factual evidence like one would champion any other business issue, and be inspirational.

Susan Anderson, Director of Portland's Bureau of Planning and Sustainability, talked about how within any city organization having local champions is essential "whether its council members or the major who have the power or capacity to deliver the bureaus and departments and to give them direction." In terms of communication, Susan recommends champions talk about things their listeners care about (e.g., saving money, looking good to the boss), show listeners how it helps them get their job done, provide listeners with positive personal feedback, or help provide publicity for their completed projects. "So they [city employees and politicians] see a benefit of making all the parking meters solar, so they see a benefit of all these different things. Not because they care about climate change personally but because their boss or politician liked it or they get some kind of personal feedback."

To be effective change agents, champions must be seen as credible by management and peers, good collaborators, excellent communicators with both internal and external stakeholders; knowledgeable about the organization and its business and culture; knowledgeable and passionate about sustainability issues; and able to foresee triple-bottom-line benefits (Blackburn 2007). Being a champion takes courage or as Rogers noted (2003) they must be risk tolerant individuals. Champions need to be cognizant that the timing of their initiative is crucial. If the implementation of a change takes too long, interest often wanes. Also, delays in making major announcements can result in leaks which lead to increased employee stress and resentment.

Champions rely on other influential people in their organization to join them to move the idea forward (Andersson and Bateman 2000), which is consistent with Rogers' (2003) discussion of the central placement of innovators and early adopters in communication networks. Building a strong coalition and enlisting the aid or endorsement of others helps add credibility and legitimacy when trying to convince leadership to act or when it is time to implement a change. Martha Jane described how after getting top management approval to enact the HEAL program in their organization she begins working with others saying:

> We engage with the human resource and accounting departments because in our case we have to have their buy-in to get access to employees and they understand how to best communicate with them. We don't try to tell them how to communicate with their folks, we try to understand what is already working and then we mold our process to what fits with their corporate culture.

Teams are needed. Team members can be drawn from a variety of units including environmental health and safety, human resources, purchasing, supply chain, finance, governmental or public affairs, community relations, law, and communications (Blackburn 2007). A team leader must coordinate and facilitate the organization wide efforts. This person does not have to be the initial champion but must have time to commit to the endeavor and be motivated; knowledgeable about sustainability; trusted; process- and goal-oriented; and possess good communication, problem solving, collaboration, and organizational skills. Once enough champions come on board, the adoption of the innovation can begin. Martha Jane described how that process works for the HEAL project saying:

> [We have something] we call the water fountain approach. Once we have success, this breeds success and people start talking to coworkers about 'you should get involved. It is the real deal. This does really work. I am much more comfortable in my home. And I am saving money.' They invite other colleagues to come to the next presentation and the innovation spreads through the workforce.

Organizational leaders, management, and change agents face internal barriers and resistance when seeking to implement sustainability-oriented changes (Dunphy et al. 2003). "Implementing sustainability is an interesting process...because it confronts people with a new reality that influences all activities and departments of the organization. Often, people cannot rely on existing routines to make sense of new ideas" (van der Heijden et al. 2012, p. 536). Change can lead to feelings of personal uncertainty, employee stress, and resistance (Yeatts et al. 2000). Employee stress during times of change has been linked to a variety of negative reactions including low motivation and morale, reduced job satisfaction, and lowered performance and high turnover; plus an increased reluctance to accept the organizational change (Bordia et al. 2004; Swanson and Power 2001; Vakola and Nikolaou 2005). Poorly managed change communication can result in rumors and resistance. In a study of change messages as discursive constructions, Bisel and Barge (2011) found the formal organizationally sponsored change messages negatively influenced employee identities. Employees experienced feelings of violation, recitation, habituation, or reservation. Their findings have implications for the way change agents approach issues of employee sensemaking, emotionality, resistance, and materiality during planned change processes. Their study is based on discursive positioning theory. This theory helps describe how everyday discourse calls identity and subject positions into being, and how these positions are adopted or rejected. Bisel and Barge's study highlights the importance of sensemaking during change; it allows us to look at the social sensemaking processes of articulation, resistance, and reflection simultaneously; it leads us to consider the relationship between emotion, positioning, and resistance; its directs us to look at the discursive and material contexts surrounding message production; and it suggests the use of turning point methodology when investigating how people respond to change initiatives.

5.3 Communicating About the Changes

5.3.1 Change Is a Process

Santiago Gowland, Unilever's Vice President of Brand and Global Corporate Responsibility, talked about how his company arrived at their *Sustainable Living Plan*. He described it as a journey with four stages: compliance, integration, transformation, and systematic change. Compliance with environmental regulations was risk driven and focused on reputation protection and having a license to operate (i.e., legitimacy). Integration involved considering the company's social, economic, and environmental impact. Transformation included using the sustainability aspects of its brands (e.g., Dove's social mission) to relieve pressure on other elements of its business of potential concern to investors or activists (e.g., concern regarding the use of palm oil). Finally, systematic change occurred when sustainability was linked to the business development agenda (e.g., developing new sustainable sources to replace palm oil). "When you realize that this is not just a compliance agenda but is something fundamental to enhancing the equity of our product brands, protecting our brands and leading the way forward, then it receives much more investment," Gowland explained (Hannaes et al. 2011, p. 14).

Along the way, formal communication designed to promote transformational change is difficult to control. van der Heijden et al. (2012) utilized Weick's sensemaking theory in their case study of how change agents working in a Dutch subsidiary addressed the sustainability initiatives of their US-based parent company, Interface, over a period of 10 years (2000–2010). Internal change agents played important roles in the sensemaking processes by articulating (e.g., translating) and presenting (e.g., embedding) ideas in ways that influenced people and implemented change. "Implementing corporate sustainability is a complex process that requires continuous internal embedding, human interactions and understanding about the nature of sustainability" (p. 538). Sustainability implementation occurs as a process of emergent change involving communication, action, and relationship building. In terms of communication, change agents' define corporate sustainability in their company and develop company-specific terms and language to put ideas about sustainability into words. In terms of action, people do not simply seek to make sense of the environment, they also enact it (i.e., they create their own environment). In terms of relationships, the focus was primarily on the processes and formalizing mechanisms' change agents developed to involve others in the sustainability discussion.

Initially, when Interface launched their sustainability initiatives in Europe, two sustainability directors spread the new vision, encouraged people to take action, and provided them with the US developed reporting and measurement instruments. Between 2000 and 2004, the Dutch change agents used the centrally developed instruments (action) and communicated using the company's sustainability jargon (communication). In 2004, the Dutch change agents experienced a downward trend in their efforts and control over sustainability-related sensemaking. They realized they did not own the language, their actions were based on the US format and produced intangible results, the sustainability directors had left, the communication about sustainability was too complex and abstract, and their local efforts depended on the external guidance and support of the US and European directors. Between 2005 and 2010, local change agents began taking control of the sensemaking process. They adapted the existing jargon about Interface and sustainability to the local subsidiary context; developed comprehensive and multiple applications of their product life-cycle analyses; provided overviews, including lists of visible results and accounts of daily practice, and communicated them widely; and established localized and interdisciplinary training programs and management development schemes. "These initiatives stimulated connections among people from different departments and disciplines and promoted cooperation and involvement with the company's sustainability ideas" (van der Heijden et al. 2012, p. 548).

5.3.1.1   One Size Does Not Fit All

van der Heijden et al. (2012) concluded that change agents can engage others only if they can modify the sustainability approach to suit all areas of their organization. It helps if there is a clear, long-term goal, and a final date to reach the goal. Bottom-up initiatives should be encouraged. To become embedded, sustainability must be continually adjusted to the perceptions and work situations of the employees.

Different sustainability issues are important in various functional areas (e.g., manufacturing, quality control, investor relations) (Blackburn 2007). Employees need to be allowed to translate the concepts to fit different areas. Individual departments need to understand, have input into, and be able to articulate how they fit within the planned change. David Driskell, Boulder's Executive Director of Community Planning and Sustainability, discussed the challenges associated with helping city departments integrate sustainability into their planning and work processes. Speaking of his interaction with people at the department manager level, David said his office needs to "start with where people are at, understanding that better, have them articulate what sustainability means to them and how what they do in their job contributes to or doesn't contribute to sustainability... Sort of like a co-learning type of process."

5.3.2 Successful Communication of Change Initiatives

Aware of multiple challenges accompanying change efforts and the importance of clear change communication, communication researchers and practitioners are concerned with how to translate new knowledge into practice. Lewis and Seibold (1993, 1996, 1998) argue communication is important to the success of change processes, the creation and diffusion of innovations, attitudes regarding and resistance to change, behavioral coping with change, and outcomes of organizational change. In 1993, they developed a model of innovation modification and intraorganizational adoption utilizing structuration theory. They mapped out factors leading up to how planned change appears in actual practice. In *Communication Yearbook 21* (1998), they proposed a research agenda for communication scholars interested in studying organizational change. In 2007, Lewis proposed 18 theoretical propositions and offered a model of communication associated with planned organizational change based on stakeholder theory. Her model connected implementers' selection of communication strategies and stakeholders' concerns, assessments, and interactions with three features of the observable postimplementation system (i.e., fidelity, uniformity, and authenticity). Various internal stakeholders, due to their hierarchical levels, occupational communities, or prior job socialization, have different experiences and awareness, which impact how they respond to change-related messages. Implementers of planned change need to recognize key stakeholders, the relative stakes of each, and then strategically adjust their communication accordingly. The negotiation of stakes among various stakeholder groups can have a powerful influence on change outcomes. In terms of postimplementation evaluation, fidelity refers to how far the change departs from the intended design. Uniformity refers to modifications in change across adopting unit(s). Authenticity addresses whether or not stakeholder support for change initiatives is genuine. Although the model Lewis presents is linear, she acknowledges that the processes are dynamic. Practitioners will find the model provides an overview of important considerations when selecting communication strategies. It is an excellent resource for researchers, as is Lewis' (2014) book chapter.

5.3.2.1 Practical Suggestions

A number of useful suggestions appear in the literature. Frahm (2011) urges that practitioners be clear about the purpose of their communication (i.e., to generate understanding of the change's potential outcomes, to communicate information during change implementation, to codify and reinforce change, or to publicize the change to stakeholders). She discusses the need to balance messages of change and stability. Intervention tactics used by practitioners can involve persuasion, edict, intervention, and participation (Lewis and Seibold 1998). Persuasion involves the use of experts with minimal management review. Edits involve sponsor control and use of personal power, limited participation, and low expert or user power. Intervention includes a problem-solving orientation, user participation during development, and selling the change. Finally, participation involves high-level goal setting, low-level decision making, and high user involvement. Scholars have documented many benefits of participatory processes during change including lower resistance and higher compliance, increased satisfaction, greater perceptions of control participation, and reduced uncertainty. However, in one study interventions influenced final adoption more than did participation, persuasion, or edict. What do change implementers prefer? They prefer the restricted and advisory approaches to input solicitation and use (Lewis and Russ 2012). In the restricted model, implementers are very selective about whose input they solicit and they tend to disregard negative feedback. In the advisory approach, input is solicited broadly, but negative feedback is inconsistently used to make slight improvements to the change initiative. Change communicators need to encourage honest dialogue with internal stakeholders, even if it is critical. In terms of communication strategy choices, it appears that implementers must decide whether or not to emphasize the positive aspects of the change or balance the positive and negative aspects of the change (Lewis 2007). Essentially, employees dislike secrecy. Beyond that, the results are mixed, although there is little empirical research on this topic. Should you send out targeted or blanket message? Blanket strategies are often used when resources are tight and the change does not require a great deal of consensus seeking. Targeted strategies are better when consensus building is important. Should change be rule-bound involving central direction and highly programmed tasks or more autonomous in nature? Autonomous approaches allow people to redefine change during the implementation process (Lewis and Seibold 1998). At the very least, people should believe their requests for adjustments will be seriously considered, if not made. Rule bounded approaches generate more resistance. Should the innovation be changed to fit the organization (adaptive) or should the organization be changed to fit the innovation (programmatic)? Practitioner suggestions point to an adaptive and empowering process. Should change be implemented across all organizational units simultaneously or run in a demonstration location first? Political resistance appears to be less if change is implemented simultaneously throughout an organization. Change-related information should be disseminated through official rather than unofficial communication channels and in a face-to-face manner. Message designers should anticipate employee needs and seek to reduce feelings of uncertainty by sharing the reasons for and value of the change, describing what the change process will entail, addressing concerns about the change, communicating the status of the implementation, and describing how work will be altered by the change (Elving 2005; Lewis and Seibold 1998). Employees need information about new goals, priorities, and policies. They will be curious about the adequacy and availability of existing resources and the use of reward programs. Supportive work relationships contribute to the formation of positive attitudes toward, and ability to cope with change, and ultimately to the success of the change intervention (Vakola and Nikolaou 2005). People want to feel emotionally connected to their organization and its change efforts. This emotional connection is increased if employees trust their organization and its management and identify with the organization. Communicators should seek to create a sense of community within the organization before, during, and after the change.

5.4   Formal Efforts to Embed Sustainability Within an Organization

Justin Zeulner, former Senior Director of Sustainability and Public Affairs for the Portland Trail Blazers, told me:

> The organization says here's our mission, here's our vision, but that in itself really has a hard time navigating all the way down to the different departments, the employees that are actually navigating and doing the work that gets you to that vision. Inherently corporate strategy has its flaws and it always has.

The description of the process I provided at the beginning of the chapter made it appear as if the Trail Blazers went through a very logical systematic process. But in actuality, for many involved in the change, it may have felt, at times, like a fly-by-the-seat of your pants experience. I mean no disrespect to the Trail Blazers. That is simply what transformational change is like as the study of the Dutch subsidiary of Interface illustrated. In his sensemaking theory, Weick (1995) discusses the concept of retrospective sensemaking. Organizational members act. Afterwards they look at their actions in light of the situation to make sense of what they did. "People make sense of things by seeing a world on which they already imposed what they believe. In other words, people discover their own inventions" (p. 15). Often times, organizational actions are guided by things which have worked in the past, either for that organization or in the broader institutional field. In this section, some formal actions organizations routinely take as they seek to formalize their sustainability-related activities are discussed. That is followed by a discussion of several forms of formal organizational communication (i.e., mission statements, goals and plans, measurement). Finally, additional formal channels for communicating about sustainability are identified.

5.4.1 Structural Change

The adoption of any transformation change is an occasion for restructuring. Over time, scholars recognized that organizational structures are not static, nor is the actual (vs. the formal) structure always under the control of decision makers. Anthony Giddens, a British social theorist, developed an important metatheory (i.e., structuration theory). Structuration theory sought to address the relationship between human agency and organizational structure. The question was 'Do individuals construct social meaning and social order through their interactions or do existing structures determine or constrain individual's behaviors (McPhee and Poole 2009). Structuration theory helps us understand how organizational structures are created, reproduced, and changed through human behaviors and how existing structures partially channel subsequent human behavior. This is called the duality of structure.

Key to structuration theory are the ideas of system, structure, and practice. "A system is a set of normal interchange patterns connecting people, behaviors, messages, relationships, and things, both human and nonhuman elements" (McPhee and Poole 2009, p. 936). Systems are the observable outcomes of the applications of rule-resource structures (e.g., an organizational chart showing the status hierarchy) and consist of concrete social practices which systematically reoccur and are recognized as meaningful by system members. So a periodic Board of Directors meeting to review the annual budget, and potentially discuss sustainability goals, is an example of a system. Structure is made up of tacit and empowering rules and resources which interact in complex ways (Rose 2006). Rules are techniques or generalizable procedures that individuals access when they want to understand or sanction each other within a concrete interaction situation. They act as guidelines, whether official or learned through experience, that individuals use to construct meaning and plan their actions. For example, an annual meeting's agenda functions to control who speaks, when, and about what (Carrington and Johed 2007). Groups protesting an organization's environmental actions may find it difficult to address their concerns at annual meetings. Resources involve the interaction dimension of power which people draw on to influence others and might include expert knowledge, official policy, friendships, or a positive reputation. Supervisor resources might include financial incentives or disincentives, legitimate power, or knowledge of official policy. Knowledge of resources may also exist at the level of practical consciousness. Some social practices become institutionalized or routinized.

But humans have agency. They can interpret and reflect upon past and current interactions. Through meaningful practice they can participate in the reproduction or destruction of a pattern of interaction. Rose (2006, p. 176) writes:

> Social actors draw on structures of "signification" (language and other symbolic codes) to produce=reproduce "communication" (meaningful and understandable interactions) via the modality of "interpretive scheme" (background knowledge regarding the codes). When persons exercise "power" (influence) at the level of concrete practice, they do so by accessing intersubjective "domination" structures—whether allocative (control of materials like budgets) or authoritative (control of people)—through the modality of "facility" (capacity, capability). Finally, social agents are able to "sanction" each other by drawing on "legitimation" structures (moral orders associated with laws, religion) via the "norm" modality. Gibbens discusses "positioning" which involves when individuals are "situated" within interaction systems. Our positioning both "constrains" and "enables" our actions.

Individuals may not always make rational decisions but we are able to rationalize our actions and reflexively monitor them. Tacit knowledge or practical consciousness help us know how to behave in social life and explain why we engage in routine behavior. Actions promoted by tacit, taken-for-granted knowledge are not easily explained, but when we have discursive consciousness of social structures and behaviors we are capable of talking about our actions. Researchers seeking an excellent overview of organizational communication scholarship drawing on structural theory are directed to McPhee et al. (2014). The theory has been applied to various issues including organizational change (e.g., Lewis and Seibold 1993), environmental reports (e.g., Buhr 2002), and employee socialization (e.g., Scott and Myers 2010).

Where the responsibility for directing sustainability initiatives is ultimately located within an organization's structure has implications for whether or not transformational change is even possible. Change agents must have the authority and the ability to inspire others (van der Heijden et al. 2012). Many large organizations create board committees (e.g., Johnson and Johnson has a Science, Technology and Sustainability Committee) and top-level positions (e.g., Chief Sustainability Officer) to oversee sustainability efforts. A C-level position symbolizes the perceived importance of sustainability to an organization's operations. For example, at the University of Colorado, Boulder, the sustainability officer was the Assistant Director of Engineering. At the University of Arkansas, Fayetteville, sustainability was the focus of two top-level positions (i.e., the Associate Vice Chancellor for Facilities, the Executive Director of the Office for Sustainability). Jarrett Smith, Sustainability Officer, University of Colorado Denver—Anschutz Medical Campus, said:

> Somebody that is going to hire a sustainability officer needs to make sure that they are empowered to do the things they need to do, and it should be in a position, whether it is in the chancellor's office or the provost's office, to be able to go to people to ask them for something and not find any opposition.

Jarrett's university signed the American College and University Presidents' Climate Commitment, created a climate action plan to reduce its GHG emissions by 20 % from its 2006/2007 baseline level, and achieved a Silver STARS rating.

But his position was not at the level needed to give him the authority and resources necessary to make major changes in his organization. When organizations create top or even mid-level positions designed to facilitate the coordination of sustainability across an organization, they are creating a communication role charged with influencing the intraorganization flow of sustainability-related information.

Restructuring at Neil Kelly   Although structural changes involve individual positions, more extensive restructuring may be necessary in green organizations or in organizations on their way to becoming green (Biloslavo and Trnavcevic 2009). Julia Spence, Vice President of Human Resources, at the Neil Kelly Company in Portland, OR, described how she sat down with the company president, Tom Kelly, to seriously look at the family-owned company's hierarchical structure saying:

> Construction companies are not noted for being progressive and we were probably more progressive than most but the information was not shared very well. We brought everybody in and really talked about what we all wanted and how do we want to communicate? How do we want people to be involved in decision making, what kind of information do we need to share, and how are we going to share it? We ended up with a team structure that we still have 30 years later that really depends on everybody in the company, whether they are a new carpenter's apprentice or a manager who has been here for 40 years, providing input into what we think will help us grow. We do an annual planning process that has helped us with our environmental sustainability. I think that team structure presents both a challenge and a terrific advantage. Because we are flat and everybody is making decisions people can suggest things. That means that everybody has to know what is going on so they are making decisions that are in alignment with our goals. Because we are all putting them together. At the same time it means that everybody has to be educated in order to do that and sometimes that's a real challenge to get information out to everybody. But, that team structure helped us to sink sustainability deeply and broadly into the company.

This flat structure is consistent with the organizational structure most reflective of a learning organization (Senge 1990). *Best Practice*: Flexible, progressive organizations capable of learning are hampered by hierarchical and bureaucratic organizational structure (Cohen and Levinthal 1990), especially if sustainability is not the focus of a top management position.

An organization's structure does more than provide authority and allow for decision-making input. It is a way to formally link departments in a configuration which allows them to more effectively enact sustainability initiatives. Systems theory directs us to focus on how formal communication messages flow across departments, among people of different and similar rank, and between superiors and subordinates along structural lines which can be used to promote information sharing and joint problem solving (Poole 2014). Susan Anderson, Director of Portland's Bureau of Planning and Sustainability, described how her organization's structure changed in the late 2000s. Separate departments focusing on planning and sustainability were combined. Now, when the city engages in comprehensive planning, sustainability is integrated throughout the plan. "It [the changed structure] has mainstreamed it [sustainability] in this organization, both in how we work with the public but also in the organization itself." In the City of Boulder, Community Planning and Sustainability departments merged at about the same time as in Portland. David Driskell, Boulder's Executive Director of Community Planning and Sustainability, explained, "We went through a process to integrate different functions [traditional planning, climate action, zero waste, economic vitality]." He described how the Office of Environmental Affairs had one person working on climate action focused on reducing transportation emissions. When he was hired, David asked, "Why do we have a separate person sitting over there in a completely different department? We should be partnering with transportation. They have a multi-million dollar budget every year. We should be helping them think about greenhouse gas emissions and how what they do impacts our inventory."

Structural elements can reinforce routines and processes which help convert external information into useful knowledge (Zahra and George 2002). Coordination capabilities (e.g., formal participation schemes, cross-functional teams, boundary-spanning mechanisms) can support cross-functional information sharing and knowledge transfer (Eisenhardt and Martin 2000; van den Bosch et al. 1999). Organization integration is enhanced when units are formally encouraged to coordinate their activities to achieve overall organizational objectives. Cross-functional and lateral forms of communication and joint decision-making processes are important for coordinated action. The more integrated the functional areas, the larger the pool of knowledge employees can access. The more aware of other units' sustainability-related information needs, challenges, and goals employees are, the more likely they can identify and transmit useful information on to the other units. The 35-person sustainability team which designed the Portland Trail Blazers' *Sustainability Charter* was scaled back to 20 people, but Justin Zeulner, their former Senior Director of Sustainability and Public Affairs, said the team continues to meet monthly to "talk about what we do next, or if there are certain hiccups that occur, or other opportunities that might exist. Issues that we need to address, we work together."

Coordination Efforts at the City and County of Denver Based on systems theory, for decades, organizational communication researchers have studied communication networks (Poole 2014) learning how managerial efforts at designing structures promoting organizational integration often fail or modify significantly over time. However, the design of structural integration schemes is useful because it initially can help us manage uncertainty (Weick 1969). Coordination capacity issues should be discussed when planning how organizational structures can facilitate the move toward increased sustainability.

Jerry Tinianow, the first Chief Sustainability Officer for the City and County of Denver, told me that when the Mayor signed the order which created his office in 2012 it also included an entity called the Sustainability Implementation Committee. This interagency committee represents 12 of the city's 21 agencies thought to be most directly related to achieving the city's 24 different 2020 goals. They meet monthly to coordinate sustainability issues across agency lines. The Sustainability Implementation Committee appointed agencies tasked with coordinating interagency efforts to achieve the 24 goals. Although it was early in the implementation process when I spoke with Jerry, ultimately one or more agencies was to serve as the coordinating agency for each of the 24 goals. The Sustainability

Implementation Committee decided the coordinating agencies would have a facil-
itative vs. a dictatorial role. Jerry said:

> What the coordinating agency does and what their responsibility is involves making sure
> that the agencies working on aspects of any one goal are talking to each other, avoiding
> overlap, and avoiding inefficiencies. Talking to the players out in the community that need
> to be involved. . . The whole idea behind the coordinating agency is there is more than one
> agency that needs to be working on each goal.

The coordinating agencies had three tasks for 2013: quantify the baseline against
which progress can be measured, identify how future progress will be tracked, and
identify existing city initiatives that can contribute to achieving one or more of the
24 goals.

A Coordination Structure at Walmart Although the previous two examples
dealt with intraorganizational coordination structures, structures can be utilized
effectively to reach beyond one organization as is the case of Walmart's Sustainable
Value Networks (SVN). The SVNs work to find ways to achieve Walmart's three
aspirational goals (i.e., sourcing 100 % renewable energy, zero waste, and providing sustainable products). Each is tasked with identifying and developing projects
that meet short-, mid-, and long-term positive outcomes related to these goals.
SVNs focus on a product category or an operational function such as GHG,
agriculture and seafood, or chemicals. SVN team members include four or five
Walmart champions to provide company-related knowledge and leadership. But the
teams also include academics, Walmart suppliers, scientists, and members of
environmental groups and NGOs which bring their own interests and expertise to
the discussions. As of 2013 there were eight SVNs, although the SVN structure
undergoes continuous evaluation. Two councils oversee the SVNs progress, evaluate future direction, and align new goals. The Operations Council SVN oversees
renewable energy, sustainable buildings, sustainable transportation, and operations
waste and recycling. The Sustainable Products Council SVN oversees index,
packaging, food and agriculture, chemical intensive products, and product GHG.
Structurally, each SVN has a network captain (generally a Walmart director or vice
president) to guide network efforts toward the SVN's goals. Above that level is the
Sustainability Team which oversees network activities, aligns overall efforts, and
provides guidelines. Next is the SVN Council (Walmart vice president level or
higher). The Chief Executive Officer is over everything. The Walmart Sustainability Office participates with and provides support to the SVNs. The legacy of SVNs
is in decentralizing the process of input and expertise in order to provide useful and
actionable information to decision makers as they work toward Walmart's three
aspirational goals.

5.4.1.1 Structural Planning and Boundary Objects

Building on Weick's (1969) sensemaking theory, Benn et al. (2013) investigated
factors which influence organizational learning about sustainability. Their findings

illustrated how knowledge sharing and boundary objects (e.g., shared reporting tools) can promote the integration and institutionalization of sustainability across intraorganizational knowledge and disciplinary boundaries. In Sect. 3.3.2.1 you read about how actor-network theory (Cooren 2009) acknowledges that both humans and nonhumans (e.g., machines, texts) can accomplish things because they have agency (i.e., the ability to act). Boundary objects are agreed-upon artifacts of practice shared across very different roles, professional backgrounds, and groups. They satisfy the informational requirements of each person or group. They allow for coordination without consensus because actors use their local understanding and reframe the object within the context of the wider collective activity. Boundary objects can include information repositories, ideal types, standardized forms, models, discourses (e.g., a common language), and processes. In their study of 20 higher education institutions in Australia, Benn et al. identified a number of boundary objects including the concepts of stakeholder engagement and/or triple-bottom line used as a teaching tool, *Education for Sustainability* discourse, role play teaching resources, interdisciplinary teaching programs and/or research centers, a sustainability website with information archival and sharing functions, and action learning teaching approaches. Readers working for higher education organizations will find this article especially useful. However, the idea of using boundary objects to create coordination and facilitate learning within any organization is useful. Boundary objects need to be adaptable to local needs but robust enough to have a common identity across locations. In van der Heijden et al.'s (2012) 10-year investigation of Interface, boundary objects were used (e.g., common language, common reporting tools) to transmit values and processes between the US home office and the Dutch subsidiary.

A Boundary Object in Boulder Creating a successful boundary object takes effort. David Driskell, Boulder's Executive Director of Community Planning and Sustainability, told me about his effort to create such an object. Boulder city departments complete departmental master plans which guide capital investments and organizational development. These plans require City Council approval. In 2011, the City Council sent one department's plan back because it did not advance sustainability in Boulder. The city manager's office asked David's staff to write a sustainability chapter for that master plan. He thought, "How do you just write a chapter at the end of the master plan that talks about sustainability?" Such an add-on defeats sustainability's holistic nature and shows it is not fully integrated into departmental plans and goals. Instead, David offered to design a framework through which the different aspects of the master plan could be looked at using a sustainability lens. Working with that framework, the first department piloted using smaller equipment, conducted a baseline GHG inventory, realized they could cut employee commute-related GHG in half by restructuring work schedules, and created a green team which generated relevant ideas. The City Council accepted the new master plan. Based on the first department's success, the city manager's office wanted all departments to use the sustainability framework. A number of departments used it (e.g., fire department, parks and recreation, transportation).

However, the day before I spoke with David he was at a meeting where a department head told him, "I don't get it [the sustainability framework]. I don't know what this has to do with us. It is just a check-the-box exercise and now I have new categories I need to cram what we do into and it's kind of silly." David said he told the department head:

> I appreciate the perspective but tell me why you don't think that what you do is relevant to these areas? Seems to me that there are things there [that you are doing]. If you are not seeing it that way, I think it is worth having a conversation about how the skill sets and things your department does might contribute because I think it is an important part of the community and you guys could be doing a lot in that area.

The department head agreed his department was contributing in many of the areas on the framework (e.g., contribute to a safe and secure community, contribute to a thriving community). David said:

> My takeaway from the whole discussion was he had no idea where this thing came from. It was plopped on to them and they were told, put this in your master plan. I had a discussion yesterday with one of our lead planners and we decided we need to step back and rethink how we are going to do a process in the city departments that are just starting their master plans because people don't get it [using the sustainability framework]. It's seen as top down, it's check-the-box, it's meaningless, and it's a waste of time.

Key Point: David's story illustrates how different departments respond to boundary objects and how communication is needed to explain their importance. We saw a similar reaction in the article describing the Dutch subsidiary of Interface's experience.

5.4.2 Formal Organizational Communication

Most beginning organizational communication textbooks (e.g., Cheney 2011) talk about how formal organizational communication follows an organization's structure. Message content, which is largely management controlled, flows in distinct directions: downward, upward, and horizontal. Downward messages generally involve goals, strategies, and objectives; job instructions and rationales; procedures and practices; performance feedback; and socialization. Upward messages routinely focus on problems and exceptions; suggestions for improvement; performance reports; grievances and disputes; and financial and accounting information. Horizontal messages focus on intradepartmental problem solving and interdepartmental coordination. Although actual organizational communication is far more complex than this, beginning sustainability communicators will find such introductory-level concepts and lists useful. *Best Practice*: Professional associations (e.g., the International Association of Business Communicators, the Public Relations Society of America) provide resources to help manage issues including structuring a communication department (e.g., Nicholson and Aiello 2008), creating a communication plan, crafting, and disseminating internal and external messages, designing internal communication campaigns, and managing change. This section focuses on channels of internal formal communication, then formal communication documents (e.g., mission and vision statements), and finally communication-related measurement systems. Strandberg Consulting (n.d.) provides a list of questions organizations can use when focusing on designing their new vision, strategy, goals, key performance measures, and measurement indices.

5.4.2.1   Formal Communication Channels

In Chap. 3, I discussed formal communication channels and settings including CEO speeches, annual meetings, sustainability reports and websites, signage within buildings, and checklists associated with certification. Authors (e.g., Bortree 2011; Craig and Allen 2013; Linnenluecke et al. 2009; Maharaj and Herremans 2008) have identified a wide range of channels used to communicate with internal and external stakeholders about sustainability. Practitioners charged with communicating about sustainability should find information about planning internal communication (Whitworth 2011), integrating employee communication media (Crescenzo 2011), and internal branding (Grady 2011) useful.

I asked Brian Sheehan, former Sustainability Manager at Sam's Club, to describe the channels he routinely used to communicate with employees. He said he sent emails to employees around Earth Day as a reminder of the organization's aspirational goals and to congratulate them on past efforts. He wrote a sustainability feature for the monthly magazine and another for the monthly memo which went out to the Sam's Club and Walmart CEOs. He blogged. Bimonthly he brought in a supply chain partner to speak at the Sam's Club Sustainability Speaker Series. He provided content for Walmart's major meetings.

I asked Brian to describe any challenges he faced when trying to communicate company goals and objectives regarding sustainability throughout the company. He said:

> I think it is a resource-based challenge for me. It is certainly not a leadership challenge, because any time we have asked our leaders for the support and the visibility that we think we need to get on the platform to address a large audience, or leverage a channel that we have, we have gotten their support. It is more of a challenge in evaluating the right channels and understanding the world of limited resources. Where are my efforts best focused? The opportunities I have had for one-on-one communication with associates, I have found to be valuable but have the least impact. I think where we get the most impact is with the biggest scale. And for us the biggest scale happens at the big meetings on the big stage in front of a room of five thousand people.

Sustainability communicators face challenges associated with selecting channels, managing relationships, and supporting networks. It is not surprising that they also face challenges communicating about complex scientific information (Bortree 2011).

5.4.2.2 Formal Communication Documents

Organizations create mission statements (who are we in terms of products, services, customers, and purpose) and vision/value/policy statements (what do we want to be long term and where do we want to go), and goals (what are the desired organizational behaviors and our triple-bottom-line status). These statements should inform the development of an organization's strategic plan (how do we get to our desired state) which in turn shapes the tactical or operating plan (how will we implement our strategy) (Blackburn 2007). Blackburn provides a model organizations can use to guide the construction of their own documents and/or policies. The issue isn't so much which type of formal statement exists but rather that an organization has internal documents which inspire and guide the actions of its employees. As an organization embarks of its sustainability journey, such formal communication statements are an important part of the strategic planning and implementation process. Successfully designing such documents demonstrates that an organization's leadership can think reflectively, plan carefully, work collaboratively, and create a process whereby informed and guided decisions can be made. The concept of auto-communication, or self-communication, applies to how companies seek to project their identities into the future using such formal documents (Cheney et al. 2014).

Visions, Goals, and Plans in Madison Madison, WI, appeared to utilize the structure Blackburn (2007) recommended where a vision leads to goals which lead to plans. Madison is a city of over 240,000 people with 800 miles of streets, 6,000 acres of parks, and 3.7 million square feet of office space. It utilizes 54 million kWh of electricity and 2.3 million gallons of fuel and generates 60,000 tons of garbage and recycling (Hoffman 2013). City government began focusing on sustainability in 2003 when their mayor formed a committee to develop a plan to make Madison a leader in energy efficiency and renewable energy. They used the Natural Step process to develop their first plan in 2005. Work on their second sustainability plan began in 2009 and the plan was adopted in 2012. The mayor budgeted $1 million per year for implementing the plan from 2014 to 2019. When developing the plan the City gathered public input using public meetings, surveys, and evaluations from stakeholders including representatives from schools and healthcare facilities, faith-based groups, architects and engineers, and other citizens. University students conducted a study on national and international best practices related to issues like sustainable neighborhoods, transportation, and food systems. The final plan includes ten sustainability priority categories based on the ICELI and STAR index categories: natural systems; carbon and energy; planning and design; transportation; economic development; employment and workforce training; education; affordable housing; health; and arts, design, and culture. Each of the ten category subsections included a vision statement, goals and action plans, links to City Agency work plans, and short, medium, and long-range recommendations. For example, the Carbon and Energy category vision statement is, "Madison embraces sustainable approaches to fuel our economy and community, achieving an 80 %

carbon reduction by 2050. Our City government and staff set examples of reduced energy use and emissions for businesses and individuals to emulate." Goals for this category include influencing reduction in transportation-related carbon impacts; systematically upgrading existing buildings, equipment, and infrastructure; and engaging the public in energy efficiency and climate change programs. Actions related to this category include purchasing hybrid buses, passing an energy-benchmarking ordinance, and utilizing the Mpower program for schools and businesses.

The focus of mission, goal, and planning documents should spring from an organization's history and capture its unique future aspirations. Several of my interviewees highlighted this point. Others talked about how any sustainability initiatives must be directly tied to an organization's mission. Justin Zeulner, former Senior Director of Sustainability and Public Affairs, for the Portland Trail Blazers told me that although his organization is a leader in sustainability in the sports entertainment industry their focus is not on sustainability:

> I am trying to win a championship. I am trying to make a better brand. I am trying to enhance the entertainment value to our fans and guests that come here. By the way, I am going to do it in a way that creates this leadership opportunity [in sustainability]. We are trying to inspire a nation through the Green Sports Alliance and we are going to be a member of BICEP. But at the end of the day I am still trying to win a championship. Everybody has got to remember that is the focus. It's not about the sustainability criterion as much as it is about how the sustainability criteria help the organization meet its goals.

5.4.2.3 Mission Statements and Guiding Principles

Many organizations find it useful to develop a mission statement or set of guiding principles involving sustainability. A mission statement is a formal, short, written statement articulating an organization's purpose and goals. Its direction setting role can be enhanced if it includes an explicit statement of a future vision for the organization. It appears on corporate websites, employee orientation materials, brochures, posters, annual reports, business cards, and is mentioned during speeches. Ideally, this statement will guide subsequent strategic and everyday decision making and action as well as clarify the organization's philosophy and intent for employees. It may outline where an organization is headed; how it plans to get there; what its priorities, values, and beliefs are; and what makes it distinctive. Mission statements represent formal communication designed to articulate an organization's identity to internal and external stakeholders; persuade multiple stakeholders; and promote distinctiveness. On the "We are Different" portion of the Aspen/Snowmass website it reads:

> We have a collective responsibility to ensure that our company is a rewarding place to work and our community a desirable place to live. We respect and nurture the delicate balance between "resort" and "community" that makes Aspen/Snowmass unique. The combination of our values-based company with unparalleled mountain sports, community, history, culture and environment gives us a unique market niche. We are successful because we live the values and principles expressed here.

Results are mixed as to whether or not mission statements actually influence an organization's performance (Sattari et al. 2011; Williams 2008). Some mission statements are too vague, irrelevant, unrealistically inspirational, not truthful or representative, or unreadable. An organization's mission statement needs to excite and engage employees and help them see the next steps which should occur (Werbach 2009). Many are not effectively implemented (Fairhurst et al. 1997). Part of the problem is that they are under-communicated. To be effective, the mission statement content needs to be integrated into the everyday activities of an organization (e.g., performance appraisals, problem-solving discussions). Under-communication suggests that the mission statement isn't really relevant and leaves organizational participants unclear about its applicability and relevance to emerging conditions in the organization.

According to Auden Schendler, Vice President of Sustainability at Aspen Skiing Company, his organization has had guiding principles for almost 20 years. On their website it says:

> We renew the spirit by sustaining the 'Aspen Idea,' a complete life nurturing mind, body and spirit. We value humanity (we treat people the way they'd like to be treated, modeling authenticity, transparency, courtesy, respect and humility), excellence (in business, quality, craftsmanship, guest services and athletic achievement), sustainability (of people, profits, the environment and the community so that we are in business forever), and passion (we live our core values and embrace life-long learning and meaningful work). As employees we are committed to the following Principles: be fair and understanding; treat everyone with courtesy and respect; communicate honestly, openly, and often; share a passion for excellence; engage as responsible citizens of our community; and ensure our financial success.

Auden said, "So we talk about this [our guiding principles] a lot and managers and the CEO talk about it a lot, and then we play it out." On their website, their CEO Mike Kaplan writes:

> Our Guiding Principles have been stripped down to their core. But they provide more direction and transparency than ever. They encompass what we do, why we come to work every day, why we choose to live here and how we'll be successful. I ask you to own these values; to take responsibility for bringing them to life. Our dedication to the Aspen Idea of the complete life, and these core principles of humanity, excellence, sustainability and passion, is what makes us unique. It's what gives employees power and autonomy in their work, and what makes Aspen/Snowmass such a rewarding place to be.

Management often unveils and communicates new mission and vision statements without considering how to implement the ideas through their organization. How people react to a mission statement is influenced by a variety of things including an organization's information environment, work unit commitment, employee trust in management, and employee role in the organization. Many who communicate mission statements (e.g., human resource personnel, mid-level managers, corporate communicators) discuss them in banal or factual ways and therefore are not communicating an inspirational vision. Fairhurst et al. (1997) propose and test a model of the processes which influence how often employees talk about their organization's mission statement with coworkers, in relationship to their job, in comparison to past mission statements; use the formal mission statement wording; explain the advantages of working to achieve the mission; identify parts of the mission not being accomplished by their department; and encourage others to try to accomplish the mission. They called this the management of meaning of the mission statement. They found the more information employees receive about critical issues of the work environment (e.g., information about safety, job duties, and their future with the company), the more they talk about the mission statement. The higher their commitment to their work group, the more they discuss the mission statement. Trust in their supervisor was unrelated and trust in upper management was negatively related to their discussion of the mission statement. Fairhurst et al. recommend that employees be able to engage with these statements and, to some extent, modify them to fit the context of their own working environments.

What content goes into a mission statement? Williams (2008) compared the mission statements of 14 higher performing and 13 lower performing Fortune 1000 firms. She analyzed their content using a typology which identified nine components to mission statements. I provide her typology here to help guide sustainability communicators seeking to craft a mission statement:

- Who are the enterprise's customers?
- What are the firm's major products or services?
- Where does the firm compete?
- What is the firm's basic technology?
- What is the firm's commitment to economic objectives?
- What are the basic beliefs, values, aspirations, and philosophical priorities of the firm?
- What are the firm's major strengths and competitive advantages?
- What are the firm's public responsibilities, and what image is desired?
- What is the firm's attitude toward its employees?

The mission statements of higher performing firms mentioned company philosophy, targeted markets, strategies for survival, concern for public image, and concern for employees more often than did lower performing firms (Williams 2008). Williams drew on rhetorical theory, specifically the concept of corporate ethos referring to Aristotle who identified three components of ethos: intelligence, character, and good will. Aristotle stressed the importance of intelligence (i.e., knowledge, good sense, and expertise) because of its impact on credibility. The content, writing style, organization, and visual rhetoric of a mission statement influences perceptions of knowledge. Perceptions of character can be influenced by the values mentioned in the mission statement, especially those illustrated through an organization's operations. Although a mission statement's authors may strategically craft the statement, ethos relies on the audience's cooperation. It is important to carefully and honestly shape the message in keeping with the audience's values in order to appeal to similarities between the organization and the audience. Using the word *we* (or other first-person-plural pronouns) in a mission statement joins a communicator and the audience and thereby encourages stakeholders to identify with the organization. Individuals who identify with an organization are more open to persuasive efforts. Higher performing firms used many more first-person-plural pronouns in their mission statements so as to build identification. Organizations in both groups mentioned excellence, integrity, and innovation, but higher performing firms mentioned respect, leadership, diversity, citizenship, and responsibility more often.

5.4.2.4   Goals and Plans

Organizations engage in annual planning and/or reevaluation processes, formulate strategies, write procedures and protocols, and develop plans. Many of my interviewees mentioned their organization's goals and/or plans. For example, Denver has a 2020 plan, Portland has *The Portland Plan*, Fayetteville has a 2030 plan, and the University of Arkansas has a 2040 plan. Ray Anderson, former Interface CEO, worked with consultants to create Seven Fonts which set goals for 2020 to reduce the company impacts and reliance on natural resources in seven areas: eliminate waste, benign emissions, renewable energy, close-loop recycling, resource-efficient transportation, sensitize stakeholders, and redesign sustainable commerce (van der Heijden et al. 2012).

Although communication is critical to the planning and implementation process surrounding goals, most communication theories focus on intrapersonal and interpersonal communication goals (see Samp 2009). Therefore, this section draws primarily on the management literature and existing processes. Practitioners will find Blackburn's (2007) discussion of strategic planning and the creation of goals (both aspirational and tactical) and clear indicators related to sustainability to be systematic, clear, and detailed. The Harvard Business School offers an interesting case study describing Governance and Sustainability at Nike (Paine et al. 2013) which describes Nike's evolution toward sustainability in terms of leadership (e.g., Board of Directors) and the strategic planning, mission, goals, and measurement process over time. The case also illustrates how structural change can enhance movement toward sustainability.

Goals at Sam's Club and Tyson Foods   Goals may begin at the inspirational level but ultimately they must become operationalized. When Walmart first established its inspirational goals related to sustainability, this was new territory for them. Brian Sheehan, former Sustainability Manager at Sam's Club, explained:

> Walmart did not have that level of knowledge or depth of understanding of the complexities behind all these different supply chains and different questions that go into making products more sustainable. We announced that we want to make more sustainable products then along the way we figured out that in order to do that you need to create a tool like the sustainability index, so that buyers and suppliers can have a conversation around the areas that have the best improvement opportunities in those product categories. I think its [Walmart's] understanding just evolved because of a dedication and focus on wanting to achieve these broad aspirational sustainability goals. And then also it has become more strategic in developing those tools that might help it set better, more achievable, more actionable goals in the future.

On the day I spoke with Kevin Igli, Senior Vice President and Chief Environ-mental Health and Safety Officer at Tyson Foods, Inc., he'd just talked with his CEO about how Tyson Foods needed new sustainability-related goals aligned with the company's revised long-term strategy. The existing 4-year-old goals were no longer a big focus for the company. Kevin told me:

> We want to set new goals that are people, planet, profit, product-related. But we want to be very careful and we want to keep these goals really tight so that we can actually do something meaningful. We don't want to have five goals under each pillar. We want to have one and to have it mean something. We want it to be tied into our strategic plan as a company and we also want it to be aligned with management commitment. Because if you don't have that, you are wasting your time.

Kevin strongly opposes setting arbitrary goals. He shared an example he had previously shared with the McDonald Corporation:

> In 2008, we [Tyson Foods] set a goal and we said 'ok in the next two years by the end of 2010 we want to reduce our water usage by 10 percent'. So two years went by and we did not meet our goal, we got 7.8–8 percent. So somebody said to me in a public meeting – 'you did not meet your goal'. And I said 'do you think that a 7.8 percent reduction in water is a good thing?' And the person said, 'well that is fantastic'. I said 'that is right'. We set an arbitrary goal but we did achieve an outcome and we missed our goal. Is that a failure? I said 'no that is not a failure that is an achievement.'

Between 2004 and 2013, Tyson reduced its water consumption by 11 % and in 2012 Kevin created a water conservation council within the company to help identify additional ways to conserve. Kevin said people often ask him about his organization's goal for GHG reduction.

> We have not set any arbitrary goal like 'we will reduce by 10 percent this year'. The reason we did not do that is because it is arbitrary. We can throw a number on a piece of paper and say – we are going to reduce greenhouse gases by 20 percent by 2020. Well, what does that really mean? If you don't engineer a way of getting there, all you are doing is satisfying a public appetite that says – 'we want you to have a goal'. It is hard to have a goal if you don't engineer a way to get to the end point. We are working with a couple of different entities on new ways to measure our energy consumption usage in an automated way. Because the better we get at that, the more we can begin to ratchet that down.

Tools exist for helping management create a vision, set goals, and establish procedures for achieving goals. In terms of creating a vision, the Delphi technique and appreciative inquiry (see Sect. 7.1.3.2) are helpful techniques for identifying new directions and goals. During strategic planning, the balanced scorecard approach and a SWOT analysis are useful in comparing what is currently being done with potential opportunities. In terms of evaluating existing procedures, cause and effect diagrams and SIPOC diagrams are useful. Cause and effect diagrams help identify failures in meeting objectives or goals. SIPOC diagrams focus on identifying where problems occur or improvements are possible in existing processes.

5.4.2.5 Measurement and Communication

In order to form a connected whole, organizations engage in strategic planning, set goals, plan ways to meet these goals, establish relevant KPIs, compare measures based on their KPIs to their goals, and then report their findings. Jerry Tinianow, the first Chief Sustainability Officer for the City and County of Denver, explained the process saying:

> I think that one of the functions that we want to do in this office is to have the city speak with one voice on sustainability. But first we have to create the substance behind the talking points. So that is where we are now. We've created the goals. And we are creating the measurements and evaluating our current initiatives. At the end of the year, when the measurement has been done, we will rework all projections [based only on current projects]. I am confident that we will see big gaps between those projections and the 2020 goals. Next year we will develop additional strategies for meeting our goals to fill those gaps. Next year is where we really unleash the creativity of the agencies to come up with the new approaches that will close those gaps.

Reliance on the information function provided by measurement results is at the heart of the certification and standardization processes discussed earlier. When organizations have clear KPI then what to measure becomes clearer. Multiple tools exist to help managers assess the outcomes of the strategic plans and the processes put in place to enact the plans. Blackburn (2007) lists a number of big picture tools.

Organizations that embrace sustainability also spend more time and effort than cautious adopters measuring how sustainability influences brand reputation, employee productivity, the ability to attract and retain talent, and improved innovation capabilities (Hannaes et al. 2011). Duke Energy's Roberta Bowman discussed the value of resources such as water or a culture of innovation as well as the costs associated with management time engaging with concerned stakeholders. "It's hard to put a tangible number around those intangibles.... What gives me encouragement and what I feel is a sign of progress is that we're having those conversations today, whereas a few years ago we may not have" (p. 16). Increasingly organizations are seeking ways to measure tangibles and intangibles related to sustainability (Meng and Berger 2012). *Key Point*: It's important to measure the tangibles and the intangibles.

But measuring the tangibles and intangibles can be challenging on multiple levels. For example, organizations invest in energy-efficient equipment and materials to save energy, but their savings may be limited by employee behaviors (e.g., increased consumption, wasteful behaviors) (Bolderdijk et al. 2013). The concept of motivated cognition indicates employees' motivation influences how they select, process, and interpret sustainability initiative-related messages. As organizations seek to motivate employees to adopt and maintain new behaviors, they may monitor individual behaviors. Electronic monitoring technology (e.g., smart meters and in-car GPS devices) can promote employee energy conservation behaviors by providing them with individualized feedback. However, electronic monitoring may raise privacy concerns. Bolderdijk et al. conducted three studies and found employees were generally unconcerned if their employers' used technology to measure individuals' conservation behaviors, especially when the benefits of the measurement program were discussed. They recommended that organizations develop policies where they only use technology to monitor individual conservation behaviors and not worker performance. Their findings are consistent with those of Allen et al. (2007) who studied employee responses to being monitored via technology. Allen et al. based their study on communication privacy management theory. This theory explores how people seek to maintain and coordinate privacy boundaries with various communication partners depending on the perceived benefits and costs of information disclosure (Petronio 2009). Distinctions are drawn between private information and public information. When people disclose private information, they rely on a rule-based management system to control the level of accessibility. An individual's privacy boundary governs his or her self-disclosures. Once a disclosure is made, the negotiation of privacy rules between the parties is required. The theory has five core principles: (1) People believe they own and have a right to control their private information. (2) They seek to control their private information through the use of personal privacy rules. (3) When others gain access to a person's private information they become co-owners of that information. (4) Co-owners need to negotiate mutually agreeable privacy rules about telling others. (5) When co-owners don't effectively negotiate and follow mutually held privacy rules, boundary turbulence often results. Although designed to focus on two individuals, the theory has been applied to an employee's relationship with his/her organization or one of its agents. Organizations and researchers interested in electronically monitoring employees' conservation behaviors will find this theory thought provoking.

Electronic monitoring of employee compliance with sustainability initiatives is fairly uncommon. The North American Task Force of the United Nations Environment Programme Finance Initiative (2010) study participants reported measuring the success of their sustainability initiatives using number of website hits, employee feedback including surveys, and employee participation in training, events, and contests. *Best Practice*: Metrics of individual employee green behaviors need to be incorporated into regular employee surveys to inform future human resource policy. For example, Procter and Gamble regularly uses their employee survey to gauge the success of their sponsored sustainability initiatives (Ones and Dilchert 2012).

But what about measurement related to the success of communication efforts? I asked my interviewees how they measured the success of their communication efforts surrounding sustainability. Most respondents made comments about looking at the number of hits to their website, shared feedback that individuals had given them, or provided their observations (e.g., reported seeing better recycling practices). Few noted using any actual measurement efforts. There is room for improvement in measuring the success of how organizations communicate about sustainability with their internal stakeholders. Meng and Berger (2012) surveyed 264 senior communication executives located on three continents and interviewed 13 extensively seeking to identify the metrics and measurement approaches being used to assess the effectiveness of an organization's internal communication initiatives on business performance and social influence. Their focus on communication's ROI showed almost 47 % had no formal measurement, and of those which did only 17 % said more than 50 % of their internal communication initiatives were measured using business outcome metrics. Most frequently, measured initiatives focused on employee awareness and understanding, employee engagement, job performance, changed employee behaviors, and increased business performance at the organizational level. Practitioners and scholars will find this article interesting in either simulating practice or influencing future research questions.

In a practitioner-focused article, Elliott and Coley-Smith (2005) report on a systematic effort the communication staff working in British Petroleum's (BP) Lubricants business took to assess communication's impact on business performance following a major change in business strategy. The BP Lubricants communication team gathered survey data to help develop insights into employee knowledge of and engagement in the new initiatives. They measured the contribution and value communication and engagement brought to the business. They wanted to ensure employees had the right information, at the right time, to make the right decisions to support strategy and had a measurement mindset focused on improved performance. They recommended the following principles:

- Identify the relationship between communication and engagement activities (inputs) and what employees know, think, and feel (outputs); what employee do (behavior and actions); and on business performance and value creation (bottom line).
- Develop a communication value tree which links communication activities to the relevant aspects and drivers of the organization's gross margin.
- Develop measures of communication and engagement KPIs. For them, two input KPIs involved communication and engagement activities. Four output KPIs involved awareness and understanding, commitment to making the strategy a success, confidence in leadership and the future of the business, and trust in leadership and the strategy.
- Develop a survey where each question measures one of the KPIs.
- Convey the survey results in an easily understood format (e.g., a dashboard).

Other authors provide suggestions for how to measure the effectiveness of internal communication (e.g., Williams 2011), public relations programs (e.g., Weiner 2011), and marketing communication (e.g., Grensing-Pophal 2011). *Key Point*: It is important that sustainability communicators measure how they are contributing to the broader organizational efforts toward sustainability.

5.5 Concluding Thoughts

This chapter has been about transformational change, the diffusion of innovations, changing information into useable knowledge, and communicating new ways of thinking. As I think of the challenges humanity faces in light of accelerating global warming, I find the *Journey of Mankind* website, presented by the Bradshaw Foundation, meaningful and comforting. That website depicts the global journey of modern humans over the past 160,000 years. Archeologists and geneticists describe how our ancestors migrated from Africa in response to climate change, branches of our human family tree were wiped out by environmental forces, other groups pressed on through inhospitable terrains, and our bodies modified to adapt to our surroundings. Our skin colors and our physical shape changed. Near the tropics we became darker as more melanin acted as a protective biological shield against ultraviolet radiation. Our bodies elongated to dissipate heat. In far northern latitudes, because of the lower ultraviolet radiation, our skin color lightened so we could produce vitamin D and our bodies compacted to retain heat. Our ability to solve problem jumped into overdrive and our simple click language expanded into the myriad of languages being spoken across our globe today. For 160,000 years we have survived through adaptation. We are a species which moves and adapts, that works cooperatively and forms organizations, and that uses language, memory, reason, and empathy to solve challenges and create opportunities. And we need to do so now. We are living in post-normal times characterized by uncertainty, contested (scientific) knowledge, and high levels of complexity (Wals and Schwarzin 2012). Conventional routines and systems will no longer work in our organizations, our resources management, and how we approach communication, education, and science. We must rethink our routines and creatively cocreate alternative ones if we are to become a more sustainable world. Rather than incremental changes in existing processes, we need fundamental changes in how we live and work and the values we pursue.

In 2014 the Center for Naval Analysis' 16-member advisory board of retired military leaders released a report reexamining the impact of climate change on U.S. national security (CNA Military Advisory Board 2014). They write that although the potential security ramifications of global climate change should be catalysts for cooperation and change, instead we see climate change is acting as a catalyst for accelerating conflicts. The report's opening statement signed by the board members says:

> We are dismayed that discussions of climate change have become so polarizing and have receded from the arena of informed public discourse and debate. Political posturing and budgetary woes cannot be allowed to inhibit discussion and debate over what so many believe to be a salient national security concern for our nation. Each citizen must ask what he or she can do individually to mitigate climate change, and collectively what his or her local, state, and national leaders are doing to ensure that the world is sustained for future generations. Are your communities, businesses, and governments investing in the necessary resilience measures to lower the risks associated with climate change? In a world of high complex interdependence, how will climate change in the far corners of the world affect your life and those of your children and grandchildren? If the answers to any of these questions make you worried or uncomfortable, we urge you to become involved. Time and tide wait for no one.

References

Allen, M. W., Coopman, S., Hart, J., & Walker, K. (2007). Workplace surveillance and managing privacy boundaries. *Management Communication Quarterly, 21*(2), 172–200.

Andersson, L. M., & Bateman, T. S. (2000). Individual environmental initiative: Championing natural environmental issues in U.S. business organizations. *Academy of Management Journal, 43*, 548–570.

Bass, B. M. (1985). *Leadership and performance beyond expectations.* New York: The Free Press.

Benn, S., Edwards, M., & Angus-Leppan, T. (2013). Organizational learning and the sustainability community of practice: The role of boundary objects. *Organization & Environment, 26*, 184–202.

Biloslavo, R., & Trnavcevic, A. (2009). Web sites as tools of communication of a 'green' company. *Management Decision, 47*, 1158–1173.

Bisel, R. S., & Barge, J. K. (2011). Discursive positioning and planned change in organizations. *Human Relations, 64*, 257–283.

Bissing-Olson, M. J., Iyer, A., Fielding, K. S., & Zacher, H. (2012). Relationships between daily affect and pro-environmental behavior at work: The moderating role of pro-environmental attitude. *Journal of Organizational Behavior, 34*, 156–175.

Blackburn, W. R. (2007). *The sustainability handbook: The complete management guide to achieving social, economic and environmental responsibility.* London: Earthscan.

Bolderdijk, J. W., Steg, L., & Postmes, T. (2013). Fostering support for work floor energy conservation policies: Accounting for privacy concerns. *Journal of Organizational Behavior, 34*, 195–210.

Bordia, P., Hobman, E., Jones, E., Gallois, C., & Callen, V. (2004). Uncertainty during organizational change: Types, consequences, and management strategies. *Journal of Business and Psychology, 18*, 507–532.

Bortree, D. S. (2011). The state of environmental communication: A survey of PRSA members. *Public Relations Journal, 4*, 1–17.

Buhr, N. (2002). A structuration view on the initiation of environmental reports. *Critical Perspectives on Accounting, 13*, 17–38.

Burns, J. M. (1978). *Leadership.* New York, NY: Harper & Row.

Carrington, T., & Johed, G. (2007). The construction of top management as a good steward: A study of Swedish annual general meetings. *Accounting, Auditing & Accountability Journal, 20*, 702–728.

Cheney, G. (2011). *Organizational communication in an age of globalization: Issues, reflections, practice.* Long Grove, IL: Waveland.

Cheney, G., Christensen, L. T., & Dailey, S. L. (2014). Communicating identity and identification in and around organizations. In L. L. Putnam & D. K. Mumby (Eds.), *The SAGE handbook of organizational communication* (3rd ed., pp. 695–716). Thousand Oaks, CA: Sage.

CNA Military Advisory Board. (2014). *National security and the accelerating risks of climate change.* Alexandria: CNA Corporation. http://www.cna.org/sites/default/files/MAB_2014.pdf. Accessed 19 June 2014.

Cohen, W. M., & Levinthal, D. A. (1990). Absorptive capacity: A new perspective on learning and innovation. *Administrative Science Quarterly, 35*(1), 128–152.

Cooren, F. (2009). Actor-network theory. In S. W. Littlejohn & K. A. Foss (Eds.), *Encyclopedia of communication theory* (Vol. 1, pp. 16–18). Los Angeles, CA: Sage.

Covin, J. G., & Slevin, D. P. (1989). Strategic management of small firms in hostile and benign environments. *Strategic Management Journal, 10*, 75–87.

Craig, C. A., & Allen, M. W. (2013). Sustainability information sources: Employee knowledge, perceptions, and learning. *Journal of Communication Management, 17*, 292–307.

Crescenzo, S. (2011). Integrating employee communications media. In T. L. Gillis (Ed.), *The IABC handbook of organizational communication: A guide to internal communication, public relations, marketing, and leadership* (2nd ed., pp. 219–230). San Francisco, CA: Jossey-Bass.

Danet, J.-B., Liddell, N., Dobney, L., MacKenzie, D., & Allen, T. (2013). *Business is beautiful: The hard art of standing apart*. London: LID Publishing.

Delmas, M. A., & Pekovic, S. (2013). Environmental standards and labor productivity: Understanding the mechanisms that sustain sustainability. *Journal of Organizational Behavior, 34*, 230–252.

Dervin, B., & Naumer, C. M. (2009). Sense-making. In S. W. Littlejohn & K. A. Foss (Eds.), *Encyclopedia of communication theory* (Vol. 2, pp. 876–880). Los Angeles, CA: Sage.

Dunphy, D. C., Griffiths, A., & Benn, S. (2003). *Organizational change for corporate sustainability: A guide for leaders and change agents of the future*. London: Routledge.

Egri, C. P., & Herman, S. (2000). Leadership in the North American environmental sector: Values, leadership styles, and contexts of environmental leaders and their organizations. *Academy of Management Journal, 43*, 571–604.

Eisenhardt, K. M., & Martin, J. A. (2000). Dynamic capabilities: What are they? *Strategic Management Journal, 21*(10/11), 1105–1121.

Elliott, S., & Coley-Smith, H. (2005). Building a new performance management model at BP: A program to track communication's impact on business performance. *Strategic Communication Management, 9*, 24–29.

Elving, W. J. L. (2005). The role of communication in organizational change. *Corporate Communications: An International Journal, 10*, 129–138.

Fairhurst, G. T., & Connaughton, S. (2014). Leadership communication. In L. L. Putnam & D. K. Mumby (Eds.), *The SAGE handbook of organizational communication* (3rd ed., pp. 401–424). Thousand Oaks, CA: Sage.

Fairhurst, G. T., Jordan, J. M., & Neuwirth, K. (1997). Why are we here? Managing the meaning of an organizational mission statement. *Journal of Applied Communication Research, 25*, 243–263.

Ferns, B., Emelianova, O., & Sethi, S. P. (2008). In his own words: The effectiveness of CEO as spokesperson on CSR-sustainability issues – Analysis of data from the Sethi CSR Monitor. *Corporate Reputation Review, 11*, 116–129.

Frahm, J. (2011). Communicating change: When change just doesn't stop: Creating really good change communication. In T. L. Gillis (Ed.), *The IABC handbook of organizational communication: A guide to internal communication, public relations, marketing, and leadership* (2nd ed., pp. 137–150). San Francisco, CA: Jossey-Bass.

Gladwell, M. (2007). *Blink: The power of thinking without thinking*. New York: Back Bay Books.

Grady, P. (2011). Internal branding, employer branding. In T. L. Gillis (Ed.), *The IABC handbook of organizational communication: A guide to internal communication, public relations, marketing, and leadership* (2nd ed., pp. 231–240). San Francisco, CA: Jossey-Bass.

Grensing-Pophal, L. (2011). Measuring marketing communication. In T. L. Gillis (Ed.), *The IABC handbook of organizational communication: A guide to internal communication, public relations, marketing, and leadership* (2nd ed., pp. 417–430). San Francisco, CA: Jossey-Bass.

Hannaes, K., Arthur, D., Balagopal, B., Kong, M. T., Reeves, M., Velken, I. et al. (2011). *Sustainability: The 'embracers' seize advantage*. MIT Sloan Management Review and the Boston Consulting Group Research Report. http://sloanreview.mit.edu/reports/sustainability-advantage/. Accessed 20 Dec 2013.

Hawkin, P. (1994). *The ecology of commerce: A declaration of sustainability*. New York: HarperCollins.

Hoffman, J. (2013). *Madison colors Wisconsin green*. Sustainable City Network webinar. http://www.sustainablecitynetwork.com/. Accessed 3 Mar 2014.

Hunter, F. (2009). *The lost artifacts of Lewis & Clark*. Frances Hunter's American Heroes Blog. http://franceshunter.wordpress.com/2009/10/06/the-lost-artifacts-of-lewis-clark/. Accessed 3 Mar 2014.

Jones, O. (2006). Developing absorptive capacity in mature organizations: The change agent's role. *Management Learning, 37*(3), 355–376.

Kiron, D. N., Kruschwitz, K., Haanaes, M., Reeves, S., Fuisz-Kehrbach, & Kell, G. (2015). *Joining forces: Collaboration and leadership for sustainability*. MIT Sloan Management Review, The Boston Consulting Group, and the United Nations Global Compact Research Report. http://sloanreview.mit.edu/. Accessed 15 Jan 2015.

Kuhn, T. R. (2014). Knowledge and knowing in organizational communication. In L. L. Putnam & D. K. Mumby (Eds.), *The SAGE handbook of organizational communication* (3rd ed., pp. 481–502). Thousand Oaks, CA: Sage.

Lewis, L. K. (2007). An organizational stakeholder model of change implementation communication. *Communication Theory, 17*, 176–204.

Lewis, L. K. (2014). Organizational change and innovation. In L. L. Putnam & D. K. Mumby (Eds.), *The SAGE handbook of organizational communication* (3rd ed., pp. 503–524). Thousand Oaks, CA: Sage.

Lewis, L. K., & Russ, T. L. (2012). Soliciting and using input during organizational change initiatives: What are practitioners doing? *Management Communication Quarterly, 26*, 267–294.

Lewis, L. K., & Seibold, D. R. (1993). Innovation modification during intraorganizational adoption. *Academy of Management Review, 18*, 322–354.

Lewis, L. K., & Seibold, D. R. (1996). Communication during intraorganizational innovation adoption: Predicting users' behavioral coping responses to innovations in organizations. *Communication Monographs, 63*(2), 131–157.

Lewis, L. K., & Seibold, D. R. (1998). Reconceptualizing organizational change implementation as a communication problem: A review of literature and research agenda. *Communication Yearbook, 21*, 92–151.

Linnenluecke, M. K., Russell, S. V., & Griffiths, A. (2009). Subcultures and sustainability practices: The impact of understanding corporate sustainability. *Business Strategy and the Environment, 18*, 432–452.

Littlejohn, S. W. (2009). System theory. In S. W. Littlejohn & K. A. Foss (Eds.), *Encyclopedia of communication theory* (Vol. 2, pp. 950–954). Los Angeles, CA: Sage.

Maharaj, R., & Herremans, I. M. (2008). Shell Canada: Over a decade of sustainable development reporting experience. *Corporate Governance, 8*, 235–247.

McPhee, R. D., Poole, M. S., & Iverson, J. (2014). Structuration theory. In L. L. Putnam & D. K. Mumby (Eds.), *The SAGE handbook of organizational communication* (3rd ed., pp. 75–100). Thousand Oaks, CA: Sage.

McPhee, R. D., & Poole, M. W. (2009). Structuration theory. In S. W. Littlejohn & K. A. Foss (Eds.), *Encyclopedia of communication theory* (Vol. 2, pp. 936–940). Los Angeles, CA: Sage.

Meng, J., & Berger, B. K. (2012). Measuring return on investment (ROI) of organizations' internal communication efforts. *Journal of Communication Management, 16*, 332–354.

Nicholson, N., & Aiello, A. (Eds.) (2008). *Structuring a communication department*. CW Bulletin (Vol. 6, p. 1). http://www.iabc.com/cwb/archive/2008/cw_news0108.htm. Accessed 19 June 2014.

Okereke, C., Wittneben, B., & Bowen, F. (2012). Climate change: Challenging business, transforming politics. *Business & Society, 51*, 7–30.

Oncica-Sanislav, D., & Candea, D. (2010). The learning organization: A strategic dimension of the sustainable enterprise? *Proceedings of the European Conference of Management, Leadership & Governance* (pp. 263–270). www.gbv.de/dms/tib-ub-hannover/647510022.pdf. Accessed 3 Feb 2014.

Ones, D. S., & Dilchert, S. (2012). Environmental sustainability at work: A call to action. *Industrial and Organizational Psychology, 5*, 444–466.

Paddack, B. (2010). *Martha Jane Murray: Working on solutions to climate change*. Arkansas Business. http://www.arkansasbusiness.com/article/37306/martha-jane-murray-working-on-solutions-to-climate-change?page=all. Accessed 7 Feb 2014

Paine, L. S., Hsieh, N.-H., & Adamsons, L. (2013). *Governance and sustainability at Nike (A) (case 9-313-146).* http://hbsp.harvard.edu/product/cases. Accessed 4 Mar 2014.

Petronio, S. (2009). Privacy management theory. In S. W. Littlejohn & K. A. Foss (Eds.), *Encyclopedia of communication theory* (Vol. 2, pp. 796–798). Thousand Oaks, CA: Sage.

Poole, M. S. (2014). Systems theory. In L. L. Putnam & D. K. Mumby (Eds.), *The SAGE handbook of organizational communication* (3rd ed., pp. 49–74). Thousand Oaks, CA: Sage.

Quinn, L., & Norton, J. (2004). Beyond the bottom line: Practicing leadership for sustainability. *Leadership in Action, 24,* 3–7.

Robertson, J. L., & Barling, J. (2013). Greening organizations through leaders' influence on employees' pro-environmental behaviors. *Journal of Organizational Behavior, 34,* 176–194.

Robinson, O., & Griffiths, A. (2005). Coping with the stress of transformational change in a government department. *The Journal of Applied Behavioral Science, 41,* 204–221.

Rogers, E. M. (2003). *Diffusion of innovations* (5th ed.). New York, NY: Free Press.

Rose, R. A. (2006). A proposal for integrating structuration theory with coordinated management of meaning theory. *Communication Studies, 57,* 173–196.

Samp, J. A. (2009). Communication goal theories. In S. W. Littlejohn & K. A. Foss (Eds.), *Encyclopedia of communication theory* (Vol. 1, pp. 129–132). Los Angeles, CA: Sage.

Sattari, S., Pitt, L., & Caruana, A. (2011). How readable are mission statements? An exploratory study. *Corporate Communications, 16,* 282–292.

Scott, C. W., & Myers, K. K. (2010). Toward an integrative theoretical perspective of membership negotiations: Socialization, assimilation, and the duality of structure. *Communication Theory, 30,* 79–105. doi:10.1111/j.1468-2885.2009.01355.x.

Senge, P. M. (1990). *The fifth discipline: The art and practice of the learning organization.* New York: Doubleday.

Singhal, A. (2009). Diffusion of innovations. In S. W. Littlejohn & K. A. Foss (Eds.), *Encyclopedia of communication theory* (Vol. 1, pp. 307–309). Los Angeles, CA: Sage.

Smerecnik, K. R., & Andersen, P. A. (2011). The diffusion of environmental sustainability innovations in North American hotels and ski resorts. *Journal of Sustainable Tourism, 19,* 171–196.

Smircich, L., & Morgan, G. (1982). Leadership: The management of meaning. *Journal of Applied Behavioral Science, 18,* 257–273.

Strandberg Consulting. (n.d.). *Developing a sustainability vision and management system.* http://www.corostrandberg.com/pdfs/Sustainability_Vision_and_Management1.pdf. Accessed 23 June 2014.

Swanson, V., & Power, K. (2001). Employee perceptions of organizational restructuring: The role of social support. *Work & Stress, 15,* 161–178.

United Nations Environment Programme (UNEP). (2014). *Integrated governance: A new model of governance for sustainability.* http://www.unepfi.org/fileadmin/documents/UNEPFI_IntegratedGovernance.pdf. Accessed 17 Jan 2015.

Vakola, M., & Nikolaou, I. (2005). Attitudes toward organizational change: What is the role of employees' stress and commitment? *Employee Relations, 27,* 160–174.

van den Bosch, F. A. J., Volberda, H. W., & de Boer, M. (1999). Coevolution of firm absorptive capacity and knowledge environment: Organizational forms and combinative capabilities. *Organizational Science, 10,* 551–568.

van der Heijden, A., Cramer, J. M., & Driessen, P. P. J. (2012). Change agent sensemaking for sustainability in a multinational subsidiary. *Journal of Organizational Change Management, 25,* 535–559.

Walls, J. L., & Hoffman, A. J. (2013). Exceptional boards: Environmental experience and positive deviance from institutional norms. *Journal of Organizational Behavior, 34,* 253–271.

Wals, A. E. J., & Schwarzin, L. (2012). Fostering organizational sustainability through dialogic interaction. *The Learning Organization, 19,* 11–27.

Weick, K. E. (1969). *The social psychology of organizing.* Reading, MA: Addison-Wesley.

Weick, K. E. (1995). *Sensemaking in organizations: Foundations for organizational science.* Thousand Oaks, CA: Sage.

Weiner, M. (2011). Measuring public relations programs. In T. L. Gillis (Ed.), *The IABC handbook of organizational communication: A guide to internal communication, public relations, marketing, and leadership* (2nd ed., pp. 363–376). San Francisco, CA: Jossey-Bass.

Werbach, A. (2009). *Strategy for sustainability: A business manifesto.* Cambridge, MA: Harvard Business Press.

Whitworth, B. (2011). Internal communication. In T. L. Gillis (Ed.), *The IABC handbook of organizational communication: A guide to internal communication, public relations, marketing, and leadership* (2nd ed., pp. 195–206). San Francisco, CA: Jossey-Bass.

Williams, J. (2011). Measuring the effectiveness of internal communication. In T. L. Gillis (Ed.), *The IABC handbook of organizational communication: A guide to internal communication, public relations, marketing, and leadership* (2nd ed., pp. 271–284). San Francisco, CA: Jossey-Bass.

Williams, L. S. (2008). The mission statement: A corporate reporting tool with a past, present, and future. *Journal of Business Communication, 45,* 94–119.

Yeatts, D., Folts, W., & Knapp, J. (2000). Older workers' adaptation to a changing workplace: Employment issues for the 21st century. *Educational Gerontology, 26,* 565–582.

Zahra, S. A., & George, G. (2002). Absorptive capacity: A review, reconceptualization and extension. *Academy of Management Review, 27,* 185–203.

Chapter 6
Using Communication to Create Environments That Empower Employees

Abstract This chapter focuses on factors that can influence employees' pro-environmental behaviors. Sustainable organizational cultures, learning organizations, and an informed and supportive managerial subculture are discussed in this chapter. Employee hiring, socialization, and training can influence employees' pro-environmental actions. Reward systems need to link sustainability goals and measures to corporate training. If employees do not see that their organization rewards people for displaying pro-environmental behaviors, training can have less of an impact (Cantor et al., *Journal of Supply Chain Management*, *48*, 33–51, 2012). Although managers may encourage employees to meet work-related pro-environmental goals, employees balance multiple goals and react emotionally to their working environment assessing issues related to perceived organizational support and justice. Informal communication helps create, reinforce, stabilize, and challenge sustainable values and goals within an organization. Additional theories or theoretical concepts discussed in this chapter include the 4I model of organizational learning, organizational climate, affective organizational commitment, perceived organizational support, employee trust, the broaden-and-build theory of positive emotions, goal congruence, path-goal theory, socialization and assimilation, uncertainty reduction theory, social exchange theory, social identity theory, person-organization fit, organizational citizenship behaviors, green teams, personal sustainability plans, social learning theory, cognitive maps, and self-efficacy. Interview data spotlights WasteCap Nebraska, Ecotrust, the HEAL program, the University of Arkansas, Fayetteville, Aspen Skiing Company, Neal Kelly Company, the City of Portland, the City and County of Denver, the Arbor Day Foundation, the South Dakota Bureau of Administration, the State Farm processing plant in Lincoln, NE, Assurity Life Insurance, and Sam's Club.

Lewis and Clark, Teamwork and Culture In December 1803, Clark established Camp River Dubois at the confluence of the Mississippi and Missouri rivers, north of St. Louis, MO. He trained the Corps of Discovery volunteers seeking to turn them into a team. He had them build log structures, learn to march in formation, and shoot more accurately. "Most of all, he tried to get the men to respect military authority and learn how to follow orders" (National Park Service n.d.). Along their journey, the Corps encountered native people whose understandings of power and authority differed from their own. This made communication difficult, open to confusion and misunderstanding. For example, the Corps hierarchy was important and rank defined who gave orders and who obeyed. In tribal societies, kinship was important and power depended upon a network of real and symbolic relationships. When the two cultures met, Lewis and Clark attempted to impose their idea of hierarchy on native peoples by giving some men they designated as chiefs medals, printed certificates, and/or gifts. In turn, native peoples tried to impose kinship obligations on the visitors through adoption ceremonies, shared names, and ritual gifts (Library of Congress n.d.). This example illustrates how organizations have cultural expectations and seek to influence others to behave in ways matching these expectations.

6.1   Pro-environmental Employee Behaviors

Employees display task-related pro-environmental behavior when they complete their required tasks in environmentally friendly ways (Bissing-Olson et al. 2012). Employee pro-environmental behaviors may include displaying personal initiative when they suggest or make environmentally friendly changes in their workplace tasks, policies, and procedures; offering constructive suggestions; identifying problems and engaging in creative problem solving; and overcoming barriers to improve existing processes. These behaviors may involve working sustainably (e.g., creating sustainable processes), avoiding harm (e.g., preventing pollution), conserving resources (e.g., recycling), and influencing others (e.g., training others regarding sustainability, lobbying). Such behaviors are more likely if employees are committed to and identify with their organization. Employees are motivated by personal engagement rather than by job descriptions or management's requirements. Those who engage in pro-environmental behaviors often are engaging in organizational citizenship behaviors (OCB). OCB have been researched since the 1970s and involve "individual behavior that is discretionary, not directly or explicitly recognized by the formal reward system, and that in the aggregate promotes the effective functioning of the organization (Organ 1988, p. 4). Such actions and behaviors are measurable (Ones and Dilchert 2012).

But not all pro-environmental behaviors make a significant impact. An environmentally significant behavior is defined by its direct or proximal impact: does it change the availability of materials or energy from the environment (direct); does it alter ecosystem structure and dynamics or even the biosphere itself (indirect) (Stern 2000). For example, one purchasing agent's decision to buy environmentally preferable products can have a small direct impact. Other environmentally significant behaviors shape the context within which choices are made that, in turn, directly cause environmental change. For example, South Dakota's law requiring all state agencies to purchase environmentally preferable products constrains the decisions of the individual purchasing agent while having a moderately large impact due to state-level purchasing power. Focusing on impact-oriented definitions of environmentally significant behavior allows us to identify and target sustainability-initiative-related behaviors that can have a large impact. Organizations can change their cultures to stimulate employee pro-environmental OCB's and thereby significantly impact their organizations' environmental and social footprints.

6.2 Organizational Cultures, Employee Socialization, and Training

In 2011, the Society of Human Resource Managers (SHRM) (2011) surveyed U.S. human resource professionals about how their companies balanced financial performance, their employees' quality of life, the welfare of society, and the environment. A sample of 5,000 professionals was randomly selected from SHRM's membership database of 250,000. Approximately 16 % ($n = 728$) responded. Of those, over half had formal sustainability-related policies that included sustainable workplace goals and policies directly tied to their organization's strategic planning process. An additional 40 % reported having informal policies (not tied to the strategic planning process). Nearly four out of 10 organizations calculated a return on investment (ROI) for their sustainability efforts. Of those, 47 % calculated a positive ROI, 46 % reported it was still too early to determine their ROI, and 6 % calculated a break-even point. Having goals and policies does not automatically result in a positive ROI. Employees must become engaged, knowledgeable, empowered, and supported by their organization. In this section, I discuss organizational cultures, especially that of learning organizations, and organizational subcultures, specifically the managerial subculture.

6.2.1 Organizational Cultures

If organizations are going to respond to environmental challenges and/or become more sustainable, new or modified values, beliefs, and processes will need to be woven into their cultural fabric and implemented through operational and strategy changes. Therefore, it is important to discuss the concept of organizational culture.

A myriad of definitions and approaches to organizational cultures exist (see Keyton 2014; McAleese and Hargie 2004; Tracy 2009). One definition is that an organization's culture is manifested through employees' shared assumptions, values, beliefs, languages, symbols, and meaning systems (Tracy 2009). Organizations have observable cultures which include structures, processes, behaviors, and artifacts as well as espoused values and philosophies displayed through their strategies and goals (Schein 2004). At the deeper level, cultures have underlying assumptions (unconscious beliefs and perceptions from which values and actions stem). It is important to remember an organization's culture is a human creation as employees build and negotiate meaning through interaction, but it goes beyond a few individuals to become something that is often intangible and pervasive.

Organizational leaders should reflect on where sustainability fits within their existing culture and then discuss changes they might make. Strandburg Consulting (n.d.) provides a series of questions leaders can use to help reflect upon their existing culture. Questions address if and how company goods, services, or processes overlap with the interests of society; if the company sees itself clearly; and if strong leadership for sustainability is articulated in a clear and simple message, disseminated throughout the organization, and repeatedly reinforced. Snowden, who writes in the field of knowledge management, emphasizes storytelling and communication interventions to help organizational participants gain insights into their organization's assumptions and practices, as well as to generate new alternatives. He introduced Cynefin, a Welsh word meaning "our place of belonging," a place of great meaningfulness for a people. For Snowden, Cynefin is a sensemaking methodology. It sets up an environment where people can come together to jointly make sense of their situation utilizing information such as organizational stories, anecdotes, legends, alternative histories, and accounts of phases and events since their organization's birth. Teams go through a well-defined group process to create plans, prescriptions, and shared memories and experiences. Snowden sought to use this approach to improve organizational capacities when situations force people to address intractable problems and complex, uncertain situations (Dervin and Naumer 2009). Intractable problems and complex, uncertain situations face organizations seeking to adapt to impending climate change-related challenges.

A focus on organizational cultures emerged in the early 1980s as scholars sought to explain how organizational meanings and behaviors are constituted through communication. Keyton (2014) summarizes the conceptual and methodological approaches communication scholars used to study organizational culture. Two distinct approaches guided their investigations: the interpretative approach and the managerial approach. The interpretative approach focuses us on how communication (e.g., metaphors, stories, myths, rites, ceremonies, jargon, physical artifacts) shapes what the organization *is* and what shared values its employees hold. This approach recognizes that management-directed cultural change is challenging, if not uncontrollable. Individuals and groups throughout an organization create, interpret, challenge, and reinterpret elements which may or may not become part of an organization's culture. van der Heijden et al.'s (2012) 10-year investigation of the Dutch subsidiary of Interface provides an example of the nonlinear nature of the organizational culture change process. The managerial approach focuses us on how organizations can design, control, and improve their corporate culture. Much of the research currently focusing on sustainability and culture appears to illustrate this approach (e.g., Linnenluecke and Griffiths 2010; Russell et al. 2007), as does work on culture management (e.g., McAleese and Hargie 2004). As management attempts to change an organization's focus toward more environmental sustainability or more socially responsible internal and/or community relations, the assumption is that they can create a single unified corporate culture (Linnenluecke et al. 2009).

Models regarding what a sustainable culture would look like and how to create one are generally lacking. Several authors have attempted to fill this gap (e.g., Harris and Crane 2002; Linnenluecke and Griffiths 2010; Russell et al. 2007). McAleese and Hargie (2004) identified five guiding principles of culture management including formulate a guiding strategy, develop culture leaders, share the culture by communicating effectively with staff, measure performance, and communicate your culture with external groups. Linnenluecke and Griffiths (2010) sought to identify what makes up a sustainability-oriented organizational culture, whether organizations can display a unified sustainability-oriented organizational culture and whether organizations can even become more sustainable through culture change. They used the competing values framework (CVF) (Quinn 1988) to discuss the relationship between corporate sustainability and organizational culture and to identify four cultural types: the human relations model, the open systems model, the internal process model, and the rational goal model. They describe how cultural control occurs within each type. In the human relations culture, training, employee development, open communication, and participative and decentralized decision making are used. Environmental health and safety, human well-being, and employee skills, satisfaction, commitment, and productivity are emphasized. Social entrepreneurs are likely to emerge who advocate sustainability principles within the organization. Open systems cultures emphasize innovation for achieving ecological and social sustainability. They seek to operate within the carrying capacity of the natural environment by minimizing their resource use and ecological footprint. An internal process culture emphasizes economic performance, growth, and long-term profitability. These organizations focus mainly on economic sustainability and may miss opportunities to develop innovative products, services, and business models because they give their employees limited room for flexibility, learning, and change. In rational goal cultures, the emphasis is on increasing resource efficiencies. Some organizations reinvest saved costs due to efficiencies in employees so as to create human systems that support value-adding and innovation. Russell et al. (2007) developed a similar typology focusing on the extent to which an organization is working toward long-term economic performance by seeking increased profits and growth, is working toward positive ecological outcomes for the natural environment, supports people and social outcomes, or takes a holistic approach. Although useful, these models do not focus on communication as an interactive meaning-making process.

Proponents of the managerial approach to organizational cultures argue that an organization's ability to import and use information about innovations (i.e., its absorptive capacity) is influenced by its culture. Those organizations with strong functional cultures employ a cohesive workforce which can work together to implement organizational goals. In contrast, firms with disjointed cultures often face difficulties integrating change into their work practices. If employees in functional organizations share their organizations' goals and values, they will be more likely to seek out opportunities to contribute to its success. This increases the likelihood that employees will be proactive in absorbing information into the organization and successfully turn information into useable knowledge.

However, it is difficult to transform for-profit organizational cultures into sustainability-focused cultures. Any change will need to convince organizational members at all levels to moderate their driving quest to meet economic goals, embrace a longer time frame (e.g., intergenerational), and value the environment as more than a resource supplier or waste repository. In the 1990s, even in the most progressive firms, true cultural transformation was rare and pro-environmental sensibilities were generally absorbed into existing cultural assumptions and beliefs. Based on interviews with 44 executives and managers representing separate organizations, Harris and Crane (2002) developed a model of the barriers and drivers influencing how managers viewed the depth, degree, and diffusion of cultural greening within organizations. Depth involves how deeply greening is being valued by different organizational members and groups. Degree involves the extent to which green values and sensibilities are included in organizational creations and artifacts. Diffusion refers to how widely green feelings and behaviors appear throughout an organization. Diffusion is hampered by organizational barriers (e.g., behaviors, systems, and structures) and cultural fragmentation (i.e., multiple subcultures). Over 10 years after Harris and Crane's research cultural transformation within an organization seems more feasible as societal *Discourses* have shifted to account for changing environmental conditions, the business case for sustainability has grown, and isomorphism pressures are occurring within many organizational fields.

6.2.1.1   Learning Organizations

Organizations which want to fully commit to pursuing sustainability will need to change their corporate cultures to evolve from reacting (adaptive learning) to innovating (generative learning). In discussing how such a transformation can occur, Oncica-Sanislav and Candea (2010) refer to Peter Senge's concept of the learning organization as an ideal type. This concept focuses us on the systems thinking and group problem-solving abilities organizations need if they are to respond more competitively in rapidly changing environments. Learning organizations are "organizations where people continually expand their capacity to create the results they truly desire, where new and expansive patterns of thinking are nurtured, where collective aspiration is set free, and where people are continually learning to see the whole together" (Senge 1990, p. 3). Proposed benefits associated with a learning organization include innovation, competitiveness, responsiveness, knowledge of customer needs, quality outputs, and an improved corporate image.

Oncica-Sanislav and Candea (2010) developed a conceptual framework useful for understanding how the learning organization paragon can influence the design of organizations concerned with long-term triple-bottom-line success. A learning organization has five main features: systems thinking, personal mastery, mental models, shared vision, and team learning. In terms of systems thinking, individuals tend to focus on the parts rather than seeing the whole, and fail to see the organization as a dynamic process. Personal mastery refers to an individual's commitment to the process of continual learning. Those with a high level of personal mastery live in a continual learning mode, realize they have more to learn and need to grow, and are deeply self-confident. Mental models are the assumptions, theories, and theories-in-use held by individuals which guide their behaviors. Organizations also operate on assumptions and memories that preserve certain behaviors, norms, and values. "Entrenched mental models. . .thwart changes that could come from systems thinking" (Senge 1990, p. 203). In learning organizations, any unwanted values or mental models supporting confrontational attitudes are replaced by an open culture promoting inquiry and trust. Development of a vision is focused on creating shared pictures of the future that foster genuine commitment and engagement, rather than compliance, which motivate employees to continually learn. Visions spread because of a reinforcing process, "As people talk, the vision grows clearer. As it gets clearer, enthusiasm for its benefits grow" (p. 227). When there is a genuine vision that their organization is a learning organization, people want to excel and learn. The accumulation of individual learning results in team learning which increases an organization's problem-solving capacity through better access to knowledge and expertise. Team learning is "the process of aligning and developing the capacities of a team to create the results its members truly desire" (p. 236). Coordination mechanisms such as knowledge management systems, boundary crossing, and openness promote team learning. Dialogue and discussion further increase and utilize team learning (Barker and Camarata 1998). Through dialogue ideally team members will suspend their assumptions and begin thinking together. Dialogue plus systems thinking facilitates dealing with complexity and focuses communicators on important opportunities and challenges rather than on personalities and leadership styles.

Ecotrust and the Learning Organization   Oakley Brooks, Senior Media Manager at Ecotrust, discussed how his organization approaches things through a resilience lens. Staff are trying to help transform economic, social, and ecological systems based on the resilience approach. They work on specialized projects as part of marine, forest, and food teams which rely on specialized and expert knowledge. Sharing only their focus on resilience, Ecotrust does not have a tightly defined brand. In order to create an opportunity for broader team learning and to create shared understandings, they hold monthly staff meetings where they do exercises to transcend their expertise silos and learn how to talk about what the other project teams are doing. Oakley explained:

> There has been an effort to have people think about problems and solutions outside of their silos and how their approaches, say forestry, could translate to other groups in the organization. So it has been a process of how you can communicate Ecotrust's mission and its work when you are out in the world and also how can you talk more openly and work more collaboratively with people within the organization. . . . One of the ways of bringing resilience back into the organization and thinking about how we work along those lines is really being open and connected to the person right next to you and thinking about

strengthening relationships between people in the organization to create a more robust work environment.

Ecotrust is using communication to create team learnings which, in turn, can influence employee communication with internal and external stakeholders.

But organizational learning about sustainability is a complex and nonlinear process. Benn et al. (2013) investigated how to embed ideas emerging at the individual and group levels into organization-wide sustainability-related initiatives. They utilized a multilevel 4I model of organizational learning partly based on Weick's (1995) sensemaking theory to look at the role of language, context, and group interpretation during organizational change, especially as individuals attempt to interpret what is meant by and how to enact sustainability initiatives. The 4I framework begins with what has already been learnt (i.e., exploitation of previous learning and old certainties) when attempting to strategically renew an organization to explore new possibilities. The 4I framework describes four interconnected processes: intuition, interpretation, integration, and institutionalization. Intuition and interpretation occur at the individual level, integration at the group level, and institutionalization at the organizational level. Their findings illustrated how knowledge sharing and boundary objects can promote the development of communities of practice which lead to the integration and institutionalization of sustainability across intraorganizational knowledge and disciplinary boundaries. However, boundaries do exist. This leads us to a discussion of subcultures.

6.2.1.2   The Managerial Subculture

Organizations may consist of multiple subcultures which can influence how employees respond to a top–down view of sustainability. For example, in a large Australian transportation-focused organization, employees who worked in a subculture with a strong emphasis on hierarchical values and who possessed little awareness of the corporate sustainability practices only emphasized an economic understanding of corporate sustainability in comparison to their coworkers in other areas of the organization (Linnenluecke et al. 2009). Often subcultures develop around professional background, departmental membership, social group membership, demographic characteristics (e.g., gender, ethnicity), and hierarchical levels. In this section, I focus specifically on the managerial subculture.

Managers can influence how their subordinates respond to sustainability initiatives. The superior-subordinate relationship is constituted through communication and is an important location for information exchange, feedback and appraisal, and mentoring (Sias 2014). Initially, you might think managers will reflect top management's vision for an organizational culture promoting sustainability. Top management expects managers to communicate change initiatives to their subordinates, to create working environments consistent with any newly mandated changes, and then to reinforce, enforce, and potentially reward employee efforts toward new organizational goals and/or directions. Organizations often assign managers to dedicated roles focused on sustainability (e.g., chief sustainability officer) and rely on line leaders and non-leadership employees to make sustainability-related changes (Hannaes et al. 2011). Managers can create a sense among employees that their organization cares about environmental initiatives (Cantor et al. 2012) which in turn can influence employee commitment to environmental initiatives and, ultimately, influence their environmental behaviors. Those managers who perceive their organization to be an opinion leader in sustainability appear more open to adopting sustainability innovations (Smerecnik and Andersen 2011) and may engage in more eco-friendly behaviors than nonmanagers (Ones and Dilchert 2012). But sometimes managers lack sufficient information about the sustainability-related changes or are unsupportive of them. There are multiple reasons why they might be unsupportive. For example, they may fear what the change will mean to their workload or power. Perhaps they feel the change initiative is a temporary fad unworthy of more than minimal effort. In the past they may have been rewarded for managing in ways that promote their organization's bottom line or primary mission and feel a shift toward sustainability will achieve neither. The manager's own personal value set may be anthropogenic business vs. sustainable business oriented (Byrch et al. 2007) (see Sect. 4.1).

Several researchers provide glimpses into the managerial subculture as it relates to sustainability and corporate social responsibility (CSR) (e.g., Bissing-Olson et al. 2012; Cordano and Frieze 2000). For example, Hine and Preuss (2009) investigated how different managerial groups perceive and respond to their organization's CSR initiatives, looking at what impact these programs had on managerial agency. They interviewed 27 mid- to upper-level managers in England and found that often managers felt the CSR activities were undertaken to create an attractive corporate reputation and to comply with governmental preferences. Many managers agreed that good things might result from the adoption of CSR policies and procedures but felt morally mute. They felt required to comply with the overriding corporate imperatives of profitability, shareholder value, and capital market responses. Managers can only make significant contributions toward sustainability initiatives if their organizational culture supports and rewards such actions. Otherwise, managers work within tightly controlled expectations, leading some to focus primarily on their personal success within highly competitive internal corporate environments.

Depending on their organizational level, even managers charged with enacting sustainability-related initiatives face challenges. Traditionally, many managers viewed waste prevention as excessively expensive and limited their environmental performance goals to achieving regulatory compliance. Cordano and Frieze (2000) investigated the source reduction preferences of 295 environmental managers. Environmental managers typically hold staff positions with limited power over resource allocations, may lack top management support, may face the belief that preventing pollution is a nonessential expense, and have limited authority to implement unilateral changes. Despite this, they potentially can influence the adoption of pollution prevention activities. Basing their research on Ajzen's theory of planned behavior (TPB), Cordano and Frieze investigated the environmental managers' pollution prevention attitudes, perceptions of their organization's subjective norms for environmental regulation, perceived behavioral control, and their facilities' past source reduction activities. As expected, environmental managers' attitudes about pollution prevention and their assessment of subjective norms about environmental regulation were both positively related to their preference to implement source reduction activities. What an organization previously did is a powerful predictor of future behaviors and an indicator of the resistance an environmental manager may or may not experience when attempting to initiate source reduction activities. Few environmental managers felt supported to improve environmental performance beyond regulatory requirements. Such managers are unlikely to voluntarily communicate the business benefits of additional environmental initiatives to their peers. Without such efforts at information exchange, it is likely that business managers will continue to only seek regulatory compliance and liability containment. The authors recommend environmental managers' select easy-to-implement short-term wins, display project benefits, counter critics, reward change, engage supervisors, and build momentum.

It is important to build commitment for sustainability among managers. Sharma (2000) developed and tested a model which investigated the relationship between managerial interpretations of environmental issues and corporate choice of environmental strategy among 99 firms in the Canadian oil and gas industry. The study focused on the relationship between whether environmental issues were framed as opportunities or threats and the influence that framing had on managers' choice of proactive or reactive environmental strategies. Findings included that if a company's managers interpreted environmental issues as opportunities, the company was more likely to voluntarily engage in pro-environmental strategies (e.g., engage in creative problem solving, search for and adopt innovative technologies, form collaborative relationships with stakeholders). If managers interpreted environmental issues as threats, the company was more likely to exhibit a conformance environmental strategy (i.e., comply with regulations, adopt industry standard practices). The more managers saw environmental concern as central to their company's identity, the more likely they were to see environmental issues as opportunities. The more they had discretionary slack when managing the business-natural environment interface, the more likely they were to interpret environmental issues as opportunities. However, the hypothesis that the greater a company integrates environmental performance criteria into its control systems, the more likely managers are to see environmental issues as opportunities was unsupported. This study suggests messages that frame environmental issues as opportunities central to an organization's success influence managers' behaviors, as does having discretionary slack when making decisions. *Best Practice*: Intentionally bring managers on board early in the process when seeking to create an organizational culture that supports pro-environmental employee behaviors.

6.2.2   Hiring, Socializing, and Training Employees

The greening of the economy (e.g., renewable energy, increasing energy efficiency) correspondingly shapes the work of existing employees and generates unique work and worker requirements. New jobs are appearing in research, design, consulting, manufacturing, renewable energy generation, and green construction (Ones and Dilchert 2012). For some jobs, the tasks, knowledge, skills, and other characteristics are changing. Companies are adding green responsibilities to existing jobs and creating new positions. In organizations seeking to create an organizational culture focused on innovation around sustainability, employee hiring and socialization is especially important. Organizations with a reputation for sustainability find attracting like-minded employees relatively easy. However, for those seeking to create a new culture emphasizing sustainability, incoming and existing employees may all need to be socialized into the evolving cultural norms. Often this involves training, especially if the changes are occurring as part of a certification program which mandates that training occurs.

6.2.2.1   Hiring

One theory (social identity theory) and one concept (person-organization fit) which help us better understand hiring issues are discussed next. Social identity theory (Reid 2009) explains how individuals are motivated to achieve and maintain a positive social identity. Employees desire to be affiliated with organizations they feel are connected to their desired personal identity. Organizational communication and management scholars have been influenced by this theory (Cheney et al. 2014). Having a public image as an organization which is a leader in sustainability can be a powerful recruiting tool when seeking qualified people interested in developing innovative ideas for dealing with issues related to climate change challenges. The person-organization fit concept appearing in the management literature refers to the degree of congruence between the attributes (e.g., attitudes, values, interests) of an individual and his/her organization. O'Reilly et al. (1991) designed an organizational culture profile instrument which uses 54 value statements to describe individual and/or organizational values (see Fields 2002). This instrument, or a modification thereof, is useful when measuring person-organization fit. Aspects of both the individual and the job situation combine to influence an employee's work performance. Value congruence is important since organizational values are components of an organization's culture. The fit between employee expectations (e.g., need for personal recognition) and an organization's climate (e.g., support for individual vs. team accomplishments) is also important. Employee stress is lower and OCBs are higher when the employee-organization fit is stronger.

Hiring at Neil Kelly and the Arbor Day Foundation Julia Spence, Vice President of Human Resources at the Neil Kelly Company, said:

> Over time, because we are known as a green company. ... We have attracted people who come to us with shared values. So that tends to feed on itself. We have people who are really dedicated. ... One of the things I think that helped us is that we have a relationship with Oregon State University. We have had summer interns from their housing and interiors programs for almost 30 years. Their focus on environmental sustainability has grown over the years. I have been on their advisory board. So we have built that capacity over all these years...and that's really good for us.

I provide this example because it illustrates how this organization strategically builds human capital. Woodrow Nelson, Vice President of Marketing Communication, of the Arbor Day Foundation, also discussed hiring with me saying:

> It is an important part of our hiring process. If this person has an affinity for environmental soundness, great. If the person seems to have the potential or an open mind, someone who wants to embrace diversity, then that is a great thing and someone we can coach and mentor. Then there are going to be others that just don't get it. Those people have got to be really talented in a whole lot of other ways in order for them to even make it to become part of the Foundation.

When organizations decide to make a turn toward sustainability, they often hire people with different skill and value sets. These individuals may help transform their organizational culture or simply adapt to the existing culture. To counteract a tendency to adapt to the existing culture, newly hired employees need reinforcement in the direction of the desired change. Companies like Aveda Corporation have aligned their human resources chain (from recruitment to selection to performance appraisal) with their environmental mission (Ones and Dilchert 2012). Scott and Myers (2010) offer an integrative theoretical perspective based on structuration theory which focuses on how employees can potentially negotiate their own roles and change organizational processes and norms. Organizational practitioners concerned with managing the hiring and socialization practices and processes in such a way as to create an organizational culture promoting sustainability will find their article interesting.

6.2.2.2   Socialization

The absorptive capacity literature discussed earlier in the book reminds us that combination capacities help create social communities rooted in action. Socialization capacities are one type of combination capacity that can facilitate "the ability of the firm to produce a shared ideology that offers members an attractive identity as well as collective interpretations of reality" (van den Bosch et al. 1999, p. 557). Organizational socialization is the "process by which newcomers learn how to fulfill their roles, are introduced to others, and become familiar with the policies and norms of the organization" (Myers 2009, p. 722). Socialization helps employees become familiar with "how things are done around here" and facilitates coordination of activities between new and existing employees. Julia Spence, Vice President of Human Resources at the Neil Kelly Company, discussed her organization's efforts to socialize new hires saying:

> There is an orientation when people first start and we still are following the pattern of the
> Natural Step, in great part because it is not prescriptive, it really is [a process] about how
> you think something through. So it can apply to anything. So we do it with an orientation
> and [reinforce it] soon after hiring with a little more in-depth training.

Researchers have focused on socialization tactics, information seeking, social support, peer relationship development, and how new hires learn their organization's culture (Kramer and Miller 2014). Early research investigated stages of socialization (e.g., anticipatory socialization, encounter, metamorphosis, and exit). Others identified the tactics organizations use to socialize newcomers' finding that institutional socialization characterized by collective, formal, fixed, serial, and investiture tactics encourage employees to adapt to their organization and are linked to higher job satisfaction and organizational identification. However, individualized socialization tactics, which are more informal, disjunctive, random, and variable, are associated with more role innovation, role ambiguity, and increased stress (Myers 2009). At least seven processes are important during the assimilation process: creating familiarity with coworkers and supervisors, enhancing acculturation, providing recognition, reinforcing involvement, increasing job competency, and allowing for role negotiation (Gailliard et al. 2010). Individuals interested in creating an organizational culture might strategize how to embed a focus on sustainability in each of these seven processes. When organizations engage in major changes, employees may need to be resocialized if they are to assimilate. "Establishing and maintaining meaningful membership in organizations involves multiple, interrelated, dynamic processes rather than the singular, linear trajectory assumed in many assimilation studies" (p. 554). Additional research is needed investigating communication linkages associated with unmet expectations, the role negotiation process, and the communication content of socialization tactics (Kramer and Miller 2014).

Communication scholars frequently use four theories to explain socialization and assimilation: uncertainty reduction, social exchange, social identity, and sensemaking. Several of these theories have been reviewed elsewhere (i.e., social identity, sensemaking). Uncertainty reduction theory was developed by Charles Berger and his colleagues to explain how people use communication when they are unsure of their own beliefs or the beliefs of others (cognitive uncertainty), or their own actions and the actions of others (behavioral uncertainty) (Knobloch 2009). People engage in passive (e.g., observation), active (e.g., asking questions of third parties), and interactive (e.g., asking questions of the target individual) communication strategies as they seek to gather information to reduce their uncertainty. Woodrow Nelson's, Vice President of Marketing Communication, of the Arbor Day Foundation, comment illustrated the passive communication strategy:

> In our employee engagement process we focus more on the culture of human interaction
> than we do on sustainability. And I think that is not necessarily by design but by default.
> Because our people, they are on board. We practice recycling at home. We bike to work.
> We embrace public transportation. We embrace healthy food, local food. It is just part of
> who we are. When you are among peers and friends that you work with, it's natural to pick
> up from each other what we are all doing. It is remarkably effortless.

rtainty organizational newcomers' experience can be reduced through l socialization, training and mentoring, conversations with supervisors kers, and observation. Newcomers weigh the social costs they face if they au... ir lack of knowledge as they decide whether or not to seek information overtly or covertly.

Organizations interested in creating pro-environmental organizational climates will find four concepts especially important (e.g., affective organizational commitment (AOC), compliance, identification, and internalization). AOC refers to "the employee's emotional attachment to, identification with, and involvement in the organization" (Meyer and Allen 1991, p. 67). AOC is rooted in social exchange theory, a theory based on ideas derived from theories across multiple disciplines (Roloff 2009). Its central assumptions are that humans exchange resources to survive, social systems develop norms and rules guiding resource exchange, and norms of reciprocity dictate that the receiver of a resource is obligated to be respectful and supportive of the giver—if not return a resource. When these norms are violated, individuals feel unfairly treated. When the exchanges are successful, stable exchange relationships and social networks are more likely to form. AOC can be influenced by employees' socialization experiences and, ultimately, be linked to employee pro-social behaviors such as increased job performance and decreased withdrawal behaviors. Compliance occurs when employees adopt attitudes and behaviors to gain rewards whereas identification occurs when an employee perceives his or her values and beliefs are congruent with those espoused and/or enacted by the organization. Identification involves a cognitive connection with an organization. Internalization occurs when employees adopt their organization's mission as their own (Caldwell et al. 1990). This is the focus of many socialization efforts. A great deal of communication research has focused on issues related to commitment, identification, and internalization seeing each as part of a connected process. See Cheney et al. (2014) for a discussion of communicating identity and identification within and among organizations.

6.2.2.3 Training

"The many small actions and decisions that all members of an organization make in their everyday work can accumulate in large improvements in the environmental impacts of the organization" (Perron et al. 2006, p. 553). But this may require employees with limited scientific training and/or interest in environmental issues to expand their knowledge bases into new areas and to change their behaviors. Otherwise, employees may be unable or unwilling to support the initiatives (Govindarajulu and Daily 2004). Through training employees become aware of relevant issues, recognize sustainability as a key part of their organization's mission connected to its triple-bottom-line, and have the knowledge and skills necessary to engage in the new actions (Ones and Dilchert 2012). It is especially critical to the effective enactment of often highly technical environmentally focused procedures and practices. Sarkis et al. (2010) investigated the implementation of environmentally oriented reverse logistics practices in 157 firms in Spain's automotive industry sector. The success of the implementation of green practices was completely dependent on employee training in the methodologies and techniques of eco-design, life-cycle assessment, recycling/reuse material use, and the disposal of production waste. Without adequate employee training, the green practices would probably not have been implemented.

Training prepares employees to act by building human capital and tactical capability. Julia Spence, Vice President of Human Resources at the Neil Kelly Company, said her organization's flat structure means that everyone participates in decision making. "That means that everybody has to know what is going on so they are making decisions that are in alignment with our goals. . . .. Everybody has to be educated in order to do that. Sometimes it's a real challenge to get information out to everybody." One of the greatest environmental impacts of the Portland Trail Blazers involved arena operations. They trained their employees and suppliers so everyone was on the same page. This included contractors, suppliers, contracted-services, cleaning staff, and security staff. Justin Zeulner, former Senior Director of Sustainability and Public Affairs, said everybody was trained so they could understand:

> What we were doing and why we were doing it and what their role would be in this. So whether it was reporting, or it's purchasing. . ... Guest services staff being able to train and educate fans about what we are up to. 'Why do I have a compostable cup?' 'Why is this food perhaps more expensive?' Because it is local and organic. To be able to have that storytelling ability and those talking points ready was really critical. And we did a phenomenal job at that.

In order to participate effectively on teams implementing sustainability-related initiatives both within and between organizations, employees need communication and interpersonal skills training (Linnenluecke and Griffiths 2010). The empowerment of all employees through training and more inclusive communication is a goal of the ISO 14000 programs. Training and higher levels of interpersonal interactions (e.g., cross-functional teams) can result in more information exchange relationships, political capital, and knowledge transfer and innovation which can increase employee productivity (Delmas and Pekovic 2013). As The Sustainability Center (TSC) works to generate an index of key performance indicators (KPIs) for different consumer brands, Sam's Club buyers must be trained in technical knowledge and improved communication skills. Brian Sheehan, former Sustainability Manager at Sam's Club, explained how he took:

> The milk or the dairy team through training around the [sustainability] index. It introduces them to TSC in relationship to Walmart, [helps them] understand what the KPIs are and the category [milk] sustainability profile is, gives them an overview of how the tool looks and functions, and then walks them through its actual use. So they can select among the suppliers, look at their scores, look at the product category score, and compare it with other product category scores. . ... And then start to talk with them [suppliers] about their results. . ... [Ultimately the buyer will] work with suppliers to improve their individual scores so that the entire product category is coming up.

Trained employees can help their organizations make ongoing adjustments and build new organizational thought. Jerry Tinianow, Chief Sustainability Officer for the City and County of Denver, talked about the approach his office is taking in terms of training:

> Our major function is to train the other agencies in sustainability theory and practice, help them to design, implement, fund and evaluate their projects. And then eventually get out of the way. The mayor would like sustainability to become so engrained in all the city agencies that first it becomes a habit for them. And then it becomes an instinct.

Several recent, interesting studies have investigated how training creates new cognitive maps within the brain. For example, Lourdel et al. (2007) investigated the cognitive maps of engineering students before and after they completed a course in sustainable development (SD). Cognitive maps have been used in environmental education to identify students' knowledge, and how it is interconnected in long-term memory. Before the training the engineering students' perceptions of SD focused primarily on its environmental and economic aspects. After the course their understanding was richer and wider including words involving social and cultural aspects, stakeholders, principles of SD, and allusions to the concept's complexity. The authors recommend this method for anyone who wants to evaluate the impact of training. I mention it here because it illustrates how training can shape worldviews as well as convey knowledge, skills, and abilities related to sustainability.

Observational and experiential learning is important. Social learning theory (Bandura 1977) helps us understand how people learn by observing their peers. But these observations relate more to judging which actions would be beneficial to engage in rather than as a way to gain knowledge, skills, and ability. Carrie Hakenkamp, Executive Director of WasteCap Nebraska, talked about how she feels peer learning is important saying:

> I have found that the best way for businesses to learn is to go out and see what other businesses are doing. . . . So we try to use that peer-learning model, that networking model. We have a lot of businesses who offer tours of their facilities to show what they have done or they do presentations about how their green team has been successful with starting sustainability initiatives in their company.

WasteCap also sponsors the Green Team Roundtable for their member organizations. In this lunch hour meeting, they discuss community topics like the solid waste management plan or, as a group, tour the landfill. Experiential learning is an important part of what Moe Tabrizi, former Assistant Director of Engineering and Campus Sustainability Director, for the University of Colorado, Boulder, promoted. He described how each summer in their Peak-to-Peak program faculty groups tour the campus and are trained in what has and is being done in terms of campus sustainability.

Although investment in sustainability-related training involves expenses, training can add value through higher quality products and services, subsequent development of efficiencies, and savings from punitive costs associated with noncompliance (Linnenluecke and Griffiths 2010). It can also influence employee productivity and effectiveness. Multiple venues exist which provide training materials related to sustainability. For organizations that want to design and conduct their own trainings, the Association for Talent Development (ATD) provides materials on how to design trainings, conduct a training needs assessment, and discuss sustainability. The ATD, formerly the American Society for Training and Development, is the world's largest association dedicated to those who develop employee talent in organizations. Present in more than 120 countries, ATD's members work in public and private organizations.

6.3   Goal Clarity and Goal Congruence

Organizations need inspirational, meaningful, and measurable goals to guide their sustainability-related efforts. Then the goals must be communicated to employees in such a way as to shape their work-related behaviors. Sometimes this occurs formally (e.g., through formal socialization and training), but employee assessment of formal goals also can be transmitted using informal communication. In this section, I focus primarily on how employees respond to the formal communication of goals. If employees are unaware of or do not support their organization's sustainability-related goals, little will be accomplished and employees may experience emotional stress. In addition to helping to create a clearly understood and shared organization culture, the cohesive execution of an organization's goals is one component of its socialization capabilities.

Organizational goals regarding sustainability ultimately are enacted by individual employees. In order to act, employees must see how their behaviors can further the organization's goals. Therefore, individual-level goals become important. Goal clarity involves the degree to which the organization creates clear objectives and performance expectations for individual employees (van der Post et al. 1997) and whether or not employees understand their associated role. It includes employee awareness of how to work together to achieve the mission and values of the organization. A great deal of organizational research has focused on measuring issues such as goal and process clarity and role ambiguity, especially during times of organizational change. Individuals seeking to promote sustainability-related initiatives at the organizational level must design, communicate, clarify, and reward individual-level goals.

6.3.1   Goal Congruence

Building on the TPB and the value-belief-norm models, Unsworth et al. (2013) proposed a model describing the psychological conditions under which organizational interventions designed to promote pro-environmental employee behaviors are likely to succeed. Their model draws on theories of values, self-concordance, goal hierarchies, goal systems, and multiple goals. Within an organization pro-environmental behaviors are only one of many behaviors or tasks that employees may choose to engage in. Pro-environmental goals are only one of many goals (e.g., efficiency goals, service and relationship goals, family goals, career goals) toward which employees are working.

Goal Conflict at Aspen Skiing Company Auden Schendler, Vice President of Sustainability at Aspen Skiing Company, wrote about this inherent goal conflict in his book, *Getting Green Done* (2009). When we spoke he said:

> This is the same problem that every corporation faces and the problem is that organizations have a mission of doing something other than sustainability. And people in the company understand that is their focus, so they are only willing to do so much on something that they see is ancillary. So, for years, I fought many, many battles trying to do some very simple basic high return on our investment sustainability projects. And lost many of them. If you google "Little Green Lies' in *Business Week*, that article is a very good summary of some of the challenges we faced. But I'll give you an example. My classic example is trying to get a hotel to retrofit light bulbs. And having the hotel manager refuse for half a dozen really good reasons ranging from the quality of light, to a concern about AAA ratings going down due to fluorescent lighting, to the opportunity costs of capital.

Within an organization, new pro-environmental goals communicated as part of an intervention designed to change employee behaviors are more likely to be activated if an individual sees the proposed environmental goal as addressing the problem and as attractive (Unsworth et al. 2013). Attractiveness depends on the intervention's characteristics and the individual's initial self-concordance. Self-concordance involves the degree to which the pro-environmental behavior expresses any of an employee's stable interests and values. Julia Spence, Vice President of Human Resources at the Neil Kelly Company, talked about self-concordance saying:

> I think a part of it is that they [employees] really like being able to come to work and do something that makes them feel good about the value of the work that they are putting in. That it's not just, let's sell this cabinet, but let's do something that really moves things forward and matches what they would like to accomplish.

Employees do not necessarily have to have humanistic, social, or biospheric values. What is important is that the employee sees (1) the proposed behavior as expressing as many of his or her values or long-term goals as possible, even if they are egoistic values, and (2) the link between his or her own behaviors and the organization's values. I asked Susan Anderson, Director of Portland's Bureau of Planning and Sustainability, to discuss any challenges her office faced when discussing sustainability with city staff. She said:

> There is still a contingent that's like—'I've got work to do. Go away'.…. [How to respond?] Talk about stuff they care about. Show them how it helps them get their job done, saves money, looks good to the boss. [Say] we will help you tell your story, or [when] they finish the project, we write it up, submit it, and get a reporter to write about them and put their picture in the paper. That kind of stuff really goes far. They see a benefit. They see a benefit to making all the parking meters solar.… Not because they care about climate change. But because their boss or politician liked it or they got some kind of personal feedback.

Self-concordant goals activate related higher order goals (more abstract and long term such as self-esteem or self-actualization) (Unsworth et al. 2013). When a pro-environmental self-concordant goal is activated, a spillover effect is likely to occur so that other behaviors connected to the higher order goal also are activated. Employees ultimately abandon behavioral goals that are not self-concordant. However, even when self-concordant, when goal conflicts occur as employees attempt to balance their competing goals, they prioritize performance or work relationship goals because green goals are generally background goals. Intervention-related pro-environmental goals are more likely to be pursued when other important goals (e.g., performance goals) are either very close or very far away from being achieved. When the intervention-related goal is focal, there are no conflicting cues, and if the behavior is perceived as self-concordant, pro-environmental activity is more likely to occur. If an intervention is to succeed, goals should be efficacious and attractive, self-concordant, in limited conflict with other goals, able to spillover into related behaviors, and achievable. Organizational messages directed toward employees as part of an intervention should stress that the proposed action is consistent with something they value and will solve the problem, that employees are capable of engaging in the proposed behavior, and that the behavior will have the desirable outcome (Unsworth et al. 2013). Leaders can increase employees' perceptions of the self-concordance of the pro-environmental behavior. Interventions can address goal conflict by providing cues to remind and refocus employees back on the pro-environmental goal. Earlier you read about how the University of Colorado, Boulder, and the State of South Dakota utilize on-location cues to reinforce behavioral change (e.g., signs over light switches). Multiple concepts appear in Unsworth et al.'s (2013) psychologically based model addressing conditions underlying pro-environmental behavior change: intervention characteristics, an individual's initial and ongoing perceived self-concordance, goal attractiveness, goal efficacy, behavioral and higher order goal activation, short-term effects, goal conflict, equifinality, the proximity of non-green and green goal attainment, long-term spillover, rebound effects, and green fads. I encourage you to look at and think through their model.

6.3.2 Goal Activation

For over a half of a century, scholars have sought to identify factors that motivate employees to engage in behaviors which will further their organization's broader goals. In this section I mention four concepts (i.e., expectancy theory, equity, justice, and rewards) to consider as management asks employees to engage in new pro-environmental behaviors.

6.3.2.1   Expectancy Theory

This is a foundational theory of individual motivation. Vroom (1964) investigated the process that helped motivate individuals to engage in a particular voluntary activity. He reasoned that an individual will select the behavior which he or she feels is most likely to lead to the results the individual desires (e.g., raise, recognition, feeling of pride). Expectancy theory is about the mental processes involved in choosing which voluntary behaviors to perform. The theory involves three components: valence, expectancy, and instrumentality. According to the theory motivational force $=$ expectancy $\times$ instrumentality $\times$ valence. Motivation is influenced by an individual's expectancy that engaging in a certain amount of effort will allow them to do a behavior, this behavior will achieve a certain result, and the individual finds the result desirable. Expectancy is influenced by self-efficacy, goal difficulty, and perceived control. Instrumentality is influenced by trust (of supervisors, of the organization), perceived control over reward distribution, and formalized policies which associate rewards with performance. Vroom explained how valence is influenced by needs, values, goals, and preferences. For example, if I am a buyer for a large retail chain, and I believe that if I find and select more sustainable products this will help my organization meet its sustainability goals, and that I will personally be recognized by management for doing so (which is something I desire), I am more likely to purchase the sustainable products. Although the theory has been criticized as being too simplistic it does direct us to thinking about how employees might go about selecting voluntary pro-environmental behaviors.

6.3.2.2   Equity, Justice, and Rewards

Perceived fair treatment motivates people. Equity theory (Adams 1963) helps explain how employees seek to maintain the equity between their job-related inputs and the outcomes that they receive compared to their coworkers' contributions and outcomes. If people feel under-rewarded or over-rewarded for their inputs, they experience distress. Inputs include time, effort, loyalty, commitment, and skills and ability. Outcomes include job security, salary, benefits, recognition, and a sense of achievement. The theory is made up of four propositions. Individuals seek to maximize their outcomes (i.e., rewards minus costs). Systems of equity evolve, change, and are maintained within groups. When they find themselves participating in inequitable relationships, individuals feel distress. They will attempt to eliminate their distress by restoring equity (e.g., work less). This theory is useful in helping us understand why employees might expect to receive rewards if they engage in extra-role pro-environmental behaviors when their coworkers do not. The idea of organizational justice stems from equity theory. It is a multidimensional construct with distributive, procedural, interpersonal, and informational components. All four components are important but here I discuss just one. Distributive justice is a concept which focuses us on employees' perceptions of the fairness of how rewards and costs are shared. Perceived fairness has been associated with positive psychological and behavioral outcomes such as trust, job satisfaction, and OCBs. Negative outcomes include withdrawal behaviors (e.g., absenteeism) and counterproductive work behaviors (e.g., resistance).

Employees monitor equity and justice. Sones et al. (2009) interviewed 12 and surveyed 1,386 employees who worked for a global elevator company located in Finland. Their respondents said organizations should allocate resources, time, and money to translate their visions for pro-environmental cultures into practice. Feedback indicated that simply expecting employees to volunteer without providing them with extra time, resources, or rewards led to reduced enthusiasm and effort. Expectancy theory suggests that valued outcomes might include pay increases and bonuses, promotions, time off, recognition, new and interesting assignments, or the intrinsic satisfaction of helping others or the biosphere as well as validating one's skills and abilities. Blackburn (2007) discussed valued outcomes organizations can use as reinforcement after assigning employees new responsibilities tied to their sustainability initiatives. He mentioned promoting performance-based objectives using adjustments to pay, bonuses, opportunities for advancement, special awards, recognition luncheons, and articles in company publications. More subtle rewards include allowing top performers to showcase their efforts in key forums attended by top management. Eisenberger et al. (1990) developed a measure useful to measuring the extent to which employees believe higher levels of job performance will be rewarded. *Best Practice*: Organizations seeking to stimulate employees to enact new pro-environmental behaviors should be aware of the importance of rewards, perceived equity, and issues related to justice.

6.4   Organizational Climate, Perceived Support and Trust, Emotional State, and Informal Communication

Concern for people is a key part of sustainability. Earlier you read that Social Accountability International (SAI) provides standards for operating in a socially responsible way. SAI's mission is to advance the human rights of workers globally. Their SA8000 standard for decent work is used in over 3,000 factories, across 66 countries and 65 industrial sectors. Hewlett Packard is among the organizations having that certification. Many more workplaces use SA8000 and SAI programs as guides for improvement. SAI has trained over 30,000 people including employees, managers, brand compliance officers, auditors, labor inspectors, trade union representatives, and worker rights advocates. They focus on issues related to child labor, forced and compulsory labor, health and safety, freedom of association and right to collective bargaining, discrimination, disciplinary practices, working hours, and remuneration. These are all critically important issues in terms of social sustainability. In this section, I identify several academic communication-related concepts related to how employees respond to their working environments. Specifically, I discuss organizational climate, employee perceptions that their organization cares about them as individuals, employee trust in their organization, the influence of employee daily emotional state on their pro-environmental behaviors, and the power of informal communication.

6.4.1 Organizational Climate

An organization's climate is defined as the shared psychological environment in which organizational behavior occurs. It is built upon recurring patterns of behavior which influence our attitudes and feelings, and, in turn, shape how we characterize our organizational life. You can think about it as the environment (or climate) within which we work. Some employees may feel they work in a harsh environment characterized by stress, isolation, and fear. Others may feel their working environment is pleasant and characterized by respect, openness, and opportunity. How employees feel about their climate is influenced by formal and informal communication, relationships and interactions. W. Charles Redding developed a prescriptive model for managers known as the ideal managerial climate. It has five components: perceived supportiveness of employees, ability to inspire trust, openness to employee communication, emphasis on high performing goals, and participatory decision making. It also can be used to assess perceptions of an entire organization (Cheney 2011). Employee responses are aggregated to identify the average employee's perception of an organization's openness or supportiveness. The climate concept focuses us on how employees evaluate their organization similarly and how this evaluation can impact employees' emotional health and behaviors (e.g., absenteeism, commitment, stress). Climate conditions effect how employees respond to new sustainability initiatives. For example, if employees do not trust their organization or feel their input is not desired, they will be less likely to voluntarily engage in pro-environmental behaviors. Although not commonly identified as an organizational climate issue, employee perceptions of organizational support, trust, and justice do influence employees' assessments of and response to their organization. These concepts are discussed next.

6.4.2 Perceived Organizational Support and Trust

How can organizations influence employee engagement in environmental behaviors such as participating in environmental management activities, promoting environmental initiatives, and proposing innovative environmental practices? Cantor et al. (2012) offer an explanatory model based on perceived organizational support (POS), a concept developed by Eisenberger et al. (1986). POS is based on social exchange theory and the reciprocity concept. The social exchange and reciprocity foundations involve employee beliefs about their organization's commitment to them. POS asserts that employees' perceptions regarding the extent their organization values and cares about their contributions have a strong influence on how they behave and on their levels of affective organizational commitment and organizational trust (Rhodes and Eisenberger 2002). POS is "associated with trust that the organization will fulfill its exchange obligations" (Settoon et al. 1996, p. 220), recognize and reward desired employee attitudes and behaviors, be fair, and reward extra role performance.

Trust is "a psychological state comprising the intention to accept vulnerabilities based upon positive expectations of the intentions or behavior of others" (Rousseau et al. 1998, p. 394). It is created over time as employees experience repeated favorable treatment which communicates their organization's general benevolence, and by employees perceptions of issues related to procedural (i.e., fairness in the processes used to manage conflicts and allocate rewards) and distributive justice. Employee perceptions of justice influence their commitment, satisfaction, and intent to voluntarily leave their employer. See Fields (2002) for some justice measures. A climate of trust is an essential factor to the change management process. If employees trust that new sustainability initiatives are not simply fads, and that they will be recognized and rewarded for engaging in pro-environmental behaviors, their adoption and display of such behaviors are much more likely.

HEAL and POS   According to the U.S. Department of Energy, buildings consume 39 % of the energy used in the USA. Although homes are twice as efficient as they were 35 years ago, residential buildings still account for about 20 % of US energy use. Earlier I mentioned the HEAL program being operated by the Clinton Climate Initiative (see Sect. 5.2.2.4). This program helps create sustainable, local jobs and, in the long term, reduces CO_2 emissions by improving energy performance in commercial and residential buildings. In 2013, the program's website says, "By enlisting employers to offer the HEAL program to qualifying employees and to community members, the program is able to make a large-scale and rapid impact in energy usage, GHG emissions and employment opportunities." Organizations place funds, often saved from energy retrofits of company facilities, into a revolving loan fund employees can use to make energy retrofits in their own homes. Loans are paid back through payroll deductions on a "pay as you save" amortization schedule that ties payments to savings realized through lower utility bills. By 2015, building retrofits and the HEAL program reduced GHG emissions by 33,500 tons annually across the USA, a number with the potential to explode as more organizations adopt the program (Energy Efficiency Program 2015).

Martha Jane Murray, the HEAL program administrator, developed and tested the program in her own factory. The next pilot test was at L'Oreal USA's manufacturing plant in North Little Rock, AR. L'Oreal has a very strong sustainability initiative, a sustainability director, and targeted goals corporate wide to reduce waste and energy and get to a net zero. Martha Jane said HEAL was:

> A way for L'Oreal to do two things: create awareness among employees about helping them achieve those goals corporate wide but also to help them personally in their own home.

> They found that providing the [HEAL] benefit gave employees a personal insight into the
> benefits of sustainability which was reflected back at the corporate level.

Formal organizational communication can set the stage for POS to occur (Allen 1992, 1995; Allen and Brady 1997) when organizations engage in voluntary actions like HEAL. Martha Jane described how that process works for the HEAL project. When presenting the program to employees:

> We always promote the company. We let the employees know that this is a benefit that
> [your organization] has brought to you and it is unique in this market. That they care about
> you beyond the 401(K) or healthcare or whatever. It is very much about helping you find
> sustainability in your personal life.

To date, many HEAL program participants are hourly wage employees. The program saves them money on their energy bills which increases their disposable income. As participants learn about the HEAL program, they also learn about other opportunities. Martha Jane said:

> We always ask when we are giving a presentation, how many of you are aware that there is
> a rebate through your utilities? We usually get less than one percent of our audience that is
> aware of those opportunities. So I think there is a constant opportunity for us to make it
> [sustainability] real. To make it authentic for people and not just something that looks like
> [company] green-washing. I think the more authentic sustainability initiatives are, the more
> personal it becomes and it certainly will engage people in other sustainable initiatives.

Given the concept of reciprocity, one would expect HEAL participants to repay their loans, remain with their employer, and engage in more OCB, including pro-environmental behaviors if cued to do so by their organization. It would be a natural extension for organizations seeking to engage in eco-efficiencies (e.g., energy efficiencies) to create conditions whereby employees could do the same in their homes. It is likely the employees who do so will discuss such actions within their communities thus opening more citizens up to considering energy efficiency-related behaviors.

Although POS originally was developed to assess how much employees feel their organization values their contributions, Cantor et al. (2012) modified the focus to assess employee perceptions of how much their organization cares about environmental issues. They surveyed 317 logistics and operations management mid-level distribution center employees working for one of *Fortune* magazines' 2010 most admired companies. Supervisory support for employee engagement in environmental initiatives and company provided training regarding environmental issues influenced employees' perceptions regarding how much their organization cared about environmental issues. Information received from supervisors, at formal organizational meetings, and appearing in their organization's sustainability report also shape employee perceptions of their organization's interest in sustainability (Craig and Allen 2013). Knowledge of their organization's involvement with sustainability influenced employee perceptions that sustainability is important to their organization's future. This perception influenced employee commitment to environmental behaviors which, in turn, should result in more innovative and frequent pro-environmental behaviors.

6.4.3   Emotional State

Although pro-environmental behaviors often are motivated by stable individual differences (e.g., attitudes, personality characteristics, personal norms, intrinsic motivation), employee behaviors may change depending on the emotions they are experiencing. Concepts such as POS and trust indicate that researchers acknowledge that employees bring their emotions to work, and that emotional reactions can influence work-related behaviors. Emotions commonly investigated include pride and more fleeting daily emotions. For example, surveying 5,220 French firms, Delmas and Pekovic (2013) explored the relationship between ISO 14001 certification and labor productivity. They argued that if employees feel proud of their organization because it has been certified then they are more likely to perform better at work, engage in more cooperative and citizenship-type behaviors, and become ambassadors for their employer. Smerecnik and Andersen (2011) argue that if an organization is externally recognized as a sustainability leader, employees are more likely to experience pride and adopt sustainability innovations as well as engage in more sustainability-related communication with key stakeholders.

Emotions were key to Bissing-Olson et al.'s (2012) investigation. They were interested in predicting two types of pro-environmental behaviors: task-related and proactive. Task-related pro-environmental behavior involves doing required tasks in environmentally friendly ways. Proactive behaviors are broader and involve displaying personal initiative. In their study, they referred to the TPB and the link between pro-environmental attitudes and pro-environmental behavior discussed in the Bamberg and Moser's (2007) meta-analysis (see Sect. 4.3.2.1), but they base their study on the broaden-and-build theory of positive emotions (Fredrickson 2001). This theory suggests that those who experience positive affect are more likely to behave in pro-environmental ways because they have more emotional energy available which allows them to consider alternative ways of thinking and acting. They looked at two types of affect: daily unactivated positive affect (i.e., feeling contented, rested, and relaxed) and activated positive affect (e.g., feeling excited and enthusiastic).

As would be expected, those who held more pro-environmental attitudes engaged in more task-related and proactive pro-environmental behaviors (Bissing-Olson et al. 2012). But emotional state mattered. When employees were feeling content, rested, and relaxed, they were more likely to engage in daily task-related pro-environmental behaviors. Daily activated positive affect increased the daily proactive (but not task-related) pro-environmental behavior of employees with less positive pro-environmental attitudes but not of those who held more positive pro-environmental attitudes. The authors recommend creating positive (i.e., calm) workplace climates supportive of pro-environmental behaviors. Such climates reinforce employees' pro-environmental attitudes. They also suggest promoting positive affect by creating messages and events that stimulate positive emotional responses, making pro-environmental attitudes salient, paying attention to job design, and facilitating positive work events. Organizations can make task-related pro-environmental behaviors more salient by presenting pro-environmental messages. But for those with limited pro-environmental attitudes, stimulating daily positive affect can be useful. Mike Johnson, Associate Vice Chancellor for Facilities, at the University of Arkansas, Fayetteville, recognized the role of excitement saying, "Part of what we are doing is the communication necessary to let people know what is going on in order to keep them excited and engaged" in campus sustainability initiatives.

Emotional contagion theory also can provide insights into how employee emotional responses influence their pro-environmental behaviors. The theory helps us understand how something as subtle as nonverbal communication can lead two individuals to emotionally converge. Mirror neurons in our brains cause us to attune emotionally (Hatfield et al. 1993). The emotional state of people within workgroups can influence cohesiveness and morale. However, if employees share feelings of alienation, uncertainty, or fear, negative results can also occur. Years ago I read a study on voluntary turnover clusters. When one employee voluntarily left an organization due to a negative emotional response, others in his or her social network were likely to follow. Research opportunities exist for scholars interested in investigating how emotions, as well as informal communication, can influence employee responses to the initiation of new sustainability initiatives within an organization.

6.4.4 Informal Communication

Chapter 5 focused on formal written communication (e.g., mission statements, sustainability reports), planned communication disseminated by change agents or champions, and communication sent from leaders or managers. However, within any organization, informal communication can be a powerful force. Informal communication is communication which is not under management control. Informal communication is important to employees because it allows for meaningful organizational relationships to be created and employees' social and support needs to be met. Through informal communication, concertive control can be exerted which reinforces formal and informal social norms. Informal communication plays an especially important role during times of organizational change when uncertainty and ambiguity are high. Sometimes, it includes workplace gossip and rumors. Through informal communication, sustainability initiatives can be discounted, distorted, challenged, supported, or modified.

In addition to structural theory, sensemaking theory, and social learning theory, another theory helpful when thinking about informal communication is social information processing theory. Social information processing theory was developed by two management theorists (i.e., Salancik and Pfeffer 1978) to explain how people respond to job design issues. Our individual needs and perceptions of our jobs are influenced by people with whom we have work-related social and informational relationships. Salancik and Pfeffer write that the social context and our own previous actions and experiences provide guides for socially acceptable reasons for action and focus "an individual's attention on certain information, making that information more salient, and provides expectations concerning individual behavior and the logical consequences of such behavior" (p. 227). The theory has been used to explain a variety of behaviors including how employees respond to media and media use in organizations, a topic with implications for how they will respond to sustainability reports and website pages discussing sustainability. Social information processing theory proposes that if individuals are exposed to more positive, informal social cues, they will express more positive feelings.

Recently, Lee and Lee (2015) investigated how multiplexity in intraorganizational networks can stimulate creativity. Previous research looked at individual cognition or investigated how an employee's position within his or her organizational network influenced his or her creativity. But Lee and Lee were interested in the multiplexity which occurs when more than one type of relationship (e.g., friendship, information exchange, advice) exists between two or three people. Idea generation occurs within relational contexts characterized by trust, positive affect, and information sharing. Creativity is a process influenced by interaction as ideas are diffused, searched for, and adopted. They found that when significant multiplexity exists, advice ties followed by knowledge and friendship ties are strongly associated with creative interaction. Although their focus was not on enlisting employees in generating creative sustainability-focused initiatives, their study suggests that efforts that allow employees to develop and use multiplex ties might be an excellent way to empower them to create creative solutions.

6.5   Engaging and Empowering Employees

In 2007, the senior management of Deutsche Post DHL Group (DPDHL), the world's largest logistics company, committed to using environmentally sound business practices everywhere they operate (Gupta 2011). The company's 2020 goal is to improve carbon efficiency by 30 %. They developed their GoGreen initiative to raise employee awareness of the importance of the CO_2 issue. They communicated this program aggressively in their internal communication. DPDHL employees took ownership of the carbon efficiency goal seeing it as a way they could help make their communities and the planet safer and healthier. They became engaged around sustainability.

In this section, employee engagement, mechanisms for promoting employee engagement, and the circular relationship between pro-environmental home and office behaviors are discussed. Engaged employees are absorbed by and enthusiastic about their work and routinely act in support of their organizations' reputation and interests (Werbach 2009). They go beyond their minimal job requirements to give both their job and their organization extra attention, thought, and energy. Their productivity is higher and they are less likely to voluntarily leave their employer. In order to become engaged, employees must feel empowered. Empowerment is the intrinsic motivation employees experience because they believe their work role has meaning, feel competent, can engage in self-determination, and perceive they can have an impact (Spreitzer 1995). Meaning is influenced by the perceived fit between the work role and an individual's beliefs, values, and behaviors. Competence refers to work-related self-efficacy. Self-determination involves autonomy over work processes and work-related decisions. Impact refers to perceived influence over strategic, administrative, or operating outcomes at work. Organizations seeking to enlist employee support and stimulate employee innovation around sustainability initiatives should design a supportive organizational climate, provide needed training, and strategically redesign and enlarge employee roles in order to empower them.

Once our basic needs are met, our happiness increases if we can be of service to something larger than ourselves, experience full engagement (flow) in what we are doing on a regular basis, show gratitude to others, and share our life with others. You might want to complete the satisfaction with life survey (see Werbach 2009). A sample question is "If I could live my life over, I would change almost nothing." Giving employees a greater purpose at work beyond generating company profits can be motivational. This is more likely when self-concordance exists. Self-concordance occurs when "pro-environmental behavior expresses any of the employee's stable interests and values" (Unsworth et al. 2013, p. 214). If organizations can tap into this self-concordance and engage their employees in planning and implementing sustainability initiatives, this may unleash employee creativity, promote leadership talent, and drive innovations (Werbach 2009). Based on their knowledge, skills, and abilities, they may become the champions of change.

6.5.1 How Organizations Engage Employees

Generally US employees are not engaged or empowered. The Gallup (2013) report entitled *The State of the American Workplace*: *Employee Engagement Insights for U.S. Business Leaders* discussed their findings from 2010 through 2012. Of the approximately 100 million full-time American workers only 30 % are engaged and inspired at work; 20 % are actively disengaged; and the remaining 50 % are just not engaged. Active disengagement is costing the annual U.S. economy an estimated $450 to $550 billion. Gallup recommends that employee engagement can increase if organizations select the right people, develop employees' strengths, and enhance employees' well-being—all topics discussed in this chapter. So organizations seeking to engage employees around sustainability initiatives can't just assume engagement happens automatically.

Organizations increasingly are seeking to engage employee support for their sustainability-related initiatives. See Exter (2014) for a comprehensive discussion of increasing employee engagement within sustainable businesses. In 2010 The North American Task Force of the United Nations Environment Programme

Finance Initiative (UNEP 2011) surveyed 20 of their members and interviewed six senior environmental managers. Participating institutions included Citigroup, Bank of America, Merrill Lynch, TD Bank Financial Group, and UBS. Active environmental employee engagement is a relatively recent phenomenon in financial institutions and efforts primarily focus on internal environmental management (e.g., resource conservation). Respondents reported using various employee engagement approaches which I include here for other organizations to consider. Their approaches included presentations to raise awareness of relevant environmental issues and how their organization was addressing them; surveys to identify employees' environmental concerns; the creation of various groups (e.g., a dedicated team or task force responsible for environmental issues, informal grassroots teams, sustainability steering committees); the use of contests, challenges, and recognition programs to incentivize employees to take pro-environmental actions (e.g., reduce resource consumption); the provision of mailboxes where employees could direct their questions and suggestions regarding sustainability issues; the dissemination of newsletters and intranet communication to keep employees informed about initiatives; and training on the implications of sustainability to employees' business roles. For example, the TD Bank Financial Group set up two programs. At the branch level, green coordinators raised employee awareness about branch-relevant environmental issues, provided information about the bank's environmental initiatives, and informed customers about the bank's environmental credentials. Within the business units and subunits, green ambassadors championed environmental programs, involved employees in enacting the environmental strategy, convened green working committees, and created and pursued their own ideas.

The Gallup organization's "Q12" survey is used by organizations to measure factors influencing employee engagement (D'Aprix 2011). Gallup's (2013) findings from 2010 to 2012 are available along with the 12 questions. Other drivers of employee engagement involve employee perceptions that senior management cares about employee well-being; employees have opportunities to improve their skills, to provide input into departmental decision making, to set high personal standards, for excellent career advancement, and to engage in challenging work assignments that broaden employee skills; employees believe their organization has a reputation for social responsibility and responds quickly to customer concerns; the existence of a positive relationship with their supervisor; and employee perceptions that their organization encourages innovative thinking (D'Aprix 2011). Prior to seeking to engage employees, Werbach (2009) suggests surveying employees about their knowledge of supervisor and peer expectations, perceived opportunity to do what they do best each day, belief that their supervisors and others in their organization care about them as a person (POS), belief that their job is important to the mission or purpose of their organization, existence of a best friend at work, and perceived opportunity to work and grow at work. Such baseline knowledge helps change agents better understand employee perspectives. In terms of superior-subordinate communication, D'Aprix suggests line managers need to proactively answer six common questions employees have: What is my job? How am I doing? Does anyone care? How are WE doing? What's our vision, mission, and values? How can I help? The North American Task Force of the United Nations Environment Programme Finance Initiative (UNEP 2011) report listed some additional potential actions associated with engagement that include: 1) solicit employee input in the development of environmental engagement strategies to create a sense of ownership and enhance employee participation, 2) encourage top management to influence those employees who are not already concerned about environmental issues, 3) incentivize employee participation through contests and challenges, 4) provide educational presentations about the organization's environmental strategies, 5) provide role-specific training to employees on how they can help improve environmental performance, and 6) develop effective communication channels employees can use to channel their concerns about environmental issues and to offer suggestions.

6.5.2   Challenges Mobilizing Employees Around Sustainability

Employee engagement programs face multiple challenges. D'Aprix (2011) likens it to throwing rocks at the corporate rhinoceros. The North American Task Force of the United Nations Environment Programme Finance Initiative (UNEP 2011) study participants identified challenges due to inadequate resources, inconsistent top management support, employees uninterested in environmental issues, problems reaching the entire employee base, and difficulty maintaining the employee engagement programs' momentum. At work, motivational forces are often lacking since employees rarely see the consequences of their behaviors in terms of their organization's overall goals (e.g., energy conservation). This makes it difficult to influence feelings of personal responsibility and to incentivize an individual's behavior (Bolderdijk et al. 2013).

Often organizations simply overlook the opportunity to change individual behaviors. Mike Mueller, Sustainability Coordinator for the South Dakota Bureau of Administration, explained why he focuses his efforts on infrastructure rather than using communication to stimulate employees' pro-environmental behaviors:

> Getting people to change behaviors is extraordinarily difficult. Swapping out a light system or boiler, while expensive, is really easy to do and considerably cuts your consumption of the resource. We have really tried to make it easy for the employees [so that] rather than telling them, 'Please turn off the light in the room', we installed motion-sensitive lighting controls. ... It is our job to pay attention to the buildings. But the employees that are in our buildings are really there to accomplish a whole different mission. They are trying to serve the poor and the disabled, protecting us in our prisons, trying to encourage job growth and economic development, trying to help schools with curriculum, and so their focus is not 'Should I throw this away? Should I recycle? What should I put in the gas tank?' The approach has really been that if we can make the decisions for them, in a way that doesn't make them uncomfortable or disrupt their other workflow, [we do] because their mission is something entirely different than ours. While we can prompt them to make good decisions

to help us save money, really our efforts are to do as much of it as possible without them having to make a conscious effort.

In 2011, the South Dakota state government employed 12,594 full-time employees and 6,237 part-time employees.

Various interviewees talked to me about picking the low hanging fruit by reducing packaging or increasing recycling and efficiencies. Many were at a loss as to where to go next in their sustainability efforts. Quite a few were aware of the need to change proactively in the face of impending climate change challenges. But few had enlisted their employees to innovate toward sustainability within their organizations and to magnify their organization's commitment toward sustainability throughout their communities.

But increasingly organizations are involving their employees. My university, the University of Arkansas, Fayetteville, signed the College and University Presidents' Climate Commitment in 2007, pledging a long-term goal to become carbon neutral by mid-century. Its mid-term (2021) goal is to reduce its greenhouse gas emissions by 50 % from 2010 levels. In the fall of 2014, I worked with a team to conduct a 12-week energy conservation pilot program. Building on theory and research, we designed a field experiment where people working in four campus buildings received a different combination of (1) weekly building specific energy feedback delivered in an email energy report and appearing on energy charts located in strategic locations in office and classroom buildings, (2) energy pledge cards, and (3) energy champions who provided one-on-one education and reinforcement messages. The energy feedback allowed occupants to see their building's total energy consumption for the first time and how their choices could save energy. The feedback chart displayed 2013 weekly kilowatt usage for the 12-week period. A dashed red line indicated what a 15 % energy reduction goal each week would look like. When I exited the elevator in my building, an updated chart allowed me to compare our 2014 energy consumption for the previous week with that week in 2013 and to our 15 % energy reduction goal. At the end of the period, our 12-week pattern was clear. Energy conservation efforts per building ranged from 7,600 to 65,000 kilowatt-hours. Combined, these four buildings reduced their carbon emissions by 89.75 metric tons. Through communication employee efforts were mobilized in this short-term campaign. However, long-term engagement is key.

6.5.3 Employee Engagement Outlets at Work, at Home, and in the Community

The influence of an individual's personal background and pro-environmental interests can be greatly magnified if their employer creates an environment where these interests are valued and their suggestions are solicited. Many organizations have green teams. These are generally self-organized cross-functional groups that voluntarily come together to educate, inspire, and empower their coworkers around sustainability. Such teams identify and implement specific solutions to help their organization operate in a more environmentally sustainable manner. Actions that green teams promote at work can spillover into employees' personal lives.

6.5.3.1   Green Teams

Fleischer (2009) shared some best practices and lessons learned on employee engagement at Deloitte, eBay, Genentech, Intel, Stonyfield, Walmart, and Yahoo! She provides her readers with business case arguments for forming green teams, describes how to start a green team, offers some potential green team projects, and provides resources for those interested in starting a green team. Practitioners will find her report informative. *Best Practices*: Hire a consultant or create a staff position to oversee the team, embed sustainability metrics into employees' performance goals, link bonuses/compensation to sustainability goals, connect green team initiatives to the sustainability initiatives of a senior-level cross-functional team of leaders from key departments, train employees on the importance of sustainability to the business, and help employees understand the importance of their individual actions. Many of these recommendations have appeared elsewhere in this book.

My interviewees representing small businesses (e.g., WasteCap Nebraska), mid-sized businesses (e.g., Assurity Life Insurance, Neil Kelly Company), larger organizations (e.g., the Lincoln, NE, State Farm processing facility), and cities (e.g., Boulder, CO; Portland, OR) talked with me about their green teams. Their advice included make sure the team doesn't get too large, create a forum within the team where no topics are off limits, give employees nameplates that say Green Team so others can identify them, utilize Sharepoint for idea exchange, draw team members with different skill sets from across the organization, work within employee time constraints during lunch-and-learns, become a part of a larger network of green teams which meets to share ideas, tour other organizations to see what their green teams have done (peer-learning and networking), and be prepared to do a cost-benefit analysis of potential new ideas. They mentioned that their green teams communicate with other employees (and sometimes other stakeholders) using websites, bulletin boards, display cases, lunch-and-learns, educational events, action events (e.g., electronic recycling day), table tents, eco-fairs, and their company's daily newswire or intranet.

Green Teams at Assurity Life and State Farm   Tammy Rogers, Senior Information Technology Business Analyst and green team leader at Assurity Life Insurance, discussed varying informational content provided at green team-sponsored forums, and being realistic. She said:

> It really is about raising awareness and presenting on a lot of different topics because the variety of topics is what draws in different people. Not everybody is into recycling but somebody might be into biking. Or they might not be into biking but they are into gardening. It has been a learning process for us because none of us had ever done this

before. So understanding and gaining experience about how to communicate the message and how to not be political about it, how not to be in your face about it so you don't turn people off. Coming up with good topics and engaging ways to present that information, is really what our goal is. ... The apolitical part of it was something we were all adamant about in our first organizational meetings. This cannot even have a hint of political or forcing opinions on people because that is a huge turn off. We are pretty low key in how we deliver our message. We have found that to be effective. You start winning people over that you did not think would ever start recycling. ... There are always challenges when you want change. There will always be people that look at everything from a political standpoint so no matter how you present it, they go on the defensive. There are just people like that so it's a cultural change. I think the reality is not everyone wants to change. Not everyone is going to change. You just keep putting the message out there and it does have an impact. But you have to be realistic about people and what their backgrounds are, how they grew up, how that influences the choices that they make.

Sometimes an organization's sustainability initiatives occur because individuals' bring something from their private life to their organizational role that motivates and empowers them to help bring about organizational change. When I was in Lincoln, NE, I spoke with Mike Malone, the green team leader at the State Farm Insurance operating facility in Lincoln. The Lincoln facility has won multiple awards for its sustainability efforts. For example, in 2011 that facility won the Keep Nebraska Beautiful Environmental Award, and the City of Lincoln and Lancaster County Environmental Leadership Award for Business and Industry. There are 1,400 employees in the two-building operating facility who do a variety of tasks involving underwriting and processing insurance claims for six states located in the Heartland zone of the company (i.e., Minnesota, Wisconsin, the Dakotas, Nebraska, and Iowa). The State Farm Heartland Green Team has 11 members: seven in Lincoln, two in Minnesota, one in Omaha, and one in Des Moines. Mike spoke with me about his experience with the green team saying,

I own a cabin on the lake in Wisconsin. So I have always been concerned about keeping water clean, and keeping the lake alive and thriving. Once I got involved with sustainability and the committee, my efforts ramped up and my views changed on not only recycling, but reusing and reducing as well.

Apparently, his organization encourages employees to engage in volunteer activities. He explained:

It looks good to be on a committee here and have something on your employee shield. ... I did not want people [on the green team] that were going to do it because it looked good on their resume. I did ask for people who were living the green life, were engaged in environmental efforts, or were really excited about the committee and its potential and what could be done. And that is really what we got. There have been a couple of people here that were not that engaged. And you know, that is going to happen and those people basically did not ask to re-up. ... We want people who are enthusiastic, people who want to be involved.

Julie Diegel, Director of Sustainability Programs at WasteCap Nebraska, talked about some of the things she discusses with the green team representatives of the WasteCap member organizations. I provide her comments here in hopes they will be useful to other green teams. She said:

We go through the process of determining what your company's biggest impacts are. Where does this green team even begin? How do you determine what is next? We go through the whole process of discovery.... What are you doing right now? How do you decide which [project] you should focus your energy on first? When you decide on a project, how do you rank that project against other possible projects? Based on what your current needs are—do you need a big win? Do you need something publically visible? You rank those projects and focus on one thing. Figuring out a process is half the battle. How do we figure out what to do and how to do it? How to measure it and how to record what we are doing so that we have a record of that so we know where we started and where we ended up.

Increasingly, green teams are focusing on helping coworkers integrate sustainability into their personal lives. Individual employees take what they learn at work back into their nonwork lives. Julia Spence, Vice President of Human Resources at the Neal Kelly Company, talked about how her definition of sustainability changed saying:

The more I learned formally through work, the more I have gone home and looked at what I do. I have changed the way I garden. I garden organically now. I buy locally. I buy organically, which I did not do before. Just the way I think about packaging [has changed]. I am a lot more conscious all the time. And that has really come out of what we have done at work. [I ask myself] do I really need this?

Julia's story was not unique. Tammy Rogers, Senior Information Technology Business Analyst and green team leader, at Assurity Life Insurance, talked about how working in a LEED-certified building influenced her and her colleagues' habits at work and at home. Like Julia, Tammy said her own definition of sustainability broadened.

Insights gained at work that translate into employees' personal lives can result in large-scale pro-environmental outcomes for their surrounding communities, if their employer supports such actions. Brian Sheehan's, former Sustainability Manager at Sam's Club, previous background working in sustainability for a city influenced his vision for what an organization Walmart's size could do in the communities where stores are located. Local Sam's Clubs could be a hub of sustainability expertise. Brian envisioned multiple Sam's associates throughout a Club as becoming aware of the sustainable products they sell and capable of helping people understand how they could operate their own organizations in a more sustainable fashion. "So in a remote part of this country.... Sam's Club, to a certain extent, could serve that need. We have business members [in our communities] who are wanting to operate more sustainably. There would be someone in the local Sam's Club they can talk to who could help them achieve their goals." Such an idea could have a major impact.

6.5.3.2   Personal Sustainability Plans

In 2007, I was present in the Walmart corporate auditorium at their annual sustainability summit when the rollout of the voluntary personal sustainability projects (PSP) was announced. Walmart associates globally were presented with a program designed to help them integrate sustainability into their personal lives. The idea was to promote small actions that could be taken anytime and anywhere that were good for the employees, their organization, and the planet. Desired actions should be repeatable, inspirational, sustainable, and enjoyable (RISE) (Werbach 2009). These PSPs have the potential to go viral as best practices spread through employee social networks via interpersonal communication. Linda Dillman, Walmart Executive Vice President of Risk Management, Benefits and Sustainability, was quoted as saying,

> Sustainability has become part of the Wal-Mart culture, and PSPs are one way for associates to become involved—in their stores, their communities and their daily lives. PSPs are being created by and for associates to help them make choices that can have a real impact on their personal health and happiness and on their families, neighbors, communities and the environment. We're excited about what we've seen and learned so far and about what can happen as this project grows (Walmart Announces 2007).

Walmart developed a website where associates could get ideas for what to change, how to set goals, and how to make the change happen. For those employees who wanted to feel connected, they could make a public commitment on the website as well as provide and solicit social support. The Walmart 2009 annual sustainability report said that more than 500,000 US associates adopted a PSP with nearly 20,000 quitting smoking. Employees had recycled three million pounds of plastic, collectively lost more than 184,000 pounds, and walked, biked, or swam more than 1.1 million miles. With 1.4 million US employees and 2.2 million globally, Walmart associates' collective action has the potential to promote positive behavioral change for themselves, their families, their 200 million weekly customers, and their communities (Weinreb 2013). Over time the program transformed into My Sustainability Plan. Partnering with the Clinton Global Initiative, Walmart now provides a royalty-free license of the My Sustainability Plan branding, program framework, curriculum, and other resources to interested organizations (My Sustainability Plan n.d.). *Action Plan*: Go to Werbach (2009) to learn more about how to develop your own personal sustainability plan.

6.5.3.3 Preparing Employees to Communicate the Sustainability Message

Employees can become change agents spreading their organization's sustainability message in their community. For example, the Austrian Federal Forest Corporation trained 180 members of its forestry staff on how to communicate about corporate sustainability to certain publics (Signitzer and Prexl 2008). The forestry staff made about 520,000 direct contacts with members of the general public annually. Because employees are the producers and users of environmental knowledge within their organizations, other stakeholders often see them as credible information sources about an organization's true activities.

If employees are to speak on behalf of their organization, there is work to be done. Although research on how employees respond to their organization's CSR-related communication is rare, Sones et al. (2009) found employees desire more concrete messages about what they personally can do for the environment. Many prefer clear, short, practical messages. They recommended sending messages stressing the benefits of optimization (e.g., eco-efficiencies) for the company and assigning an environmental contact person in each department. Others said they were too busy to attend informal meetings or indicated they were only willing to do things that did not require too much effort. People found it hard to censor coworkers for doing things that aren't pro-environmental. Some were not willing to make suggestions for fear of being seen as green. Organizations can use at least three different CSR communication strategies in messages directed toward stakeholders (e.g., employees): information, response, and/or involvement (Sones et al. 2009). The information model relies on one-way communication. Although the two-way response strategy does gather information from stakeholders, its main purpose is to influence stakeholder attitudes and it does not adapt in response to stakeholder feedback. In contrast, stakeholder involvement strategies seek to create and maintain iterative and progressive two-way symmetric communication (i.e., dialogue).

Sones' et al. (2009) study makes it clear that employees need to be part of the dialogue (i.e., informed, engaged, and empowered) if they are to communicate their organization's sustainability message. Employee understanding of their organization's view of sustainability increases if they see sustainability as part of their organization's identity and are aware of the corporate sustainability policy, mission, vision, and goals (Allen et al. 2012; Craig and Allen 2013; Linnenluecke et al. 2009). Increased awareness of their organization's sustainability practices provides employees with the holistic understanding of corporate sustainability (i.e., more concern for planet, people, and profits) they need if they are to communicate with external stakeholders.

Communication plans developed around preparing employees to discuss sustainability with external publics (e.g., community, supply chain members) are important. For people unfamiliar with constructing communication plans, templates are available online from docs.google.com. Information to share with employees includes relevant facts, key talking points, and memorable stories they can share. Interface, the world's largest manufacturer of commercial carpets and floor coverings, developed a speaker's bureau where knowledgeable employees share their organization's best practices with interested organizations. Employees are more credible speakers when they can talk about their own direct experiences. They are more likely to particulate when they are recognized and praised for doing so.

6.6   Concluding Thoughts

As of 2013 there were 45,508 companies listed in stock exchanges around the world. The number of formal unlisted companies is unknown because that's impossible to track. Then, there are other types of organizations (e.g., NGO's, small businesses, nonprofits, governmental entities). In 2012, 26 % of the world's adult population worked full time for an employer. Although organizations contribute significantly to resource degradation, global warming, and worker exploitation, they also create the contexts within which we individuals can be part of something larger than ourselves. It is through our organizations that we can most effectively manage climate change challenges. Some organizational cultures have already changed to promote pro-environmental planning and behaviors. Yet many others must change in order for meaningful proactive action to occur on the societal and global levels. How can you be one of the visionaries leading your organization to empower its employees to either seek a different path or tread more firmly along your present path to creating a more sustainable future for all humanity?

References

Adams, J. S. (1963). Towards an understanding of inequality. *Journal of Abnormal and Normal Social Psychology, 67*, 422–436.

Allen, M. W. (1992). Communication and organizational commitment: Perceived organizational support as a mediating factor. *Communication Quarterly, 40*, 357–367.

Allen, M. W. (1995). Communication variables shaping perceived organizational support. *Western Journal of Communication, 59*, 326–346.

Allen, M. W., & Brady, R. M. (1997). Total quality management, organizational commitment, perceived organizational support, and intraorganizational communication. *Management Communication Quarterly, 10*, 316–341.

Allen, M. W., Walker, K. L., & Brady, R. (2012). Sustainability discourse within a supply chain relationship: Mapping convergence and divergence. *Journal of Business Communication, 49*, 210–236.

Bamberg, S., & Moser, G. (2007). Twenty years after Hines, Hungerford, and Tomera: A new meta-analysis of psycho-social determinants of pro-environmental behavior. *Journal of Environmental Psychology, 27*, 14–25.

Bandura, A. (1977). Self-efficacy: Toward a unifying theory of behavioral change. *Psychological Review, 8*, 191–215.

Barker, R. T., & Camarata, M. R. (1998). The role of communication in creating and maintaining a learning organization: Preconditions, indicators, and disciplines. *Journal of Business Communication, 35*, 443–467.

Benn, S., Edwards, M., & Angus-Leppan, T. (2013). Organizational learning and the sustainability community of practice: The role of boundary objects. *Organization & Environment, 26*, 184–202.

Bissing-Olson, M. J., Iyer, A., Fielding, K. S., & Zacher, H. (2012). Relationships between daily affect and pro-environmental behavior at work: The moderating role of pro-environmental attitude. *Journal of Organizational Behavior, 34*, 156–175.

Blackburn, W. R. (2007). *The sustainability handbook: The complete management guide to achieving social, economic and environmental responsibility*. London: Earthscan.

Bolderdijk, J. W., Steg, L., & Postmes, T. (2013). Fostering support for work floor energy conservation policies: Accounting for privacy concerns. *Journal of Organizational Behavior, 34*, 195–210.

Byrch, C., Kearnins, K., Milne, M., & Morgan, R. (2007). Sustainable "what"? A cognitive approach to understanding sustainable development. *Qualitative Research in Accounting & Management, 4*, 26–52.

Caldwell, D. F., Chatman, J. A., & O'Reilly, C. A., III. (1990). Building organizational commitment: A multi-firm study. *Journal of Occupational and Organizational Psychology, 63*, 245–261.

Cantor, D. E., Morrow, P. C., & Montabon, F. (2012). Engagement in environmental behaviors among supply chain management employees: An organizational support theoretical perspective. *Journal of Supply Chain Management, 48*, 33–51.

Cheney, G. (2011). *Organizational communication in an age of globalization: Issues, reflections, practice*. Long Grove, IL: Waveland.

Cheney, G., Christensen, L. T., & Dailey, S. L. (2014). Communicating identity and identification in and around organizations. In L. L. Putnam & D. K. Mumby (Eds.), *The SAGE handbook of organizational communication* (3rd ed., pp. 695–716). Thousand Oaks, CA: Sage.

Cordano, M., & Frieze, I. H. (2000). Pollution reduction preferences of U.S. environmental managers: Applying Ajzen's theory of planned behavior. *Academy of Management Journal, 43*, 627–641.

Craig, C. A., & Allen, M. W. (2013). Sustainability information sources: Employee knowledge, perceptions, and learning. *Journal of Communication Management, 17*, 292–307.

D'Aprix, R. (2011). The challenges of employee engagement: Throwing rocks at the corporate rhinoceros. In T. L. Gillis (Ed.), *The IABC handbook of organizational communication: A guide to internal communication, public relations, marketing, and leadership* (2nd ed., pp. 157–270). San Francisco, CA: Jossey-Bass.

Delmas, M. A., & Pekovic, S. (2013). Environmental standards and labor productivity: Understanding the mechanisms that sustain sustainability. *Journal of Organizational Behavior, 34*, 230–252.

Dervin, B., & Naumer, C. M. (2009). Sense-making. In S. W. Littlejohn & K. A. Foss (Eds.), *Encyclopedia of communication theory* (Vol. 2, pp. 876–880). Los Angeles: Sage.

Eisenberger, R., Fasolo, P., & Davis-LaMastro, V. (1990). Perceived organizational support and employee diligence, commitment, and innovation. *Journal of Applied Psychology, 75*, 51–59.

Eisenberger, R., Huntington, R., Hutchison, S., & Sowa, D. (1986). Perceived organizational support. *Journal of Applied Psychology, 71*, 500–507.

Energy Efficiency Program. (2015). https://www.clintonfoundation.org/our-work/clinton-climate-initiative/programs/energy-efficiency-program. Accessed 28 Jan 2015.

Exter, N. (2014). *Employee engagement with sustainable business: How to change the world whilst keeping your day job*. Abingdon: Routledge.

Fields, D. L. (2002). *Taking the measure of work: A guide to validated scales for organizational research and diagnosis*. Thousand Oaks: Sage.

Fleischer, D. (2009). *Green teams: Engaging employees in sustainability. Green impact: Helping companies go green*. www.GreenImpact.com. Accessed 26 Mar 2014.

Fredrickson, B. L. (2001). The role of positive emotions in positive psychology: The broaden-and-build theory of positive emotions. *American Psychologist, 56*, 218–226.

Gailliard, B., Myers, K. K., & Seibold, D. R. (2010). Organizational assimilation: A multidimensional reconceptualization and measure. *Management Communication Quarterly, 24*, 552–578.

Gallup. (2013). *State of the American Workforce*. http://www.gallup.com/services/178514/state-american-workplace.aspx. Accessed 2 Jan 2014.

Govindarajulu, N., & Daily, B. F. (2004). Motivating employees for environmental improvement. *Industrial Management and Data Systems, 104*, 364–372.

Gupta, A. (2011). Championing sustainability programs through internal communication. *Strategic Communication Management, 15*, 6–18.

Hannaes, K., Arthur, D., Balagopal, B., Kong, M. T., Reeves, M., Velken, I. et al. (2011). *Sustainability: The 'embracers' seize advantage*. MIT Sloan Management Review and The Boston Consulting Group Research Report. http://sloanreview.mit.edu/reports/sustainability-advantage/. Accessed 20 Dec 2013.

Harris, L. C., & Crane, A. (2002). The greening of organizational culture: Management views on the depth, degree and diffusion of change. *Journal of Organizational Change Management, 15*, 214–234.

Hatfield, E., Cacioppo, J. T., & Rapson, R. L. (1993). Emotional contagion. *Current Directions in Psychological Science, 2*, 96–99.

Hine, J. A. J. S., & Preuss, L. (2009). Society is out there, organisation is in here: On the perceptions of corporate social responsibility held by different managerial groups. *Journal of Business Ethics, 88*, 381–393.

Keyton, J. (2014). Organizational culture: Creating meaning and influence. In L. L. Putnam & D. K. Mumby (Eds.), *The SAGE handbook of organizational communication* (3rd ed., pp. 549–568). Thousand Oaks, CA: Sage.

Knobloch, L. K. (2009). Uncertainty reduction theory. In S. W. Littlejohn & K. A. Foss (Eds.), *Encyclopedia of communication theory* (Vol. 2, pp. 976–978). Los Angeles, CA: Sage.

Kramer, M. W., & Miller, V. D. (2014). Socialization and assimilation: Theories, processes, and outcomes. In L. L. Putnam & D. K. Mumby (Eds.), *The Sage handbook of organizational communication* (pp. 525–547). Thousand Oaks, CA: Sage.

Lee, S., & Lee, C. (2015). Creative interaction and multiplexity in intraorganizational networks. *Management Communication Quarterly, 29*, 56–83.

Library of Congress. (n.d.). *Rivers, Edens, Empires: Lewis and Clark and the revealing of America.* http://www.loc.gov/exhibits/lewisandclark/lewis-landc.html. Accessed 2 July 2014.

Linnenluecke, M. K., & Griffiths, A. (2010). Corporate sustainability and organizational culture. *Journal of World Business, 45*, 357–366.

Linnenluecke, M. K., Russell, S. V., & Griffiths, A. (2009). Subcultures and sustainability practices: The impact of understanding corporate sustainability. *Business Strategy and the Environment, 18*, 432–452.

Lourdel, N., Gondran, N., Laforest, V., Debray, B., & Brodhag, C. (2007). Sustainable development cognitive map: A new method of evaluating student understanding. *International Journal of Sustainability in Higher Education, 8*, 170–182.

McAleese, D., & Hargie, O. (2004). Five guiding principles of culture management: A synthesis of best practice. *Journal of Communication Management, 9*, 155–170.

Meyer, J. P., & Allen, N. J. (1991). A three-component conceptualization of organizational commitment. *Human Resource Management Review, 1*, 61–89.

My Sustainability Plan. (n.d.). Retrieved from https://us.walmart.mysustainabilityplan.com/. Accessed 15 Nov 2014.

Myers, K. K. (2009). Organizational socialization and assimilation. In S. W. Littlejohn & K. A. Foss (Eds.), *Encyclopedia of communication theory* (Vol. 2, pp. 722–724). Los Angeles, CA: Sage.

National Park Service. (n.d.). *The journey. Lewis and Clark expedition: A national register of historic places travel itinerary.* http://www.nps.gov/nr/travel/lewisandclark/journey.htm. Accessed 2 July 2014.

O'Reilly, C. A., III, Chatman, J., & Caldwell, D. F. (1991). Person and organizational culture: A profile comparison approach to assessing person-organization fit. *Academy of Management Journal, 34*, 487–516.

Oncica-Sanislav, D., & Candea, D. (2010). The learning organization: A strategic dimension of the sustainable enterprise? *Proceedings of the European Conference of Management, Leadership & Governance* (pp. 263–270). www.gbv.de/dms/tib-ub-hannover/647510022.pdf. Accessed 3 Feb 2014.

Ones, D. S., & Dilchert, S. (2012). Environmental sustainability at work: A call to action. *Industrial and Organizational Psychology, 5*, 444–466.

Organ, D. W. (1988). *Organizational citizenship behavior: The good soldier syndrome.* Lexington, MA: Lexington Books.

Perron, G. M., Cote, R. P., & Duffy, J. F. (2006). Improving environmental awareness training in business. *Journal of Cleaner Production, 14*, 551–562.

Quinn, R. E. (1988). *Beyond rational management: Mastering the paradoxes and competing demands of high performance*. San Francisco: Jossey-Bass.

Reid, S. (2009). Social identity theory. In S. W. Littlejohn & K. A. Foss (Eds.), *Encyclopedia of communication theory* (Vol. 2, pp. 896–897). Los Angeles, CA: Sage.

Rhodes, L., & Eisenberger, R. (2002). Perceived organizational support: A review of the literature. *Journal of Applied Psychology, 87*, 698–714.

Roloff, M. (2009). Social exchange theory. In S. W. Littlejohn & K. A. Foss (Eds.), *Encyclopedia of communication theory* (Vol. 2, pp. 894–896). Los Angeles, CA: Sage.

Rousseau, D. M., Sitkin, S. B., Burt, R. S., & Camerer, C. (1998). Not so different after all: A cross-discipline view of trust. *Academy of Management Review, 23*, 393–404.

Russell, S. V., Haigh, N., & Griffiths, A. (2007). Understanding corporate sustainability: Recognizing the impact of different governance systems. In S. Benn & D. C. Dunphy (Eds.), *Corporate governance and sustainability: Challenges for theory and practice* (pp. 36–56). London: Routledge.

Salancik, G. R., & Pfeffer, J. (1978). A social information processing approach to job attitudes and task design. *Administrative Science Quarterly, 23*, 224–253.

Sarkis, J., Gonzalez-Torre, P., & Adenson-Diaz, B. (2010). Stakeholder pressure and the adoption of environmental practices: The mediating effect of training. *Journal of Operations Management, 28*, 163–176.

Schein, E. H. (2004). *Organizational culture and leadership* (3rd ed.). San Francisco: Jossey-Bass.

Schendler, A. (2009). *Getting green done: Hard truths from the front line of the sustainability revolution*. New York: PublicAffairs.

Scott, C. W., & Myers, K. K. (2010). Toward an integrative theoretical perspective of membership negotiations: Socialization, assimilation, and the duality of structure. *Communication Theory, 30*, 79–105. doi:10.1111/j.1468-2885.2009.01355.x.

Senge, P. M. (1990). *The fifth discipline: The art and practice of the learning organization*. New York: Doubleday.

Settoon, R. P., Bennett, N., & Liden, R. C. (1996). Social exchange in organizations: Perceived organizational support, leader-member exchange, and employee reciprocity. *Journal of Applied Psychology, 81*, 219–227.

Sharma, S. (2000). Managerial interpretations and organizational context as predictors of corporate choice of environmental strategy. *Academy of Management Journal, 43*, 681–697.

Sias, P. M. (2014). Workplace relationships. In L. L. Putnam & D. K. Mumby (Eds.), *The SAGE handbook of organizational communication* (3rd ed., pp. 375–399). Thousand Oaks, CA: Sage.

Signitzer, B., & Prexl, A. (2008). Corporate sustainability communications: Aspects of theory and professionalization. *Journal of Public Relations Research, 20*, 1–19.

Smerecnik, K. R., & Andersen, P. A. (2011). The diffusion of environmental sustainability innovations in North American hotels and ski resorts. *Journal of Sustainable Tourism, 19*, 171–196.

Society of Human Resource Managers. (2011). *Advancing sustainability: HR's role survey report*. http://www.shrm.org/research/surveyfindings/articles/pages/advancingsustainabilityhr%E2%80%99srole.aspx. Accessed 20 Dec 2013.

Sones, M., Grantham, S., & Vieira, E. T. (2009). Communicating CSR via pharmaceutical company web sites: evaluating message frameworks for external and internal stakeholders. *Corporate Communications: An International Journal, 12*, 144–157.

Spreitzer, G. M. (1995). Psychological empowerment in the workplace: Dimensions, measurement, and validation. *Academy of Management Journal, 38*, 1442–1465.

Stern, P. C. (2000). Toward a coherent theory of environmentally significant behavior. *Journal of Social Issues, 56*, 407–424.

Strandburg Consulting. (n.d.) *Developing a sustainability vision and management system*. http://www.corostrandberg.com/pdfs/Sustainability_Vision_and_Management1.pdf. Accessed 23 June 23 2014.

Tracy, S. J. (2009). Organizational culture. In S. W. Littlejohn & K. A. Foss (Eds.), *Encyclopedia of communication theory* (Vol. 2, pp. 713–716). Los Angeles, CA: Sage.

UNEP Finance Initiative. (2011). *If you ask us: Making environmental employee engagement happen*. http://www.unepfi.org/fileadmin/documents/ifyouaskus_engagement.pdf. Accessed 3 Sept 2014.

Unsworth, K. L., Dmitrieva, A., & Adriasola, E. (2013). Changing behavior: Increasing the effectiveness of workplace interventions in creating pro-environmental behaviour change. *Journal of Organizational Behavior, 34*, 211–229.

van den Bosch, F. A. J., Volberda, H. W., & de Boer, M. (1999). Coevolution of firm absorptive capacity and knowledge environment: Organizational forms and combinative capabilities. *Organizational Science, 10*, 551–568.

van der Heijden, A., Cramer, J. M., & Driessen, P. P. J. (2012). Change agent sensemaking for sustainability in a multinational subsidiary. *Journal of Organizational Change Management, 25*, 535–559.

Van der Post, W. Z., de Coning, T. J., & Smith, E. M. (1997). An instrument to measure organizational culture. *South African Journal of Business Management, 28*, 147–168.

Vroom, V. H. (1964). *Work and motivation*. New York: McGraw Hill.

Walmart Announces. (2007). Retrieved from http://news.walmart.com/news-archive/2007/04/05/wal-mart-announces-expansion-of-associate-driven-personal-sustainability-projects. Accessed 1 Nov 2014.

Weick, K. E. (1995). *Sensemaking in organizations: Foundations for organizational science*. Thousand Oaks, CA: Sage.

Weinreb, E. (2013). *How Walmart associates put the 'U' and 'I' into sustainability*. GreenBiz. com. Retrieved http://www.greenbiz.com/blog/2013/01/09/walmart-associates-u-i-sustainability. Accessed 20 Dec 2013

Werbach, A. (2009). *Strategy for sustainability: A business manifesto*. Cambridge, CA: Harvard Business Press.

Chapter 7
Facilitating Group Collaboration and Enhancing Supply Chain Conversations

Abstract This chapter is about groups and group communication at the organizational and interorganizational levels. The enactment of sustainability initiatives often requires group collaboration. Characteristics of strong teams are identified. Theory and research related to small group decision-making and problem-solving processes are reviewed. Group-level techniques for frame breaking and building so as to identify creative alternatives are identified. Then, the focus shifts to interorganizational collaboration (IOC) efforts. Complex environmental issues accompanying global climate change require large collaborative efforts. Theory related to the role of communication in the creation of emergent collaborative structures is identified. Theories and research related to learning in IOCs are reviewed. Techniques used to build communities of practice and enhance group learning are discussed. The ways language and texts can contribute to IOC effectiveness are reviewed. A specific case of IOC interaction, the nongovernmental organization (NGO)–corporate alliance, is discussed in terms of their joint external communication efforts. Finally, the focus is on sustainable supply chains. Why do organizations form them, how are they governed, and what challenges exist? Internal and external communication issues are identified and best practices are offered. Interview data spotlights the City of Denver, Aspen Skiing Company, Tyson Foods, Sam's Club, and Bayern Brewing.

Lewis and Clark and the Role of Groups Large journeys often require the talents of multiple people working toward the same goal. William Clark recruited 44 men of diverse backgrounds. They were good hunters, healthy, unmarried, accustomed to the woods, and strong. Two had blacksmithing experience, one knew carpentry, and a few knew Indian languages. Thirty intended to make the entire journey. Six returned at the midpoint taking maps, notes, and specimens of plants, animals, and minerals back to President Jefferson. During their 2-year journey, the Corps of Discovery split into smaller groups to accomplish parts of their overall mission. For example, on their way home, they broke into two groups. Clark and his group traveled down the Yellowstone River, while Lewis and his group headed toward the Great Falls of the Missouri River in Montana and then north along the Marias River. Both groups faced challenges to the achievement of their goals before they reunited near Sanish, ND. This example illustrates how people working together bring

© Springer International Publishing Switzerland 2016   231

M. Allen, *Strategic Communication for Sustainable Organizations*, CSR,
Sustainability, Ethics & Governance, DOI 10.1007/978-3-319-18005-2_7

multiple resources into play when facing insurmountable tasks for individuals to accomplish and how group structures shift over time.

7.1 Intraorganizational Groups

Organizations which embrace sustainability are moving from relying on experts to having internal group members plan and implement sustainability initiatives within and between organizations (Hannaes et al. 2011). In Chap. 5, you read how the Portland Trail Blazers created an internal sustainability team of 35 individuals representing every department and all organizational levels. That group established the vision, developed a set of sustainability goals, and designed actions to minimize the Trail Blazers' triple-bottom-line (TBL) impacts and to benefit the community. You also read how the City and County of Denver created an interagency committee representing 12 of the city's 21 agencies charged with coordinating efforts to achieve the city's 24 different 2020 goals. Blackburn (2007) describes the role of a core team in strategic planning; the usefulness of teams in government, small businesses, and colleges and universities interested in addressing sustainability; and teams as a way to engage employees. At this moment, across our world, groups are working toward sustainability within and between organizations of various types (e.g., NGOs, governmental entities, for-profit organizations). In this section, I identify some key group-related theories, review relevant research involving groups seeking to promote sustainability initiatives, and offer suggestions for effective group functioning.

7.1.1 *Small Group Decision-Making and Problem-Solving Processes*

For decades, scholars and practitioners have sought to identify the characteristics of highly functioning teams. For example, Larson and LaFasto (1989) conducted a 3-year study of teams and team achievement. They interviewed multiple and varied teams and found eight common characteristics of effective teams: a clear, elevating goal; a results-driven structure; competent team members; unified commitment; a collaborative climate; standards of excellence; external support and recognition; and principled leadership. These same characteristics appear in the literature discussing effective organizational and interorganizational groups.

Effective small groups begin by focusing on four questions: (1) Does the situation require some kind of choice? (2) What do we want to achieve in our decision? (3) What choices do we have? (4) What are the positive and negative aspects of each choice? (Hirokawa and Salazar 1999). How groups answer these four questions shape the social context within which their final decisions are made.

High-quality decisions are more likely if group members publically commit to seeking the best possible decision, are willing to expend the needed effort, can identify potential obstacles and needed resources, possess needed knowledge and skills, utilize strategies appropriate to the work and the setting, develop interventions to overcome any constraints which limit their decision-making effectiveness, review their process, and reevaluate their judgments and decisions. Finally, a decision-making group's effectiveness is influenced by conditions within the group (e.g., leadership) as well as by the larger organizational context (e.g., external support for the group's success).

7.1.1.1 Functional Group Decision-Making Theory

Group communication scholars began developing theory in the 1970s (Poole 1999). In the 1980s, Gouran and Hirokawa developed one of the most influential group communication theories, the functional theory of group decision making (Salazar 2009). This theory promotes rational decision making and critical, reflective discussion. Its rational systematic approach is based on the work of philosopher John Dewey (Schultz 1999). Functional decision making involves understanding the problem, identifying criteria for judging solutions, generating a relevant set of alternatives, examining the alternatives in light of the criterion, and selecting the alternative that best meets the desired characteristics. Initially, group members must correctly understand the issue or problem to be resolved. Problems may be made up of questions of fact (e.g., How much CO_2 are we omitting?), questions of conjecture (e.g., Is it cost-effective to change our processes so as to reduce our CO_2 emissions?), and questions of value (e.g., Do we have a responsibility to reduce our CO_2 emissions?). Which types of questions are your group, or the groups you are researching, focusing on? Do group members share a common understanding of the questions they need to address? After a group has identified the problem, members should identify the minimal characteristics any acceptable alternative must have (i.e., criterion) (e.g., be technologically and financially feasible, have a demonstrated record of success). How well groups discuss and evaluate their alternative choices in light of the criteria influences group decision-making performance.

While groups attempt to generate solutions and make decisions, they are simultaneously undergoing internal orientation processes. In the 1970s, Fisher developed a theory of decision emergence (Littlejohn and Foss 2005) which discussed how new groups go through four phases: orientation, conflict, emergence, and reinforcement. During orientation, group members become acquainted, begin to focus on the task, and start sharing their insights. During the conflict phase, debate occurs as people solidify their attitudes. Debate involves breaking issues/problems into parts; distinguishing between the parts; justifying/defending assumptions; persuading, selling, and telling; and finally gaining agreement. Polarization is likely. Conflict is a normal part of the process when groups create collaborative decisions around sustainability-related challenges (Livesey et al. 2009). The emergence phase is characterized by the beginning of cooperation, more ambiguous comments, and the emergence of a possible decision. During reinforcement, the decision solidifies and group comments reinforce the decision.

The functional model of decision making and the phase model are only generalizations. There is no single blueprint for how a group should work through a problem or arrive at a decision. Many groups do not follow a rational approach in their decision making (Schultz 1999). Groups move back and forth between issues and possible solutions amid shifting group dynamics. Some groups discuss, argue, and reconsider the same issues multiple times before reaching a decision. However, the functional model of decision making *can* help groups learn to make more high-quality decisions. *Best Practice*: Vigilance to rational decision making is especially important when groups face unfamiliar, ambiguous, or difficult decisions and/or the group is heterogeneous (i.e., members represent different cultures, interests, value sets, or organizations). Unfamiliarity, ambiguity, difficulty, and heterogeneous participants are likely to characterize large groups facing complex sustainability-related challenges.

7.1.1.2 Resources for Improving Decision Making

Groups need a structure that promotes competent work, an environmental context that supports and reinforces excellence in decision making, and effective coaching in and assistance during the decision-making process (Hirokawa and Salazar 1999). Diagnosis and intervention techniques (e.g., training) can improve group communication performance (Schultz 1999). Facilitators can observe what is occurring during group discussion, provide feedback as to the group's strengths and weaknesses, suggest changes to help the group become more effective, and help group members practice new behaviors. The coding system developed by Bales, the interaction process analysis (see Bonito 2009), is useful when evaluating task and socioemotional group dynamics. Coders identify the occurrence of various communication behaviors (e.g., shows solidarity, gives suggestions, asks for suggestions, or displays antagonism). If group members learn to recognize problems and change their processes, this can result in better decisions.

Various tools exist to help groups structure their discussions (e.g., create an agenda), analyze the problems faced (e.g., reflective thinking), create new alternatives (e.g., consensus mapping), and agree on a decision (e.g., straw poll) (see SunWolf and Seibold 1999). During the problem analysis phase, useful procedures are reflective thinking, devil's advocacy, the Delphi method, dialectical inquiry, nominal group technique, flowchart, multiattribute decision analysis, problem census, single-question format, fishbone diagram, 6M analysis, cognitive map, journalist's six questions, Pareto analysis, ideal-solution format, is/is not analysis, force-field analysis, stepladder technique, risk procedure, program evaluation and review technique, multidimensional scaling, focus groups, interpretive structure modeling, and expert approach. Ways to create options include brainstorming, reverse brainstorming, idea writing, consensus mapping, lateral thinking, buzz groups, morphological analysis, role storming, ideals method, synectics, brain writing, object stimulation, excursion, visioning, collective notebook, lotus blossom, semantic intuition, and creative problem solving. *Best Practice*: I provide this list here to remind readers of the tools they can use to improve their decision-making processes.

7.1.2 Dialogue and Learning

> The nature of sustainability challenges seems to be such that a routine problem-solving approach falls short, as transitions towards a more sustainable world require more than attempts to reduce the world around us into manageable and solvable problems. Instead, such transitions require a more systemic and reflexive way of thinking and acting, bearing in mind that our world is one of continuous change and ever-present uncertainty (Wals and Schwarzin 2012, p. 13).

Functional group decision-making theory can't generate the kind of transformational changes we need in the face of global climate change, especially when the problem solvers represent multiple organizations and interest groups. In a world where people strongly disagree along cultural and ideological lines, debate-based decision-making techniques may only increase polarization. Dialogue offers us another alternative. Dialogue is an interactive effort to co-create novel ideas and understandings which rests on using inquiry, advocacy, and reflection. It can be a catalyst for generating new understanding, insight, and action, for bringing about individual and collective shifts in mindsets and behaviors, and for encouraging collaboration and collective action. Wals and Schwarzin (2012, p. 15) explore the:

> practice of 'dialogic interaction', which is defined as reflexive conversation and engagement among a heterogeneous group of people, who attempt to explore a diversity of potentially incompatible perspectives in a mutually respectful, trusting and collaborative way, and which can be considered as a type of interaction that carries considerable potential in community and organizational learning towards sustainability.

Dialogue is key to many of the decision-making models and studies discussed later in this chapter (e.g., Brulle 2010; Livesey et al. 2009; Manring 2007; Mitchell et al. 2012; Reed et al. 2014; Wals and Schwarzin 2012).

Dialogue requires we abandon the culturally entrenched directive and/or adversarial decision-making patterns which often characterize decisions about resource allocation and the natural environment. Its success is influenced by the personal characteristics of participants, interaction dynamics within a group, and contextual factors in a group's interaction environment. Supportive individual-level attitudes include a commitment to adhere to dialogic interaction principles and a willingness to reduce competitive and egotistic drives. Supportive group dynamics include an inspirational joint vision or goal and the practice of symmetric conversation. Group dynamics that inhibit dialogic interaction include conflict avoidance tendencies and hidden power aspirations and hierarchies. Groups overcome tendencies destructive to dialogue as they move from either/or comparisons to yes/and thinking, develop group cohesiveness, and form a joint vision. A facilitator can assist participants in the use of dialogic interaction principles using various communication tools (e.g., talking sticks, visualization aids, and improvisation games). A facilitator might suggest group members read *Getting to Yes*: *Negotiating Agreement Without Giving In* (Fisher et al. 2011) in preparation for working past initial conflicts. Facilitators and mediators bring important skill sets which can enhance the design and implementation of any collaborative process. But, ultimately, groups must learn to interact dialogically without a facilitator's assistance.

Dialogic interaction is a key mechanism for supporting group learning processes (Wals and Schwarzin 2012). Since the early 1990s, researchers have discussed how organizations can improve their sustainability by learning cooperatively with their various stakeholders. Learning theories and heuristics undergird much action-oriented sustainability research. Dialogue opens up a space for learning where issues can be seen holistically as people move from single-loop to double-loop learning. Several of the articles mentioned in this chapter discuss single-, double-, and even triple-loop learning (e.g., Manring 2007; Mitchell et al. 2012; Wals and Schwarzin 2012). All reference Argyris and Schön (1978). Single-loop learning occurs as groups attempt to correct their behavior so as to not deviate from system norms. It involves comparing current conditions to existing theories-in-use—the assumptions, values, rules, and norms guiding routine behaviors. It often results in incremental operational solutions. When organizations face a rapidly changing environment and critical situations or want to follow a new vision, a different type of learning is needed. Double-loop learning involves questioning the underlying assumptions and utility of existing theories-in-use. Problems are solved in such a way as to modify underlying norms, policies, and objectives. For example, the TBL concept may stimulate double-loop learning as companies which previously only focused on financial performance begin to include social and environmental performance indicators in their decision making (Mitchell et al. 2012). Double-loop learning provides an opportunity for dialogue and reflection which can restructure thought processes and result in the potential for radical and/or strategic change. Triple-loop learning involves learning how to learn by reflecting on how we think about the rules, not only about whether a rule should be changed. Learning how to learn involves uncovering tacit knowledge and processes and making organizational processes open and participatory. *Best Practice*: In Sect. 6.2.1.1, you read about how to create organizational cultures focused on learning. Those same ideas apply here.

Several studies (i.e., Mitchell et al. 2012; Wals and Schwarzin 2012) drawing on the concepts of dialogue and double-loop learning will be discussed next. Building on two case studies of sustainable community development, a Dutch case study of a sustainable neighborhood and an Indian case study of an international sustainable city, Wals and Schwarzin (2012) identify sustainability competences necessary to dialogic interactions. Sustainability competences refer to the capacities and qualities needed if we, and the organizations and communities we represent, are to adequately deal with the five features which characterize sustainability-related problems (i.e., their indeterminacy, value-ladenness, controversy, uncertainty, and complexity) identified in Sect. 1.1. The competences include empathetic listening, suspension, slowing down, and assertiveness. Empathetic listening includes perception checking, active listening, feedback aimed at clarifying the action effects, and nonaggressive and nonevaluative assertions. Suspension involves participants learning how to put their judgments and emotional responses on hold and only make helpful and relevant comments. Participants need to slow their conversations down, interact calmly, allow others time to reflect, avoid interrupting, and pause before speaking. Group members voice their opinions and perspectives and critically reflect on their own and others' contributions. They learn to use advocacy to share their perspectives to enhance group learning rather than to simply support their entrenched positions. *Best Practice*: Consider providing internal teams with training in effective group communication skills.

In a second study, Mitchell et al. (2012) drew on work discussing learning cycles and organizational learning theory to investigate whether or not double-loop learning, change agency, and radical change could occur based on collaboration generated around the process of TBL reporting over two consecutive reporting cycles. Their research questions were as follows: Can the TBL reporting process help organizations learn about sustainability and how can they enhance it? Can the TBL reporting process help organizations engage and build their capacity and those of their stakeholders to respond to the sustainability challenges being faced? If learning and capacity building evolves, does the organization implement changes that respond to their sustainability challenges? If not, why? If yes, will those changes enhance sustainability? Over a 2-year period, Mitchell et al. worked with Murrumbidgee Irrigation (MI), a private company with around 200 employees, which manages the Murrumbidgee Irrigation Area (MIA) in central New South Wales, Australia. MIA faces a critical sustainability issue: severe water shortages, partly due to the over-allocation of irrigation water. The water management crisis led to calls for radical reform in how water use is viewed and in the structures and governance of irrigation water allocation. Social attitudes are shifting as more Australians believe irrigation policies and practices are too focused on production outcomes at the expense of environmental resources. Organizations such as MI face challenges to their social license to operate and are adopting TBL reporting partly as a defensive strategy.

At the end of the 2 years, Mitchell et al. (2012) found that organizational participants were beginning to learn how to learn. But they were not seriously questioning the organization's assumptions about the sustainability of MI's core activities. The researchers saw few examples of double-loop learning and few practical changes. Any changes were small, isolated, and ineffectual in the face of severe and enduring drought. The authors offer insights regarding how TBL reporting processes might be improved so as to promote more radical agendas for change. They argue that small changes in the aggregate moving in a similar direction can result in radical transformation. However, the transformational change research discussed in Sect. 5.1 suggests this is unlikely. Second, learning, commitment, and change are more likely if workplace proponents, rather than external change agents, initiate changes. Third, if done correctly and communicated widely within an organization, the reflection which occurs as part of a TBL reporting process can develop into an iterative cycle of experientially based and action-oriented learning. However, my own research suggests organizations do not devote enough attention to ensuring this occurs. Report content will be overlooked unless it is systematically and repeatedly communicated and linked to changed task and reward structures. Finally, multistakeholder interactions inspired by TBL reporting activities have the potential to result in greater collaborative activity. In the MIA, for example, regional collaboration on how to improve TBL outcomes would provide organizations with an opportunity to investigate their own theories-in-use and begin questioning societal-level governance values and arrangements.

7.1.3   Group-Level Techniques for Frame Breaking and Building

> We cannot think about sustainability in terms of problems that are out there to be solved or in terms of 'inconvenient truths' that need to be addressed. Instead, we need to think in terms of challenges to be taken on in the full realization that, as soon as we appear to have met the challenge, things will have changed and the horizon will have shifted once again (Wals and Schwarzin 2012, p. 13).

As our climate shifts forcing our communities and organizations to adapt, more creative ideas will be needed ever more quickly. Creativity involves "the generation, application, combination, and extension of ideas" (SunWolf 2002, p. 205). Communication scholars are uniquely able to assess communication's role in facilitating creative group interactions (Jarboe 1999). In this section, I discuss creativity and chaos, techniques including appreciative inquiry and symbolic convergence, skill-building strategies, and organizational conditions that hamper creativity.

7.1.3.1   Creating Chaos to Allow for Creativity

Salazar developed an untested theory relating communication to group creativity (Salazar 2009). A group's ability to be creative is unlikely if the group continues to function as it always has. Too much imposed order (through overt action or due to existing norms) and too much chaos both hamper creativity. Creative groups function somewhere between order and chaos. Salazar writes:

> Whenever we see group members changing their interaction patterns, whether by changing the frequency and directional flow of communication, using new procedures to solve problems or make decisions, displaying little or no discussion typical of group member roles, or questioning or supplanting existing assumptions about what counts as good or bad information, the group is showing evidence of operating in a complex state (p. 212).

Complex states allow creativity to emerge. Researchers built on Salazar's ideas to describe a three-step process groups might use to become more creative. First, take inventory of the group's structural (i.e., who talks to whom), technical (i.e., the procedures used to accomplish tasks), relational (i.e., group member roles), and information (i.e., assumptions regarding what is right/wrong or good/bad in how members communicate about information) systems. Second, introduce changes to disturb one or more of the four systems. Finally, use these changed features in the group's everyday functioning, and recognize how that change can influence other aspects of group work such as changed roles and power differences. *Best Practice*: Pattern breaking creates space where creativity can emerge.

7.1.3.2 Stimulating Appreciative Inquiry Around Sustainability

Appreciative inquiry (AI) focuses a group's discussion on what works well within a particular society, organization, or group. The hope is that this focus on what is working well can help create a blueprint for future action. The process of AI begins during interviews as multiple people in an organization (or other social system) identify what they believe to be moments of excellence, high points, core values, proud moments, and life-giving forces. Then, the process shifts to envisioning what might be. Provocative proposals that remind members of *what is* best about their organization (e.g., "Our citizens volunteer to build a healthy community," "Our organization seeks innovative ways to address challenging problems") emerge from the stories elicited during the interviews. These provocative propositions are discussed in a series of small or large open group meetings. Next, people dialogue over *what should be* in light of the interview information and the provocative propositions. The final step concerns *what will be* as members decide on the next steps needed to create the kind of desired future captured by the provocative propositions. *Best Practice*: This seems like a useful process as organizations create new visions, set new goals, and plan sustainability initiatives.

Several recent studies reported on the use of AI to promote a shared focus on enhanced sustainability. For example, in 2009, the City of Cleveland, OH, held an AI summit to envision a new sustainability-focused local economy (Meyer-Emerick 2012). The AI process helped the community create a preferred future. The resulting plan, *Sustainable Cleveland 2019*, developed from the shared vision of Cleveland becoming a Green City on a Blue Lake by 2019, the 50th anniversary of the Cuyahoga River fire. In 1969, the Cuyahoga was one of the most polluted rivers in the USA. Publicity surrounding the fire raised public interest in the US environmental movement. In another study, Hinrichs (2010) described the success of combining AI with social constructionism in the SOAR (strengths, opportunities, aspirations, and results) program utilized by a global equipment manufacturer with 60,000 employees. The SOAR program sought to increase employee collaboration, shared understanding, and commitment to action. SOAR conversations resulted in initiatives and projects which were translated into individual performance goals. *Best Practice*: Practitioners will find the SOAR questions useful in guiding similar discussions. Sample questions include the following:

- Strengths: How do our strengths fit with the realities of current environmental demands or align with our enterprise strategy?
- Opportunities: How can we learn to reframe challenges to be seen as opportunities?
- Aspirations: Reflecting on our strengths and opportunities, who are we, who should we become, and where should we go?
- Results: What would a dashboard look like that could provide useful and continuous feedback to stakeholders?

7.1.3.3   Symbolic Convergence Theory and Shared Interpretations

Bormann developed symbolic convergence theory to explain how groups create a common consciousness (i.e., shared emotions, motives, and meanings) that can help them form into a coherent unit (Poole 1999). This theory explains one way communicators might create a vision of and motivation around sustainability initiatives that involve issues related to ethics or an ideal future. Symbolic convergence begins with the creation and sharing of group fantasies. A fantasy is any message that does not involve a group's present condition (e.g., an imagined future). These fantasies refer to characters and events in another time and place. If group members talk about and build on these dramatizations, they may come to share similar interpretations, emotions, and common experiences. This is called fantasy chaining. Fantasy chaining may include identifying positive or negative actions, heroes and villains, a common plot, and/or a reinterpretation of the group's past. Ultimately, fantasy themes are repeated enough to become a fantasy type (e.g., battle against good and evil) or create a rhetorical vision describing a group's place in the world and its vision for the future. For example, Interface, the world's largest carpet manufacturer, developed its' Mission Zero goal at the small group level. This idea came to symbolize change and ultimately help motivate and guide the decision making of multiple Interface groups as they sought to create a more ideal future. *Best Practice*: Group creativity may be stimulated by fantasy chains.

7.1.3.4   Limiting or Unleashing Creativity

Organizational factors that can limit group creativity include time pressures, risk-aversive organizational cultures, production blocking norms, a hierarchical network structure, and a collective information sampling bias. Time constraints limit group members from adequately expressing their ideas or searching for new alternatives. Production blocking occurs when idea generation is limited by the sequential flow of communication messages (i.e., when one person speaks, others are silenced). Network structure can be a problem if group members direct most, if not all, messages to their group leader. A collective information sampling bias occurs because decision-making groups prefer to talk about information everyone in the group already knows rather than information individual members uniquely know

(Bowman 2009). People enter groups with an already formed preference for a particular decision. This becomes a problem if group members do not reveal enough new or uniquely known information to counteract preexisting biases. This theory has implications for groups seeking to creatively address topics related to sustainability. It points to the need to use discussion mechanisms to solicit and share uniquely known information.

The communication process is essential to group creativity. Group members may find a new set of nonlinear communication and decision-making skills useful. SunWolf (2002) identified and defined 36 techniques for enhancing group creativity. Half of them overlap the techniques previously identified for generating alternatives (SunWolf and Seibold 1999). Those which do not include analogy storm, bug lists, crystal ball, ideal needles, ideals method, imaginary world, left–right brain alternation, lion's den, manipulative verbs, mess finding, mind mapping, object stimulation, organized random search, picture tour, problem reversal, progressive abstraction, wildest idea, and wishful thinking. *Best Practice*: I provide this list here to illustrate how groups have techniques at their disposal which may be useful in stimulating creativity. Unleashed creativity can spark profound changes (see Senge 2005). Researchers should investigate how groups seeking to deal with large environmental challenges stimulate creative decision making.

7.2 Interorganizational Collaboration (IOC) Efforts

> The global scientific community has made it clear that human activity is already changing the world's climate system. Accelerating climate change has caused serious impacts. Higher temperatures and extreme weather events are damaging food production, rising sea levels and more damaging storms are putting our coastal cities increasingly at risk and the impacts of climate change are already harming economies around the world, including those of the United States and China. These developments urgently require enhanced actions to tackle the challenge (The White House 2014).

Climate change presents the largest challenge our species has ever attempted to understand, influence, and/or adapt to. We are facing some wicked problems (Rittel 1973) which are hard to articulate, made up of unique but interlinked problems, and open to multiple causal explanations. Minimal opportunities exist for us to engage in trial and error. In terms of solutions, it is difficult to claim success (e.g., mitigation strategies lack a definitive scientific test). It is almost impossible for us to foresee the best course of action, predict the effectiveness of our chosen strategies or actions, or even understand the complexity of the problems as elements in our natural environment interact to create chaotic cycles. But act we must.

> Determining the meaning of sustainability is a process involving all kinds of stakeholders in many contexts, i.e. people who may not agree with one another. In dealing with conflicts about how to organize, consume and produce in responsible ways, learning does not take place in a vacuum but rather in rich social contexts with innumerable vantage points, interests, values, power positions, beliefs, existential needs, and inequities (Wals and Schwarzin 2012, p. 13).

The most powerful tool we have is our ability to collaborate as we attempt to problem solve, plan, implement, assess, and redesign in an ongoing process. Collaborative efforts have grown along with our knowledge of the challenges we face. IOC efforts cross governments, scientific disciplines, geographic boundaries, and communities and seek to build on the strengths of various stakeholders working together (i.e., businesses, governments, NGOs, communities) to plan and implement interventions and responses. For example, in 2014, the USA and the People's Republic of China jointly announced their commitment to strengthen bilateral cooperation on climate change and to work with other countries to adopt a document with legal force at the 2015 UN Climate Conference in Paris (The White House 2014). Both presidents announced post-2020 goals in order to limit the global temperature rise to 2 °C. The USA said it will reduce emissions by 26–28 % by 2025. China said it will increase the use of non-fossil fuels in primary energy consumption to around 20 % by 2030. Dare we hope our leaders actually enact these goals and these goals are sufficient?

In order to accomplish complex goals and in situations characterized by tight time constraints and/or cost concerns, organizations come together temporarily to find solutions (Stohl and Walker 2002). The hope is that through collaboration they can complete difficult, complex projects relatively quickly, pool and leverage their financial and material resources, and increase innovation due to the increased available strength, knowledge, and skills contributed by each partner. Several recent reports have focused on collaboration. The Network for Business Sustainability published a review of more than 275 relevant articles appearing in management and public policy academic journals between 2000 and 2012, plus materials drawn from books and recent practitioner-oriented reports (Gray and Stites 2013). Their report synthesized some of the growing literature addressing partnerships for sustainability. They provide links to their article summaries and descriptions of almost 150 past and current partnerships from across the globe. Each description identifies the issues addressed and some key outcomes or learning. Several years later, the MIT Sloan Management Review, The Boston Consulting Group, and the United Nations Global Compact surveyed nearly 3,800 managers and interviewed sustainability leaders from around the world to investigate the growing importance of corporate collaboration and boards of directors to sustainable business (Kiron et al. 2015). Practitioners and scholars will find both reports enlightening. Neither focus specifically on communication, although Gray and Stites do mention communication briefly as one of the important process-related factors they identify.

This section explores communication-related characteristics of IOCs. IOCs are distinct organizational forms made up of members representing multiple organizations concerned with addressing a focal problem none can solve alone (Koschmann 2012). I identify their key characteristics and relevant theories. Several interesting studies are summarized. Suggestions for how to improve IOC effectiveness are offered.

7.2.1   Common Characteristics of and Theories Explaining IOCs

IOCs share common characteristics. They have permeable boundaries, are interdependent with their environment, and are designed to be temporary. Membership fluctuates as different organizations or their representatives join or leave the IOC over time. Structures emerge but are fluid and temporary. Often, the hope is that the IOC can help change conditions within the broader institutional fields of which the organizational-level members are a part. But IOCs lack institutional power, many of the material artifacts (e.g., physical resources), structural constraints (e.g., chains of command), and legal formalities (e.g., employment contracts) that characterize other organizations and enable their taken-for-granted existence. IOCs depend on their members' willingness to work together voluntarily, share resources, and take action in the absence of formal authority or market incentives. Within an IOC, conflict is common due to value and goal differences. Individual members must resolve differences between what two groups (i.e., the IOC and their home organization) expect of them. Outside interests influence what IOC members say and do. Agreements emerge from advocacy and persuasion and rely on the fragile social infrastructure of interpersonal relationships, as well as the power and resources IOC members' import from their home organizations.

In order to proceed, shared understanding must occur. Munshi and Kurian (2015) offer a framework of sustainable citizenship that looks at the deliberative processes stakeholders use to identity and buy into shared values when power inequities exist. They argue that our traditional ideas of dialogue privilege the powerful. Sustainable citizenship is about working with the dialectical nature of complex issues. It involves efforts through deliberative democracy to create inclusive processes that highlight marginalized voices as central to decision making. Their study focused on a project to publically engage various New Zealand stakeholders around a discussion of controversial technologies (e.g., nanotechnology, synthetic biology, gene mapping, and assisted reproductive technologies). Stakeholders included technoscientists, entrepreneurs, venture capitalists, environmental and social activists, young adults, Maori groups, artists, trade unionists, journalists, science communicators, policy practitioners, and people representing nonprofit organizations. What resulted was a clash of ideas that resulted in the creation of shared values diverse publics could connect to. The process they used to identify and share differing values relied on Q-sort surveys. Participants ranked two sets of 41 statements involving their sustainability-related beliefs and values and their perspectives on new and emerging technologies. They sorted the statements, placing four statements each in the columns of extreme agreement and disagreement. The rest were spread across the columns for minor agreements and disagreements and a neutral column. The diverse perspectives of the various groups emerged and were shared. Originally, groups were uncomfortable with the positions of other groups. But the process enabled them to identify what each group *could* be comfortable with.

7.2.1.1   Theories Applied to IOC

There is no grand theory of IOC. It has been investigated from many theoretical perspectives including resource dependency theory, corporate social performance theory, institutional economics theory, strategic management theory, social ecology theory, microeconomics theory, institutional theory, negotiated order theory, and political theory (Walker and Stohl 2012). Gray and Stites (2013) identified and briefly described six additional theories: environmental justice, network theory, critical theory, actor–network theory, deliberative democracy, and dialogue. Other theories you will see mentioned in this section are sensemaking theory, organizational learning theory, and the communication model of organizational communication. Typically, collaboration is viewed as a nonemergent structure, and communication is seen as one of the many variables or simply a vehicle for information transfer. That is not the perspective of the theories and articles reviewed here. Communication constitutes IOCs and is necessary to their effective functioning. The bona fide group collaboration model (BFGCM) is one of the communication theories best suited to understanding the emergent structure of IOCs. It will be discussed next. Then, the focus shifts to the role of learning and language in IOCs.

7.2.2   Communication and Emergent Structure

Bona fide group theory was developed by Putnam and Stohl to explain naturally occurring temporary groups (Littlejohn and Foss 2005; Stohl 2009). Bona fide groups share three characteristics: permeable boundaries, interdependence among the individual group members, and links between boundaries and context. Permeable boundaries define individuals as being part of the group while also allowing them to move in and out of the group. Groups are interdependent with the physical and social environments within which they are embedded. A group's borders are unstable and ambiguous as groups change, redefine, and renegotiate boundaries to alter their identities and the context within which they are embedded. Differences exist between regular groups and a bona fide collaborating group in terms of "What is relevant? Where do loyalties reside? When does trust develop? Why do group members do what they do? How do collaborative groups acquire the resources needed to accomplish their goals?" (Stohl and Walker 2002, p. 244).

A natural extension of the bona fide group theory was to focus on communication within IOCs. Walker, Craig, and Stohl expanded the theory to create the bona fide group collaboration model (BFGCM) (Stohl and Walker 2002). An IOC is influenced by environmental exigencies, collaborative partners, relational boundaries, negotiated temporary systems (NTS), innovation outcomes, mutually accountable ends, and individual goals. An environmental exigency results in the formation of the IOC, influences its structure and processes, and directly shapes the results of the collaboration. At the model's heart is the NTS that is quickly negotiated both formally and informally by group members and *is* the collaborating group. The NTS is organized around a reciprocal interdependency with minimal initial structure or hierarchy. Within the NTS decision-making process, commitment, trust, power, and knowledge management are important to group success. Collaboration partners may vary in project involvement and commitment, but ultimately, mutually accountable ends are required of any outcome. *Best Practice*: BFGCM is a useful model for researchers when investigating collaboration among members of a supply chain, among multiple organizations within or across industry sectors, and among organizations of varying types. It is also useful in understanding IOCs focusing on eco-collaborations and sustainable ecosystem management.

Parts of the BFGCM were tested in two engineering IOCs utilizing a longitudinal approach to network analysis (i.e., SIENA). Walker and Stohl (2012) found the IOC to be volatile and nonhierarchical. Participants changed task and resource linkages frequently and quickly, resulting in continuous network restructuring. They communicated with whoever they needed to do their jobs. They were more concerned with tasks and less concerned with developing relationships. Who knew what was more important than who knew whom or who did what for whom. Unexpectedly, based on the theory, the researchers found that the actual number of communication links utilized remained fairly stable at a moderate level of density. They concluded that there may be an upper limit to the number of communication links individuals and groups can effectively manage. Moderate density levels allow participants to take advantage of the information richness of weak ties while protecting them from receiving too much information from too many people. The authors note that flexibility and the lack of procedural routines are important if IOCs are to work effectively.

7.2.3 *Learning Within an IOC*

Communities of practice can be designed to enhance learning among partners within a collaborating group. But power dynamics can hamper results. Best practices for facilitating learning in IOCs are offered.

7.2.3.1 Communities of Practice and Sustainability

Organizations interested in advancing sustainability must engage in deliberation, dialogue, and systemic learning (Reed et al. 2014). The communities of practice concept (Wenger 1999) emphasizes how shared learning, knowledge, and identity creation occur through social interaction. The authors (e.g., Attwater and Derry 2005; Benn et al. 2013; Reed et al. 2014) have discussed the utility of communities of practice as organizations work together to respond to complex issues involving sustainability-initiative creation and implementation. The theory rests on four premises: humans are social creatures, knowledge involves situated competence, knowledge requires active engagement, and learning produces meaning (Swieringa 2009).

Communities of practice create domain-based knowledge among groups of professionals—inside and across organizations. It is within communities of practice that group-level learnings about sustainability can occur. People build relationships where they help one another learn how to do something and clarify what is meaningful. Participating together over time, people develop a shared repertoire of resources (e.g., practices, tools). What they develop may appear stable but must be capable of modification as new concerns emerge. Most research looking at communities of practice has focused on small-scale case studies rather than network partnerships among organizations that "span spatial scales, governance responsibility, and scales of influence" (Reed et al. 2014, p. 230). For example, communities of practice helped promote more efficient urban water management practices through the Hawkesbury Water Recycling Scheme near Sydney, Australia (Attwater and Derry 2005). Coupled with the use of action research strategies, that community of practice developed a pragmatic approach for co-constructing and communicating more effective risk management strategies. Also, in their study of 20 Australian higher education institutions, Benn et al. (2013) explored how communities of practice influenced organizational learning around sustainability.

But can communities of practice help larger groups create alternatives to deal with systemically based environmental challenges? Reed et al. (2014) investigated if communities of practice created among organizations of different types at different locations were capable of facilitating learning and action. They designed a framework for instilling collective learning and action strategies across a 3-year multilevel, multi-partner network made up of 70 individuals representing community resident–practitioners from Canada's 16 UNESCO biosphere reserves, government scientists and policy practitioners, and academics located across five Canadian time zones. Biosphere reserves are geographic areas which function as living laboratories and sites of excellence to conserve biological and cultural diversity, advance sustainability, and support scientific research, learning, and public education. Reed et al.'s framework focused on helping IOC members use dialogue to learn through deliberation, networking, and experimentation. It consists of seven action steps beginning and ending with reflection and evaluation. Other steps include defining the project, participating in an action, sharing and communication, integrating and co-creating, and negotiating and deciding. These actions are supported by seven characteristics of collaborative environmental management: trust building, common interests and a shared vision, incentives, perceived value in sharing information, willingness to engage in collaborative learning and decision making, effective information flow, and effective leadership. They judged the project's success by asking participants to evaluate the IOC on these seven characteristics. Their study does an excellent job describing a participatory action research process which included an initial meeting where a visioning exercise was used, working groups formed to focus on one of the three agreed-upon themes (i.e., management and governance of sustainable tourism, land management and ecological goods and services, and education for sustainable development), and a bilingual facilitator. Working groups met for a year to complete their projects. The bilingual facilitator maintained contact with each working group to ensure their information/organizational needs and targets were met, facilitate their meetings, and help them plan activities and projects. *Best Practice*: If you are involved in creating a large community of practice, read this article.

How can members in an ecosystem management network learn to co-create a shared conceptual infrastructure for generative learning, consensus building, and collaborative decision making? An ecosystem management network is a strategic partnership or alliance among stakeholders who work together to improve sustainable resource management in a complex ecosystem. The stakeholders may include local private and institutional interests; local governments, planning commissions, or boards; school districts; county governments, planning boards, and quasi-governmental authorities; environmental nonprofit agencies; and NGOs and government entities at the state, interstate, and federal levels. Like many other collaborative groups, they are a loosely coupled, permeable, dynamic political system where communication, influence, and negotiation skills are important. Manring (2007) modeled ecosystem management networks as emerging learning organizations using two case studies as evidence: county land-use planning in Monroe, PA, and the management of the lower Roanoke River in North Carolina. She discussed the role of net brokers, multiple leaders, and web cultures. Net brokers manage the network and play a variety of roles including facilitator, coordinator, environmental scanner, and policy entrepreneur. They identify and connect all stakeholders with vested interests and complementary resources and help stakeholders create a common bond that promotes mutual trust. Each person or group has something unique to contribute making such networks leaderful rather than leaderless. Having more than one leader creates resilience in the network. Web cultures form among stakeholder organizations which allow for generative or double-loop learning to emerge. Stakeholders create new ways of looking at ecosystem resource issues and identifying solutions that transcend individual stakeholder boundaries and views. *Best Practice*: Generative learning and consensus building are critical to the success of any IOC.

7.2.3.2 Power Dynamics Within IOCs

Communication scholars have a long history of studying influence processes within groups (Meyers and Brashers 1999). It would be naïve not to note that power dynamics play important roles in IOCs. Those with greater formal authority (e.g., government agencies), who control scarce or critical resources (e.g., capital, expertise), and who have discursive legitimacy (i.e., the perceived right to speak legitimately for issues or other organizations) are more likely to assume initial leadership roles (Manring 2007). Having more power, their rules and resources often influence the early IOC structure, goals, and processes. For example, government agencies regulating water quality initially influenced the Roanoke IOC to incorporate their existing water quality rules, practices, and resources into the IOC's decision making. The power dimension influences, but does not determine, decisions surrounding problem definition, membership, and collaborative practices. Persuasion and coalition formation also effect problem definition and solution generation especially when the problem being faced has not been addressed in the institutional fields of a collaboration's dominant members.

Changes recommended by an IOC may require that members influence their own organization and their institutional field. An IOC member organization's ability to do so depends on their motivation, capability, and strategic position in their institutional field. Organizations in low-status positions are more likely to want to change the rules of the field. But through collaboration, high-status organizations may be influenced by low-status collaborators. For example, an environmental activist group might influence the development of national environmental policy through its collaborative efforts with major corporations and key government agencies. Also, coercive isomorphism may force members in an institutional field facing similar environmental circumstances to adopt shared structures and processes regardless of the recommendation of an IOC.

7.2.3.3   Learning-Related Best Practices for IOCs

Knowledge sharing and generation tools in the form of selected boundary objects (e.g., shared reporting tools) help promote the development of communities of practice across institutions (Benn et al. 2013). These tools reinforce the integration and institutionalization processes. Focusing specifically on designing strong communities of practice, other good suggestions exist. When developing a community of practice, ask individuals to recommend others who should participate so as to pull together a diverse group. Make sure the value gained by participating is immediately apparent to each member. Clarify who is participating, what each person brings to the table, and what roles each person plays. Design ways to allow for various levels of contribution: from reading, to engaging, to creating content. Allow communities to evolve by letting them define themselves and manage their own structure and growth. Promote regular meetings but change up the format or location (e.g., face to face, monthly webinars, speaker series). Provide new participants with onboarding experiences. Create a workable balance of new and expert members. Make sure the group has and measures their key performance indicators (KPI). Design the community so it doesn't need any single individual to operate effectively. Multiple resources exist to help groups develop communities of practice.

Participants in any IOC must be willing and able to build trust and respect; establish common goals, shared norms, common interests, a unifying goal, and a transcendent vision; demonstrate effective information flow; participate in collaborative decision making; generate value through information sharing; be willing to engage; provide leadership and facilitation; and co-create knowledge. Multiple authors discuss the role of a strong facilitator (e.g., Manring 2007; Reed et al. 2014; Wals and Schwarzin 2012). In light of these suggestions, IOCs can use a series of diagnostic questions to evaluate themselves (Manring 2007, p. 342):

> Does the network have a unifying purpose based on the value and goal of consensus building through collaboration? What is the nature of the voluntary links and relationships between independent yet interdependent network members? Are individual members of the network honing their own skills of clarifying and deepening their personal visions? Does the network have in place communication pathways that facilitate shared learning? Is the network building a shared vision for sustainable management of the ecosystem? Are network members increasing their abilities to scrutinize their own mental models and internal pictures, to make them open to the influence of others, and to discard old ways of thinking and problem solving? Is the quality of stakeholders' generative learning and capacity to think together improving? Is a spiral of trust evolving within the network, as evidenced by members acknowledging the legitimacy of each other's goals and committing to the collaborative partnership? Does the network show evidence of systems-thinking capability to integrate multiple perspectives and fuse them into a coherent body of knowledge that transcends stakeholders' original point of view and boundaries? Are there effective multilevel leaders and net brokers in place to manage the various leadership needs of the network? Are members of the network aware of their webbed culture and able to articulate the values and shared beliefs of their interorganizational learning network? Is the network demonstrating consensus building and collaborative decision making?

7.2.4   The Role of Language and Texts Within IOCs

In this section, I discuss two sustainability-related studies which focus on language and text. The first, based on a communicative model of organizational constitution, looks at how language and texts influence the formation of an IOC's collective identity. The second, based on performativity theory, investigates how language and storytelling can enhance creative problem solving and create changes in how sustainability is viewed in the broader community an IOC represents. The section concludes with a discussion of the role strategic ambiguity can play in an IOC's early development.

7.2.4.1   The Creation of a Collective IOC Identity

Since widely different groups often work together in an IOC, it is important that the group develop a collective identity. This concept refers to a shared understanding of the "we-ness" of a group. The collective identity of an IOC is a discursive resource shaped and reshaped through communication and drawn on for strategic ends. Having a collective identity can motivate and guide internal activity as well as create external legitimacy, provide social capital, and serve as a rationale for group action. A group's collective identity may appear stable, but it is a function of sustained interaction patterns, not an inherent property of the IOC that exists apart from its current membership and organizing practices.

An IOC's collective identity can be theorized as an authoritative text created through the communication processes of co-orientation, abstraction, and reification (Koschmann 2012). An authoritative text is an abstract textual representation of a collective that portrays its structure and direction, shows how activities are to be coordinated, and indicates authority relations. Co-orientation occurs as people use conversations to experience, accomplish, and align their actions in relation to common objectives. Conversations sometimes become texts (e.g., written standards, guiding metaphors). Textual representations gain power through abstraction and reification and go on to guide the voluntary actions of diverse stakeholders. In order for localized interactions to have an impact, they need to be extended in a process called distanciation. One of the most developed constitutive models of organizational communication comes from the Montreal School of James Taylor, François Cooren, and colleagues at the Universitè de Montrèal (Cooren et al. 2011). Their theorizing is particularly interested in the emergence of distinct organizational forms (e.g., IOCs) that transcend and eclipse their individual members.

Koschmann (2012) traced the emergence of a new collective identity for one IOC. A new director took over the guidance of an IOC which was having difficulty identifying and articulating its purpose. At one meeting, the new director described the IOC as operating like a dashboard for the community. That simple comment became a metaphor which extended through time and space to guide subsequent interactions. The director's dashboard comment began to circulate in meetings and written documents. The comment was used by IOC members to describe "who we are" as a group. The dashboard metaphor became a concise way to speak about the IOC and helped reinforce its existence. It guided the IOC's discussion of who should win their community award. What the IOC stood for was discussed in light of the qualities its members wanted to see displayed by their award recipients. The dashboard-inspired requirements were written in the award nomination handout which, in turn, shaped subsequent conversations and behaviors. *Best Practice*: The article focuses researchers and practitioners on the importance of conversations regarding who we are and what we do which then go on to become texts capable of guiding subsequent interactions.

7.2.4.2 Using Language to Create New Possibilities

Performativity theory provides another framework for understanding the role organizational discourse plays in creating new possibilities within collaborating groups. Butler developed the theory which has been used by researchers in economics, economic geography, and sociology. Its value to management and communication rests in its ability to illuminate the micro processes by which subjects (individuals or organizations) reconstitute themselves and the way social change occurs (Livesey et al. 2009). Performativity theory provides a framework for looking at the link between individual- or organizational-level transformations and societal-level changes. Language and storytelling play central roles, as do repetition, naming, and recounting as social understandings are brought into being, (re) enacted, and institutionalized. Through discourse, values, beliefs, practices, policies, and laws of societies, organizations, and identities are constituted and performed and become seen as real.

Sustainability is a contested space of political action and ethical choice. It is a meaning always in the making, the product of social negotiation and ethnical decision making shaped by factors tied to particular places and times. Sustainable development (like other abstract concepts) becomes real for businesses and for society through local enactment. Livesey et al. (2009) applied performativity theory to the Ricelands Habitat Partnership (RHP), an eco-collaboration between the rice industry and environmental advocates in California's Sacramento Valley. By the mid-1980s, duck and geese populations in North America were alarmingly low due to habitat loss in Canada and California. The researchers used the theory to explain the transformation of the rice farmers and their industry toward more sustainable agriculture in the face of ambiguities and contradictions due to changing social discourses.

The RHP appeared to be an odd alliance, but actually, its members shared some similarities. RHP members were farmers, ornithologists, nature enthusiasts, hunters, university scientists, and government agency employees. Many members shared conservation-based, rather than preservation-based, values. Many were duck hunters or came from duck hunting families. Common values were reinforced by organizational overlaps, social networks, and personal familiarity. But still joining together in a collaborative effort was challenging due to the very different missions, goals, ideals, and political positions of their employing organizations and by distinct individual perspectives. Points of tension emerged around participants' core identities, their views regarding appropriate knowledge practices and decision strategies, and the decision whether or not to communicate publically about the collaboration. RHP interactions were characterized by conflict and cooperation, resistance and acceptance, and debate and dialogue. Each side had to overcome skepticism about the other and become open to finding common ground. Conflict and difference became the ground for discovery. The RHP members reflected critically on their own relationships with nature. Alternative visions began to emerge. Participants formed new understandings about how to be good farmers, how to protect the Sacramento Valley, and what an expanded sense of community and common good might mean. They created a new vision for how the health of business and the natural environment might coexist. They developed innovations in beliefs and practices of farming and environmental protection. The last step of Livesey et al.'s (2009) analysis emphasized the performative effects of storying the RHP. The stories the RHP generated helped institutionalize eco-collaboration as a viable form of sustainable agricultural development and created new replicable sustainability practices (i.e., winter flooding of rice fields allowing for waterfowl habitat, dam deconstruction). By promulgating the RHP story, media, industry, NGO, and academic accounts showed how a new era of collaborative conservation and sustainable farming might occur. *Best Practice*: This article illustrates how dialogue within an IOC and stories about what they accomplished transformed stakeholder behaviors, lead them to enact more sustainable practices, and ultimately created new societal alternatives.

7.2.4.3 Strategic Ambiguity Within IOCs

The concept of strategic ambiguity is important to emerging IOC coalitions, especially in the early stages of relationship formation (Wexler 2009). Many of the IOCs described thus far had members representing different discourse communities. Wexler discusses three discourse communities in terms of those who focus on profits, people, or planet. Representatives of each discourse community attempt to influence the others during problem conceptualization and solution generation. *Best Practice*: An IOC will have greater success initially when potential goal-oriented actions are discussed in ways open to multiple interpretations and sufficiently malleable to facilitate dialogue. The TBL's strategic ambiguity provides each discourse community a way to show others how and why staying in the IOC is strategically useful. Wexler identified seven uses for such strategic ambiguity (p. 65):

1. SA is used when disseminating texts or information that is tentative (early plans) or require interpretation (mission statements). It enhances latitude by suggesting the issues are open for discussion and invites dialogue.
2. SA is used to buy time when pressed for a decision or specific information. The decision is that the best answer is not to be specific at this time.
3. SA is used when addressing issues that are controversial. It allows people to focus on general and abstract concepts where agreement is possible, not on specific points where disagreement often exists.
4. SA is used to foster organizational change by requiring buy-in from stakeholders. It works well when discussing ideas about the future because it allows for multiple meanings and remains open to revision.
5. SA is used to seal or buffer parts of an organization from closer scrutiny. It decreases a group's need to be closely accountable for its earlier ambiguous positions.
6. SA is used by organizations seeking deniability rooted in earlier ambiguous communication. It can be used to obfuscate and protect those with privilege in the corporation from close scrutiny.
7. SA is used in communities of practice with escalating demands for action in the midst of organizational crises or scandal. It helps diffuse responsibility.

7.2.4.4 The External Communication of IOC Alliances

Although some members of an IOC may hesitate to speak publically about the collaborative effort and their role in it (Livesey et al. 2009), corporations and NGOs are forming and communicating about their strategic alliances at unprecedented levels. Cross-sector alliances can provide a way to solve complex environmental challenges, improve economic production, address issues disruptive to the business environment, respond to stakeholder pressures, and address the TBL (Shumate and

O'Connor 2010a). Corporations and NGOs differ in how much the public trusts them about issues related to the environment, human rights, and health. NGOs have greater public trust than do governments, the media, or corporations and are often asked to provide a social safety net when environmental and/or social conditions deteriorate.

Several theories have been applied to these NGO–corporate alliances, but none really focused on communication's role in the creation, valuation, and existence of an alliance until Shumate and O'Connor (2010b) introduced their macro-level symbiotic sustainability model (SSM). Community ecology theory suggests that relationships between organizational populations, as in the case of NGO–corporate alliances, are different than relationships between organizations in the same population (i.e., alliances among NGOs in an issue industry and alliances among corporations in an economic industry). "Participants in such alliances were likely to have different performance measures, competitive dynamics, organization cultures, decision-making styles, personnel competencies, professional languages, incentives and motivational structures, and emotional content" (Austin 2000, p. 14). At least three other management theories guided previous research on NGO–corporate alliances: transaction cost economics, stakeholder management theory, and collaboration theory. Shumate and O'Connor (2010b) describe each theory and review relevant research emerging from that theoretical perspective.

From the SSM perceptive, NGO–corporate alliances are an effort to use communication to influence institutional positioning, i.e., how an organization's identity is established and recognized as having a certain place in the larger social system (Shumate and O'Connor 2010b). NGO–corporate alliances are symbiotic (i.e., characterized by mutual dependence). Generally, corporations and NGOs do not have overlapping resource or identity niches. The communication of an alliance with organizations in another identity niche allows both organizations to share a desirable identity with stakeholders whom they might not reach otherwise. Through cross-sector alliances, organizational actors use communication to co-construct a relationship in hopes of mobilizing and/or creating economic, social, cultural, and political capital. Both partners spend resources (e.g., money, staff time) to communicate their cross-sector alliances to stakeholders through interactive websites, advertising, public relations, and special events. Such communication allows alliance partners to enter the public dialogue, offer legitimacy claims, and create positive associations that can influence their institutional operating environments.

Shumate and O'Connor (2010b) offer six propositions concerning the role of communication, capital mobilization resulting from NGO–corporate alliances, NGOs and corporations' choice(s) of alliance partner(s), the number of partners with whom organizations are likely to communicate, and potential risks and rewards of the alliance. They illustrate their propositions by discussing the Rainforest Alliance and Chiquita Better Banana program. The lack of existing relationships with other organizations within that industry made both organizations attractive alliance partners. Both organizations had sufficient capital and a limited number of other NGO–corporate alliances. Both received criticism from their own stakeholders. However, both reaped economic, social, political, and cultural capital from stakeholders in response to the alliance. They (Shumate and O'Connor 2010a) tested several of their propositions by looking at the websites of 155 US Fortune 500 corporations representing 11 economic industries. They examined (a) the number of NGOs with which corporations communicated alliances, (b) the patterns of the communicated alliances between corporations in economic industries and NGOs in social issue industries, and (c) the relationship between corporate stakeholders and the communication of alliances with NGOs surrounding particular social issues. The SSM proposes that when corporations form enduring alliances with a few NGOs, the communication of such alliances sends messages about the legitimacy of the corporation and the corporation's character. Alliances with many NGOs or alliances with multiple NGOs within an issue industry diminish the communicative value of such alliances. In order to mobilize stakeholder capital, each NGO–corporate alliance needs a unique message to communicate to stakeholders. When corporations have multiple alliances with NGOs, communication becomes less memorable. They found corporations limited the communication of alliances they reported on to fewer than five NGOs. Most only communicated about one alliance with an NGO in a single issue industry. The SSM proposes that stakeholder expectations lead corporations in the same economic industry to engage around similar social issues. They found corporations in the same economic industry tended to report alliances with different NGOs in the same issue industry. Even if multiple NGOs receive corporate support, only a few are communicatively linked to the corporation. Only a small set of social issues are included in NGO–corporate representational communication, leaving many issues marginalized. Some issue industries were preferred across economic industries. The most popularly cited issue industry was generalist foundations and funds (e.g., the United Way, Cherish the Children Foundation). The number and type of NGO alliances that corporations in various economic industries reported did not differ by industry but were due to other differences between industries (e.g., potential for public disapproval). This study directs us to think about the strategic reasons and implications for how cross-sector alliances and other forms of IOCs communicate their relationships to the broader public and to other stakeholders. *Best Practice*: As we seek to shift societal-level debates so as to promote wider acceptance of sustainability efforts, the public communication of joint ventures is important. As we saw in Livesey et al.'s (2009) discussion of performativity theory, websites tell stories which can shape the perceived availability of new options as we face environmental challenges.

7.2.5 Additional Best Practices for IOCs

It is important to understand what the scholarly literature is saying about collaborations generally. The Gray and Stites (2013) review defines and categorizes the various types of partnerships for sustainability and the various sectors involved (i.e., business, NGOs, government, and community). They offer a model of four factors that influence partnership outcomes: external drivers, partner motivation, partner and partnership characteristics, and process issues. The key process issues they identified as important for achieving optimal partnership outcomes are the willingness to explore differences, create a shared vision, agree on norms, build trust, manage conflict, reach consensus, devise accountability criteria, share power and ensure voice, and cultivate effective leadership. They offered recommendations for all partners and specific recommendations for businesses, NGOs, governments, communities, and indigenous peoples. Their general recommendations include as follows: adopt a problem-centric model of stakeholders, frame the partnership as a learning process, construct fair processes and manage conflicts, don't expect to reach quick solutions, ensure voice for all participants, set evaluation criteria, allow time for representatives' constituencies to review and ratify agreements, and develop leaders competent in partnership skills.

The more sustainability-related collaborations an organization engages in, the more successful its collaborations are. Organizations learn how to collaborate through collaborating. Extensive knowledge sharing, both formal and informal (e.g., immersive learning experiences in key locales), strengthens collaborations. Based on survey data gathered from almost 3,800 managers from around the world, Kiron et al. (2015) offer six recommendations on how to improve collaborations: build internal collaborations, develop a shared language among members of the collaboration who speak different sustainability dialects (e.g., business language vs. NGO language), engage in due diligence, plan entrance and exit strategies, remember people matter, and develop board engagement. They quote Wood Turner, former Vice President of Sustainability Innovation at Stonyfield, as saying, "Internal collaborations can be very successful in keeping people excited and aligned with big picture sustainability goals. Internal collaborations also create bridges inside the organization" (p. 12). Due diligence involves beginning with a structured discussion among partners representing different interests in terms of "Is it a good opportunity?" "What is the best solution from a purely business or NGO perspective?" Controversial issues such as the ability and willingness to commit to and monitor certain standards and the potential for reputational risk need to be discussed. As we learned from the BFGCM, collaboration partners may enter and exit discussions throughout the process. Kiron et al. recommend letting partners focus on the parts of the process for which they are best suited and discussing timing issues during initial meetings. They quote Shelly Esque, Vice President of Legal and Corporate Affairs at Intel and Chair of the Board of the Intel Foundation, as saying, "It is important to focus up front on the ending. It avoids the dilemma of walking away feeling the job is not done, or feeling that we left too early or stayed too long" (p. 13). Collaboration is about relationships and trust. Therefore, having the right partners, both in terms of individuals and organizational actors, is important. Board engagement around sustainability also is critical, but relatively rare (see Sect. 5.2.2.4). Understanding emergent structure, the importance of learning opportunities, the role of language and texts, and the best practices offered by Gray and Stites (2013) and Kiron et al. (2015) can help organizations develop more successful collaborations.

7.3   Supply Chains

Several years ago, Sam's Club periodically hosted in-house lectures at their Bentonville, AR, headquarters where representatives of their vendor companies spoke about their sustainability initiatives. I attended when Michael Kobori, Vice President of Sustainability at Levi Strauss and Company, spoke. In 1991, Levi Strauss became the first in their industry to establish a code of conduct for their direct contractors. Over the years, the company created sustainability initiatives related to water quality standards, developed a restricted substance list for certain chemicals, and made 100 % organic cotton jeans. Their 2007 environmental product life-cycle assessment indicated most of their environmental impact involved their raw materials (i.e., cotton), water used to make jeans, and how consumers cared for their jeans. Cotton is the third worst chemically intensive crop and over 30 million farmers grow it. The company partnered to start the Better Cotton Initiative to help farmers reduce the amount of water and chemicals they need to grow cotton. More than half of the water used associated with the company's products stems from their supply chain. In 2011, they began using techniques to reduce the water used in garment finishing by up to 96 %. In 2013, they developed the first standard for water recycling and reuse in the apparel industry. Levi Strauss is influencing its' supply chain. Almost half of the climate impact of a pair of Levi's® 501® jeans comes from consumers (http://www. levistrauss.com/sustainability). The average US consumer throws away 68 pounds of clothing and textiles annually. As a result, 23.8 billion pounds of clothing end up in landfills. In 2013, the company started reaching out to consumers through education, marketing, and labeling efforts. On their website, it says, "We're committed to do our part to reduce energy and material use throughout our operations and supply chain and to continue to engage consumers on how they can also reduce their environmental footprint." When I heard Kobori speak, he talked about the importance of the Levi–Walmart relationship, reaching out to other brands and their supply chain members and working with NGOs. He said, "It's not just us. We're all going to need to collaborate on these things." Increasingly, organizations (both public and private) are working with their supply chain members to enact their sustainability goals and initiatives. This section discusses sustainable supply chains, their growing importance, the major theories appearing in the scholarly literature, the challenges and barriers, and communication.

7.3.1   *What Is a Sustainable Supply Chain?*

A supply chain includes the different companies involved in producing, handling, and/or distributing a specific product. As organizations become more sustainable, their attention extends to the sustainability-related practices and concerns of organizations upstream (e.g., suppliers, producers) and downstream (e.g., distributors, customers) in their supply chain. Over time, sustainable supply chain management (SSCM) evolved to be a strategic process which involves:

> the management of material, information and capital flows as well as cooperation among companies along the supply chain while taking goals from all three dimensions of sustainable development, i.e., economic, environmental and social, into account which are derived from customer and stakeholder requirements (Seuring and Muelle 2008, p. 1670, as cited in Gimenez and Sierra 2013).

SSCM issues influence decisions about suppliers and products, deciding how to dispose of waste and pollutants (Cantor et al. 2012), procurement, and logistics. SSCM initiatives can involve designing and implementing frameworks, methods, and tools that suggest, recommend, or prescribe certain modes of action. They may be process or product oriented, produced internally (e.g., an internal environmental management system) or externally (e.g., a code of conduct for the supply chain). SSCM initiatives occur at the intraorganizational level (e.g., waste reduction, energy consumption) and the interorganizational level (e.g., green purchasing). They may involve specific supply chain activities (e.g., green reverse logistics) or the entire supply chain. Attempts have been made to generalize the components of SSCM across industries, and sustainability-based policies and practices in specific industries have been analyzed (Vurro et al. 2009).

Several organizations you read about (e.g., Levi Strauss, the Portland Trail Blazers) have sought to holistically identify their environmental impact. Since the 2002 World Summit on Sustainable Development, there have been calls globally for wider adoption of an integrated life-cycle approach. Life-cycle management (LCM) was developed to create partnerships and procedures to holistically minimize impacts. It is a "product management system aiming to minimize environmental and socioeconomic burdens...during the entire life-cycle...[relying on]... collaboration and communication with all stakeholders in the value chain" (United Nations Environment Programme 2009). LCM takes into account upstream and downstream impacts. Upstream activities include resource extraction and product production. Downstream activities include product distribution, use, and disposal. Two widely known examples of LCM partnerships for sustainable resource use are the Forest Stewardship Council (FSC) and the Marine Stewardship Council (Balkau and Sonnemann 2009). Both bind various partners (e.g., suppliers such as timber farms and fisher people, distributors, retailers) along the supply chain under common objectives and procedures. Procedures include audits, product specification, and certifications. Successful LCM partnerships include a shared belief in the need for action on sustainability issues; a neutral forum where partners can initially gather; a formal process (e.g., contractual arrangements, membership on joint initiatives) linking the stakeholders; a mechanism members can use to agree on issues, priorities, and processes (e.g., a board or steering group, a code); the ability to gather and interpret technical data and develop plans, protocols, contracts, or regulations; an implementation, auditing, and monitoring process; and a communication system to keep stakeholders informed.

I asked all my interviewees to discuss their supply chain relationships. Not all had focused on making them more sustainable. Susan Anderson, former Administrative Coordinator for the Missoula, MT, Sustainable Business Council, told me that in their Strive Toward Sustainability certification program, they asked their business members, "Do you actually talk to your supply chain? Because we really think one of the key ways that we can get the supply chains to change is to make certain people know what you are looking for." Also, along my travels, I learned that the State of South Dakota and the City of Denver have green purchasing policies. As societies' largest consumer of goods and services, governments need to incorporate life-cycle performance criteria into their public purchasing. Doing so shows suppliers that LCM has commercial advantages and can improve communities' long-term performance in line with environmental and social public policy priorities. Sustainable public procurement can be improved by consolidating information about which suppliers offer sustainable products and processes, improving procurement manager training, providing guidance on how to meet both financial and sustainability targets, and setting mandatory targets (Balkau and Sonnemann 2009; Preuss and Walker 2011). Public sector organizations need to share best practices and collaborate, as well as incorporate sustainable procurement criteria in outsourcing contracts.

7.3.2 How and Why Does the Idea of a SSC Spread?

An overall descriptive model for SSCM diffusion at the intraorganizational and interorganizational levels is lacking. But the diffusion of innovation theory (see Sect. 5.2) has been influential. Carbone et al. (2012) draw on theories of the diffusion of business ideas and practices and on the SSCM literature to outline diffusion mechanisms for SSC decisions and actions. Mechanisms supporting the diffusion of SSC include nonmarket stakeholder pressures, the internal company perspective, and stakeholder pressures at the interorganizational level. Nonmarket stakeholder pressures come from regulatory bodies, NGOs, media, the civil society, and the environment. Isomorphic pressures within an institutional environment are important. For example, if respected organizations within an industry discuss using SSCM, other organizations in that industry are likely to adopt such practices. However, major differences exist among industries and countries due to varying regulations, uncertainty in their institutional environments, and varying coercive, normative, and mimetic pressures. Internal company influences include top management's vision, customer or consumer pressures, and the type of innovation-driven or process-driven SSCM initiatives (e.g., TQM, lean organizations). An organization's adoption of a SSCM initiative is influenced by their position in a supply chain, their proximity to the sustainable initiative's trigger, and their bargaining power relative to customers and suppliers. Adoption can be reactive or proactive (Wolf 2014). First, I will discuss reactive responses.

7.3.2.1   SSCM as a Reactive Response

Researchers have pursued three main lines of inquiry (i.e., motivation driven, performance driven, stakeholder driven) to show how each influences SSCM diffusion (Carbone et al. 2012). SSCM adoption decisions may be motivated by resource-, transaction cost-, or capability-related concerns (Vurro et al. 2009). The motivation-driven approach is the oldest and provides extrinsic reasons including in response to regulations, to preempt legal sanctions, and in anticipation of structural changes in sourcing regulations. In terms of performance, researchers investigated the impact of SSCM initiatives on financial and sustainability-related performance. Wolf (2014) designed a resource dependence theory-based study utilizing a dataset of 1,621 organizations in 32 countries. In terms of SSCM, she looked at supply chain standards, supply chain monitoring systems, and green procurement. She found stakeholder pressure and SSCM both contribute to an organization's sustainability performance—however she concluded that the benefits to supply chain managers seen from SSCM go beyond risk reduction. Organizations build a reputation as being a good citizen by promoting environmental and social sustainability in their supply chains. This reputation improves legitimacy and access to key resources. Finally, the stakeholder-driven research views SSCM as occurring in response to stakeholder demands (e.g., customers, suppliers, competitors, or companies sharing the distribution network), NGOs, and regulatory agencies (Carbone et al. 2012). Auden Schendler, Vice President of Sustainability at Aspen Skiing Company, illustrated this when he talked about their supply chain. He mentioned a time when his organization used its position in a supply chain to pressure for pro-environmental change.

> If we have an opportunity to pull a supply chain lever that has disproportionate impact, we will and we have done that. One example is our work with Kimberly-Clark, we joined a boycott of Kimberly-Clark because of their forestry practices. We stopped using Kleenex.… Because Aspen is a pretty high-profile name, we got Kimberly-Clark's attention. Kimberly-Clark at the time was a $30 billion dollar company, bigger than half the economies in the world. And the result was that we ended up in conversations with the CEO. We were able to broker conversations with the Natural Resources and Defense Council and we were part of this broader boycott. So we had a role, it was probably a small role, in pushing Kimberly-Clark in a more sustainable direction. They actually moved a little bit and then moved a lot. So the change that we wanted to be part of with that organization was probably more high-impact than anything we have ever done.

Large for-profit organizations increasingly realize their brand competitiveness, and legitimacy may be influenced by their suppliers' environmental and worker rights practices. With the expansion of supply chains into developing nations, stakeholder criticisms of perceived social and environmental deficiencies occurring upstream in an organization's supply chain have increased (Wolf 2014). Stakeholders hold organizations accountable for actions and decisions regarding product design, sourcing, production, and distribution both in their home country and in their global supply chain. For example, in 2009, following a 3-year investigation of Brazil's cattle industry, Greenpeace released a report alleging Adidas, Clarks, Nike,

Reebok, and Timberland were sourcing leather from illegally deforested areas and that the Brazilian government was complicit (Vurro et al. 2009). The story dominated newspaper headlines worldwide. In response to public pressure, supply chain relationships shifted toward extensive stakeholder collaboration in corporate decision making. In collaboration with Greenpeace, Timberland called for environmentally and socially responsible leather sourcing policies in the Amazon and required its leather suppliers to not purchase cattle raised in newly deforested areas. Nike announced it would stop using leather from the Amazon until deforestation due to expanding cattle herds ceased. SSCM provides large organizations with a way to reduce reputation and brand risk while managing the consistency and reliability of materials integral to their operations.

7.3.2.2   SSCM as a Proactive Response

SSCM formation can be proactive (Wolf 2014) when organizations recognize that promoting social welfare and environmental protection can enhance their economic performance (e.g., serve as a catalyst of interfirm resources and competitive advantage) while ensuring suppliers are able to deliver products over the long run. For example, Walmart decided to make its supply chain for fish products sustainable, partly so they could stabilize their future supply. In the 1990s, Walmart faced fish shortages and realized overfishing, degradation of oceanic wildlife, and pollution were all increasing. Today, they only buy fish from Marine Stewardship Council-certified suppliers. Also, SSCM is a way organizations can enact their core values (Vurro et al. 2009), although intrinsic reasons for implementing SSCM (e.g., ethics, philosophical motivations) rarely appear in the SSCM research (Carbone et al. 2012). Reasons which do appear include reduced costs, increased operational efficiencies (Morali and Searcy 2013), improved reputation, and competitive advantage (Carbone et al. 2012). Differentiation strategy and awareness building also are important issues. Cruz and Boehe (2008) discussed these issues in relationship to JOBEK, an organization serving approximately 80 % of the European market for sustainable hammocks, hanging air chairs, and hammock stands. Located in Brazil, JOBEK implemented internationally certified sustainable business practices and utilized FSC-certified wood. Differentiation strategies are an integral part of global sustainable supply chains. Differentiation occurs as an organization argues its products have unique characteristics. Sustainability only works as a differentiation strategy if consumers recognize and value it as a unique characteristic. Awareness building involves activities (e.g., training, advertising, persuasion) that promote the creation of shared visions or standards along a SSC. Global SSCs rely on awareness building regarding sustainability standards all along the chain. Certification agencies can help create a more symmetric, network-type relationship between global buyers (distributors and retailers) and producers from developing countries. Coherence in practices along the supply chain is a prerequisite for the credibility of certification (e.g., FSC's chain-of-custody concept). Consumers must recognize and trust the certification.

7.3.3 Major Theories Applied to SSCM

SSCM research lacks a theoretical foundation and efforts to introduce theoretical frameworks are just beginning (Morali and Searcy 2013). Multiple organizational theories are referred to in the SSCM literature (Sarkis et al. 2010) including complexity theory, ecological modernization theory, information theory, institutional theory, the resource-based view (RBV), resource dependence theory (RDT), social network theory, stakeholder theory, and transaction cost economics (TCT). Morali and Searcy (2013) include a brief description of contingency theory, institutional theory, the RBV, RDT, and stakeholder theory and discuss each theory's implications for supply chain relationships. In terms of the articles reviewed in this section, Gimenez and Sierra (2013) referred to TCT and RBV and Wolf (2014) based her study on RDT. Readers interested in understanding how these three theories apply to SSCs will find these two articles informative. Institutional theory, stakeholder theory, ecological modernization theory, and the resource dependency theory have been discussed elsewhere in this book. Morali and Searcy's (2013) summary of contingency theory, the resource-based view, and transactional cost economics is provided next followed by a theoretical model proposed by Vurro et al. (2009). In terms of communication, Allen et al. (2012) investigated supply chain relationships as discursive constructions.

Contingency theory argues that the optimal design and leadership style of an organization is influenced by internal and external constraints. An effective organization, its structures, subsystems, and strategy must fit the environment in which it operates. An organization fits its structure to its strategy to increase its bottom-line results. Size, structure, and strategy are related. For example, large corporations are more likely to adopt and implement corporate sustainability practices, publish sustainability reports, and utilize codes and certifications. Organizations operating in the same environment or institutional fields (e.g., petrochemical companies) face similar constraints which result in similar contingencies leading to similar organizational structures and processes (Morali and Searcy 2013). Different units within an organization appear to be influenced by varying SSCM reasons (Carbone et al. 2012). For example, production is influenced by process-driven reasons, marketing and sales by market-driven reasons, purchasing by purchasing-driven reasons, and research and development by innovation-driven reasons.

RBV focuses on how organizations seek valuable, rare, imperfectly imitable, and non-substitutable (VRIN) resources in order to achieve and maintain competitive advantage. The learning that occurs between buyers and suppliers when they work together to improve environmental performance is a VRIN (Gimenez and Sierra 2013), as is the ability to form collaborative sustainability-focused relationships. Collaborative relationships increase trust and understanding of what the buyer wants which contribute to improved supplier environmental and social performance. Improved supplier environmental and social performance influences the buyer's environmental and social performance. Assessment (e.g., monitoring) enables collaboration by identifying areas for supplier development (e.g., training).

With TCT, the focus is firmly on controlling the relationship. Organizations seek to limit the transaction costs they experience with their suppliers (Gimenez and Sierra 2013). Transaction costs include both the direct costs of managing relationships (e.g., monitoring supplier activities) and the potential opportunity costs resulting from poor governance decisions. The assumption is that without controls, organizational actors will engage in opportunism. Assuming some suppliers will act unethically or illegally, monitoring is used to reduce excessive opportunism. Companies use governance mechanisms (e.g., audits) to improve their suppliers' environmental and social performance and better manage their own reputation and performance.

Building on stakeholder network theory, social network theory, and the network determinants of supply chain configurations, Vurro et al. (2009) proposed a theoretical model discussing a supply chain network's density and the centrality of the focal organization. Network density, defined as the degree or completeness of the ties between network actors, influences corporate responsiveness because it influences the ease of communication and the efficiency of information flow among network members. Centrality facilitates information access and allows organizations to be information gatekeepers and liaisons among network members. In their model, they describe four governance mechanisms used in supply chains depending on network density and network centrality. These mechanisms are discussed in the next section.

7.3.4 The Governance of Supply Chain Behaviors

Governance involves "the relations through which key actors create, maintain, and potentially transform network activities" (Raynolds 2004, p. 728, as cited in Gimenez and Sierra 2013). Governance mechanisms are those practices firms use to manage relationships with their suppliers, especially upstream members, and to stimulate improved sustainability-related performance. Gimenez and Sierra describe various governance mechanisms and identify related research streams based on SC, TCT, social network theory, and global value chain analysis.

Different governance mechanisms used to increase supplier performance and capabilities include rewards, coercion, legal instruments (Carbone et al. 2012), supplier assessment (e.g., questionnaires, company visits), supplier incentives, supplier competition, and collaboration (e.g., providing training and support) (Gimenez and Sierra 2013). Assessment reduces the risk that suppliers will act illegally and/or unethically. Training allows a buyer to increase its supplier's environmental capabilities and influence the delivery of more environmentally friendly products or services that can influence the buyer organization's reputation and performance. Organizations engage in differing levels of supplier assessment and collaboration. Both approaches can have a positive and synergistic effect on the environmental performance of an organization's supply chain. Firms which assess and collaborate outperform other firms. In 2008, nearly all Global 250 companies had a supply chain code of conduct to guide their suppliers' ethical and socially responsible practices. Yet, only half of the Global 250 disclosed details of the processes and mechanisms they used to activate and monitor their codes and standards (Vurro et al. 2009).

Many organizations may want to measure performance on sustainability initiatives within their supply chain, but lack the necessary quantitative performance measures. Less than half of the companies Morali and Searcy (2013) contacted said they measured the success of their supply chain sustainability initiatives. There appears to be less emphasis on measuring supplier performance than on measuring a company's own success. Few organizations measure the effects throughout a product's life cycle or the social dimension of sustainability. But that is not the case many large organizations are facing. For example, Kevin Igli, Senior Vice President and Chief Environmental Health and Safety Officer at Tyson Foods, Inc., talked about their supply chain and measurement saying:

> Where and how a company sources its supplies and services says a lot about its commitment to sustainability—you are as good as the company you keep. As with all other aspects of business, you should fully understand why your company engages in, and what it expects to gain from, certain activities. Establishing the business case for evaluating the sustainable practices of supply partners is no exception. It's important to know up front why you are evaluating the sustainable practices of your supply partners and what you hope to achieve. We interact with a number of significant companies in our supply chain. These include large retailers, restaurant companies, food service providers and others. We have a lot of engagement with them. The engagement process involves surveys, benchmarking and discussion on how to get better together. With respect to surveys, we receive many. We try to encourage people to take a look at the information we provide on our website. Then we will engage in a conversation about further details. If your company is considering the use of a survey to evaluate the sustainability performance of your supply chain, keep the survey straight-forward, relevant and, most importantly, manageable. For example, it's both time-consuming and frustrating to respond to sustainability surveys containing 100 or more questions that, for the most part, are 'not applicable.' This is especially true when a supplier makes a sustainability report available. Tailor your survey so that it asks about the issues most important to your company and stakeholders.

"Best practices now clearly indicate how the global pursuit of improved social and environmental standards further down the often complex supply chain have transformed cooperation from being an option into an indispensable necessity" (Vurro et al. 2009, p. 608). So there is a growing trend to form collaborative sustainability-focused relationships that allow for joint learning. Organizations with strong SSCs build and maintain integrated supply chain management approaches using long-term cooperation, shared knowledge, and jointly developed competence with upstream and downstream supply chain members. Brian Sheehan, former Sustainability Manager at Sam's Club, talked about this joint learning saying:

> We will work with those buyer and supplier groups in different communication forums like a summit. We recently had a dairy summit and the purpose of that was to communicate broadly to the supplier community in these product categories that our Sustainability Index is real. It is important for a variety of reasons, but mostly because it can help improve the sustainability-related performance of products in those categories. And that moving

forward there is the expectation of suppliers completing a score card, but also partnering on
initiatives that will improve the sustainability performance and address the KPIs. And then
spelling out what those KPIs are that Walmart is going to go after in milk, yogurt and
cheese.

Section 3.2.3 mentioned how relevant KPIs are identified by the Sustainability
Consortium and discussed by the relevant Walmart Sustainable Value Network
(which is an IOC) before they are broadly communicated at these summits.

Collaboration opportunities differ for a variety of reasons including the
bargaining power (i.e., asymmetrical, symmetrical) supply chain members have
(Cruz and Boehe 2008), member willingness (Vurro et al. 2009), and where
organizations are located within their stakeholder networks. Increased centrality
allows organizations to influence their network and coordinate integrated
approaches along their supply chain. Central actors can gather knowledge from
across their network, gain influence and control, resist adaptation requests, shape
the interpretation of sustainability, and influence how sustainability issues are
implemented throughout the supply chain. Improved centrality attracts partners
and improves the opportunities for cooperation, knowledge sharing, and expertise.
Increased density means a focal organization is likely to receive increased attention
and pressure to comply with the expectations of both upstream and downstream
partners. Centrality and network density influence relational attitudes promoting
collaboration.

Mapping an industry's network structure will help identify which sustainable
supply chain governance (SSCG) model (i.e., transactional, dictatorial, acquiescent,
and participative) is most likely to be successful. Vurro et al. (2009) identify,
describe, and illustrate these four SSCG models. A transactional SSCG model
emerges in industry network conditions of low centrality and low density. The
focal organization lacks influence and the supply chain structure is dispersed.
Incentives supporting the meaningful integration of sustainability along the supply
chain are lacking, monitoring costs are high, and the sustainability-focused actions
of a supply chain member may be overlooked. Buyers select suppliers for short-
term commitments and limited information sharing. A dictatorial SSCG model
emerges where low supply chain density combines with concentrated power held
by centrally located firms. The focal organization can resist pressures to conform to
sustainability expectations and impose its own practices, norms, behaviors, and
sustainability standards both upstream and downstream. This style requires an
organization to use its power to enforce the practices it has imposed and to monitor
competitors. A competing orchestrator (e.g., an NGO, corporate social watchdog)
can mobilize opposition to overthrow the dictator. The acquiescent SSCG model
applies when an actor occupies a peripheral position in a densely connected supply
chain. Although network density facilitates the flow of information, the peripheral
organization must comply in order to remain in the network. Commitment to
sustainability is compliance based. If peripheral actors have the needed resources
and competences, they are likely to comply. If not, they may attempt to conceal
any questionable practices and/or be forced out of the network. Finally, the participative SSCG model is more likely in highly dense supply chains where

sustainability-oriented organizations are centrally located. The central organization establishes the basis for upstream and downstream sustainability initiatives but is influenced by its partners. If it is open to the involvement of its partners, joint activities and cooperation are likely. Those organizations often stress compromise and seek to balance, pacify, and bargain with influential stakeholders, build multistakeholder collaborations, and develop joint rules with their supply chain partners (e.g., certifications, environmental management schemes). Mutual understanding and a shared culture stimulate joint innovation processes and improve network flexibility. Vurro et al.'s model raises some provocative issues worthy of future investigation. Communication scholars, with a background in network analysis and/or bona fide group theory, could provide additional insight into message flow and content in these four SSCG types.

7.3.5   Challenges/Barriers Facing Sustainable Supply Chains

Organizations seeking to establish sustainable supply chains face challenges in integration and implementation (Morali and Searcy 2013; Vurro et al. 2009. Supply chain members may not understand what sustainability efforts entail or be unaware of how the three pillars of sustainability interact. Capital investment commitments are required. Risk management, supplier monitoring programs, and measurement processes are needed. Transparent information and knowledge sharing are important. Corporate strategies and corporate cultures need to be aligned. SSCM can be costly, require complex and difficult-to-achieve coordination efforts, and be hampered by insufficient communication (Carbone et al. 2012). Increasingly, supply chain researchers are focusing on how to promote employee participation in pro-environmental behaviors (Cantor et al. 2012). Overcoming employee resistance and promoting engagement present challenges. Individual-level challenges are related to cognition, information processing, information dissemination, reflective learning, motivation, perceptions, knowledge, skills, attitudes, limited control over work tasks, and values.

Procurement officers are among those tasked with interacting with sustainable supply chain members. In an effort to create a framework of the psychological barriers hampering sustainable development through public procurement, Preuss and Walker (2011) synthesized research from social, organizational, and environmental psychology, sustainable purchasing, and supply management to generate a midrange theory. They interviewed 72 procurement managers in local government organizations and the National Health Service in the UK to identify factors negatively influencing procurement officers' decisions. They found multiple barriers at the individual, small group, organizational, and external levels. Thorsten Geuer, brew master, at Bayern Brewing in Missoula, MT, shared an example which illustrates the role individual attitudes (e.g., of the sales person, of the owner) can have on supply chain relationships. Bayern Brewing has been the only German microbrewery in the Rockies for over 25 years (see Sect. 5.2.2.4). The company actively promotes recycling and utilizes recycled bottles and cardboard carriers as part of their sustainability efforts. Thorsten said:

> We have sales people come in here at least once a month, trying to sell us boxes for our beer, because that is what is done in the usual market. If you challenge them saying 'I need a box that can come back to us [i.e., is reuseable], that is refoldable and stable enough, it will be able to hold a six pack,' most of them are already checked out because there is no longer a box like that. [They think] I am a sales person—that is not my job. Then we have other people who will actively come in here and think 'how can we solve this'? We are lucky enough that we work with a packaging company that does a lot for the fishing industry in Alaska. They fill the boxes up in Alaska and ship them to the U.S. When empty, they fold flat and are shipped back to Alaska. The concept that we have is not new. It is just not out there in the brewing industry.... Our president found an old Miller box, because they did that same thing back in the day. It was a wax box with a foldable top which they collected and reused. So having the technology or idea that was out there before, we transferred that into something that will be usable for us. It was a combination of the Alaska example and the Miller beer box. We patented the design.

The organizational-level factors Preuss and Walker (2011) identified include issues of managerial control (e.g., policy decisions), organizational structure (e.g., organizational roles), and organizational norms, routines, and cultures. Role specialization, impersonal norms of behavior, and structural control devices can present barriers. Procurement priorities emphasizing cost savings and inventory availability can influence a procurement manager's motivation to address sustainable development challenges. Senior management priorities and performance measurement criteria de-emphasizing sustainable development present barriers. Other organizational barriers exist due to the strategic planning process, limited trust and cooperation, resistance, the autonomy needs of functions or units, resource availability, and the degree of organizational centralization. Finally, external barriers focus on boundary spanning, interpersonal trust, and cooperation. Readers will find that Cantor et al.'s (2012) model does a thorough job identifying barriers.

7.3.6   Communicating Within and About Supply Chains

Kevin Igli, Senior Vice President and Chief Environmental Health and Safety Officer at Tyson Foods, Inc., recommends that organizations:

> Take action on the information you receive from your supply partners. Ensure sustainability is part of your regular discussions with suppliers. Actively seek out forums to share sustainability advice, tips, ideas, and lessons learned with your suppliers. Lead by example and work with your supply partners to seize opportunities that generate joint success.

Limited communication research has focused on supply chains. One exception was the work by Allen et al. (2012). Consistent with recent research concerning the nature of organizations and organizational discourse, they conceived of organizations (and supply chains) as discursive constructions. Individuals who represent their organization in a sustainably focused supply chain relationship have an important role to play in that relationship's ongoing (re)construction. Allen et al. focused on two levels of sustainability discourse—organizational and individual. They analyzed the content of the corporate training materials used by two multinational corporations (MNC) linked in a supply chain relationship to see if common themes and language appeared. They also interviewed representatives of the two organizations. Three research questions guided their study: How do two large MNCs operating within a supply chain relationship define and/or discuss sustainability in their training materials? To what extent does the sustainability-related content appearing in corporate training materials overlap with how employees discuss sustainability? To what extent do sustainability-related comments made by representatives of these two organizations overlap? Although content differences emerged between the two training documents, some shared terms appeared including corporate mission, corporate sustainability, corporate performance, sustainable packaging, reduced waste, product price, healthy product, and sustainable product. Both organizations used training to emphasize the importance of sustainability to their employees. Some overlaps appeared between the training documents and the interviewee comments. Shared sustainability-related issues and values emerged within the interview comments of respondents representing the two organizations such as it is the "right thing to do," is a "good way to do business," and has the potential for making a "big impact." It is important to study overlaps and disconnects in talk and text because expressed differences (e.g., in goals, values, approaches) appearing in supply chain partner conversations surrounding joint sustainability-related initiatives do not bode well for their accomplishment.

Allen et al. (2012) discussed the situation facing communicators within a supply chain relationship as being similar to the four embedded narrative phases (manipulation, competence, performance, sanction) Cooren (2000) discussed. Regardless of organization, individuals were given a "having to do" (manipulation: embed sustainability into your conversation with your supply chain partner) and a "being able to do" (competence: you have been trained in what to say or what to look for). Each supply chain representative was expected to perform the "having to do" (performance: embed sustainability in your buyer–supplier conversations so as to reach sustainability goals) and was told their performance would be linked to either punishment or reward (sanction: those who successfully drive sustainability will be rewarded and those who do not will not succeed either in the company or within the buyer–supplier relationship). The outcomes of the competence–performance link can influence individual employees' success and the achievement of corporate supply chain goals and ultimately reinforce any global-level shift toward sustainability we are able to achieve through supply chains. It would be interesting to see additional research comparing how supply chain partners discuss sustainability.

7.3.6.1 Communicating About Supply Chains in Sustainability Reports

When attempting to understand how organizations are working with their supply chains to promote their environmental and social initiatives, one easily accessible place where the topic is discussed is in corporate reports (e.g., sustainability reports, CSR reports). In this section, I review several studies (e.g., Maharaj and Herremans 2008; Morali and Searcy 2013; Tate et al. 2010) that investigated the content of sustainability reports in order to better understand changes in supply chain expectations.

In 2002, Shell Canada, one of the first companies to publish a sustainable development report instead of a corporate social responsibility report, created a booklet, *Business Principles: Expectations for Shell Canada Suppliers and Contractors*, that clearly communicated their business conduct expectations to their vendors. Shell Canada sought to require their suppliers to protect the environment and work to prevent and reduce pollution. Suppliers were asked to establish safety and sustainable development policies, implement environmental management programs, provide training and certification, have emergency response procedures, describe their storage and disposal capabilities, and commit to adhering to Shell Canada's internal corporate safety and sustainable development requirements (Maharaj and Herremans 2008).

More recently, Tate et al. (2010) investigated the CSR reports of 100 global companies to see how they referred to their supply chain. Tate et al.'s content analysis identified ten themes including supply chain, institutional pressure, community focus, consumer orientation, external environment, risk management, measures, energy, healthcare, and green buildings. For each theme, they pointed out how SSCM was relevant. They concluded that while institutional pressures were the driving force behind strategy development in all the industries studied, different companies emphasized different facets of social, environmental, and economic responsibility upstream and downstream in their supply chain depending on their industry, size, and geographic location.

Finally, Morali and Searcy (2013) assessed the content of the sustainability reports of 100 Canadian firms. They found only 13 % of the reports mentioned having a management mechanism that tied sustainability to their procurement practices. This suggests that a governance structure for SSCM may be a marginal practice. Seventy-two percent reported having a SSCM strategy or program. But most only addressed environmental sustainability leaving the social and economic aspects unaddressed. Most statements on supply chain strategy focused on local purchasing practices. Forty-five percent of the companies reported at least one procurement-related indicator and many listed at least one KPI. Most companies cited policy, practices, and proportion of spending on locally based suppliers as an indicator. Fifty percent mentioned one relevant standard for supplier behavior. Almost every corporate report mentioned a code of business conduct. Other standards involved product-/process-related certifications and management systems and initiatives. In terms of collaboration within their supply chain, 42 % reported they had such a partnership, mainly focusing on upstream initiatives with suppliers. Finally, 33 % mentioned having a supplier management monitoring system. These three studies show that supply chain issues are being addressed in some sustainability reports. But what is said, if anything, is often vague. Certainly, that does not mean supply chain documents do not exist or discussions are not occurring, but they are not making their way into sustainability reports. Descriptive and historical analyses such as these are useful in providing context. However, what we need now is a concerted effort to shape (e.g., create shared standards), enact, and discuss (between partners and within society) supply chain relationships within and across industries and continents.

7.4   Concluding Thoughts

Global warming results in rising seas. Globally, 60 % of cities with populations over five million are within 60 miles of the sea. In the USA, over 123 million people live in coastal counties. By 2100, sea-level rise could put up to 7.4 million US coastal residents at risk and cut the country's GDP by as much as $289 billion (Stephens and Kille 2014). On January 30, 2015, President Obama issued an executive order establishing a federal flood risk management standard and a process for soliciting and considering stakeholder input (The White House 2015). Collaboration was at the heart of the executive order. The National Security Council staff coordinated an interagency effort and solicited the views of governors, mayors, and other stakeholders as they sought to develop a new flood risk reduction standard for *federally funded* projects. Meanwhile, the Georgetown Climate Center (2015) has a webpage listing state and local adaptation plans. Fifteen states have plans, but implementation is limited. For example, California has 345 state-level goals of which 48 are completed and 251 are in progress. Massachusetts has 373 state-level goals (24 completed, 191 in progress). New York has 121 goals (17 completed, 63 in progress), while New York City has nine plans beginning in 2008. We need large-scale urban adaptation strategies as we face the effects of long-term global climate change. Such strategies require creative problem solving and collaboration. They will be complex and expensive to execute as millions of people are mobilized. Planning is essential, but communication is critical.

References

Allen, M. W., Walker, K. L., & Brady, R. (2012). Sustainability discourse within a supply chain relationship: Mapping convergence and divergence. *Journal of Business Communication, 49*, 210–236.

Argyris, C., & Schön, D. A. (1978). *Organizational learning: A theory of action perspective*. Reading, MA: Addison-Wesley.

Attwater, R., & Derry, C. (2005). Engaging communities of practice for risk communication in the Hawkesbury Water Recycling Scheme. *Action Research, 3*, 193–209.

Austin, J. E. (2000). *The collaboration challenge: How nonprofits and businesses succeed through strategic alliances*. San Francisco: Jossey-Bass.

Balkau, F., & Sonnemann, G. (2009). Managing sustainability performance through the value-chain. *Corporate Governance, 10*, 46–58.

Benn, S., Edwards, M., & Angus-Leppan, T. (2013). Organizational learning and the sustainability community of practice: The role of boundary objects. *Organization & Environment, 26*, 184–202.

Blackburn, W. R. (2007). *The sustainability handbook: The complete management guide to achieving social, economic and environmental responsibility*. London: Earthscan.

Bonito, J. A. (2009). Interaction process analysis. In S. W. Littlejohn & K. A. Foss (Eds.), *Encyclopedia of communication theory* (pp. 529–530). Thousand Oaks, CA: Sage.

Bowman, J. (2009). Collective information sampling. In S. Littlejohn & K. Foss (Eds.), *Encyclopedia of communication theory* (pp. 117–118). Thousand Oaks, CA: Sage.

Brulle, R. J. (2010). From environmental campaigns to advancing the public dialog: Environmental communication for civic engagement. *Environmental Communication: A Journal of Nature and Culture, 4*, 82–98.

Cantor, D. E., Morrow, P. C., & Montabon, F. (2012). Engagement in environmental behaviors among supply chain management employees: An organizational support theoretical perspective. *Journal of Supply Chain Management, 48*, 33–51.

Carbone, V., Moatti, V., & Wood, C. H. (2012). Diffusion of sustainable supply chain management: Toward a conceptual framework. *Supply Chain Forum, 13*, 26–39.

Cooren, F. (2000). *The organizing property of communication*. Philadelphia: John Benjamins.

Cooren, F. A., Kuhn, T., Cornelissen, J. P., & Clark, T. (2011). Communication, organizing and organization: An overview and introduction to the special issue. *Organization Studies, 32*, 1149–1170.

Cruz, L. B., & Boehe, D. M. (2008). CSR in the global marketplace: Towards sustainable global value chains. *Management Decision, 46*, 1187–1209.

Fisher, R., Ury, W. L., & Patton, B. (2011). *Getting to yes: Negotiating agreement without giving in*. New York: Penguin.

Georgetown Climate Center. (2015). *How prepared are U.S. states for climate change*? http://www.georgetownclimate.org/. Accessed 15 Jan 2015.

Gimenez, C., & Sierra, V. (2013). Sustainable supply chains: Governance mechanisms to greening suppliers. *Journal of Business Ethics, 116*, 189–203.

Gray, B., & Stites, J. P. (2013). *Sustainability through partnerships: Capitalizing on collaboration*. Network for Business Sustainability. http://nbs.net/wp-content/uploads/NBS-Systematic-Review-Partnerships.pdf. Accessed 15 Jan 2015.

Hannaes, K., Arthur, D., Balagopal, B., Kong, M. T., Reeves, M., Velken, I. et al. (2011). *Sustainability: The 'embracers' seize advantage*. MIT Sloan Management Review and The Boston Consulting Group Research Report. http://sloanreview.mit.edu/reports/sustainability-advantage/. Accessed 20 Dec 2013.

Hinrichs, G. (2010). SOARing for sustainability: Longitudinal organizational efforts applying appreciative inquiry. *AI Practitioner, 12*, 31–36.

Hirokawa, R. Y., & Salazar, A. J. (1999). Task-group communication and decision-making performance. In L. R. Frey (Ed.), *The handbook of group communication theory and research* (pp. 167–191). Thousand Oaks, CA: Sage.

Jarboe, S. (1999). Group communication and creativity processes. In L. R. Frey (Ed.), *The handbook of group communication theory and research* (pp. 335–368). Thousand Oaks, CA: Sage.

Kiron, D. N., Kruschwitz, K., Haanaes, M., Reeves, S., Fuisz-Kehrbach, & Kell, G. (2015). *Joining forces: Collaboration and leadership for sustainability*. MIT Sloan Management Review, The Boston Consulting Group, and the United Nations Global Compact Research Report. http://sloanreview.mit.edu/. Accessed 15 Jan 2015.

Koschmann, M. A. (2012). The communicative constitution of collective identity in interorganizational collaboration. *Management Communication Quarterly, 27*, 61–89.

Larson, C., & LaFasto, F. M. J. (1989). *Teamwork: What might go right/what can go wrong*. Newbury Park, CA: Sage.

Littlejohn, S. W., & Foss, K. A. (2005). *Theories of human communication* (8th ed.). Belmont, CA: Thomson Wadsworth.

Livesey, S. M., Hartman, C. L., Stafford, E. R., & Shearer, M. (2009). Performing sustainable development through eco-collaboration: The Ricelands Habitat Partnership. *Journal of Business Communication, 46*, 423–454.

Maharaj, R., & Herremans, I. M. (2008). Shell Canada: Over a decade of sustainable development reporting experience. *Corporate Governance, 8*, 235–247.

Manring, S. L. (2007). Creating and managing interorganizational learning networks to achieve sustainable ecosystem management. *Organization & Environment, 20*, 325–346.

Meyer-Emerick, M. (2012). Sustainable Cleveland 2019: Designing a green economic future using the appreciative inquiry summit process. *Public Works Management & Policy, 17*, 52–67.

Meyers, R. A., & Brashers, D. E. (1999). Influence processes in group interaction. In L. R. Frey (Ed.), *The handbook of group communication theory and research* (pp. 288–312). Thousand Oaks, CA: Sage.

Mitchell, M., Curtis, A., & Davidson, P. (2012). Can triple bottom line reporting become a cycle for "double loop" learning and radical change? *Accounting, Auditing & Accountability Journal, 25*, 1048–1068.

Morali, O., & Searcy, C. (2013). A review of sustainable supply chain management practices in Canada. *Journal of Business Ethics, 117*, 635–658.

Munshi, D., & Kurian, P. A. (2015). Imagine organizational communication as sustainable citizenship. *Management Communication Quarterly, 29*, 153–159.

Poole, M. S. (1999). Group communication theory. In L. R. Frey (Ed.), *The handbook of group communication theory and research* (pp. 37–70). Thousand Oaks, CA: Sage.

Preuss, L., & Walker, H. (2011). Psychological barriers in the road to sustainable development: Evidence from public sector procurement. *Public Administration, 89*, 493–521.

Reed, M. G., Godmaire, H., Abernethy, P., & Guertin, M.-A. (2014). Building a community of practice for sustainability: Strengthening learning and collective action of Canadian biosphere reserves through a national partnership. *Journal of Environmental Management, 145*, 230–239.

Rittel, H. (1973). Dilemmas in a general theory of planning. *Policy Sciences, 4*, 155–169.

Salazar, A. (2009). Functional group communication theory. In S. Littlejohn & K. Foss (Eds.), *Encyclopedia of communication theory* (pp. 417–421). Thousand Oaks, CA: Sage.

Sarkis, J., Gonzalez-Torre, P., & Adenson-Diaz, B. (2010). Stakeholder pressure and the adoption of environmental practices: The mediating effect of training. *Journal of Operations Management, 28*, 163–176.

Schultz, B. G. (1999). Improving group communication performance. In L. R. Frey (Ed.), *The handbook of group communication theory and research* (pp. 371–394). Thousand Oaks, CA: Sage.

Senge, P. M., Scharmer, O., Jaworski, J., & Flowers, B. S. (2005). *Presence: An exploration of profound change in people, organizations, and society*. New York: Random House.

Shumate, M., & O'Connor, A. (2010a). Corporate reporting of cross-sector alliances: The portfolio of NGO partners communicated on corporate websites. *Communication Monographs, 77*, 207–230.

Shumate, M., & O'Connor, A. (2010b). The symbiotic sustainability model: Conceptualizing NGO-corporate alliance communication. *Journal of Communication, 60*, 577–609.

Stephens, R., & Kille, L. W. (2014). *Global warming, rising seas and coastal cities: Trends, impacts and adaptation strategies*. http://journalistsresource.org/studies/environment/climate-change/impact-global-warming-rising-seas-coastal-cities#sthash.SNmwq9ta.dpuf. Accessed 18 Jan 2015.

Stohl, C. (2009). Bona fide group theory. In S. Littlejohn & K. Foss (Eds.), *Encyclopedia of communication theory* (pp. 78–80). Thousand Oaks, CA: Sage.

Stohl, C., & Walker, K. (2002). A bona fide perspective for the future of groups: Understanding collaborating groups. In L. R. Frey (Ed.), *New directions in group communication* (pp. 237–252). Thousand Oaks, CA: Sage.

SunWolf. (2002). Getting to GroupAha!: Provoking creative processes in task groups. In L. R. Frey (Ed.), *New directions in group communication* (pp. 203–217). Thousand Oaks, CA: Sage.

SunWolf, & Seibold, D. R. (1999). The impact of formal procedures on group processes, members, and task outcomes. In L. R. Frey (Ed.), *The handbook of group communication theory and research* (pp. 395–431). Thousand Oaks, CA: Sage.

Swieringa, R. C. (2009). Community of practice. In S. W. Littlejohn & K. A. Foss (Eds.), *Encyclopedia of communication theory* (Vol. 1, pp. 147–148). Los Angeles, CA: Sage.

Tate, W. L., Ellram, L. M., & Kirchoff, J. F. (2010). Corporate social responsibility: Thematic analysis related to supply chain management. *Supply Chain Management, 46*, 19–44.

The White House. (2014). *U.S.–China joint announcement on climate change.* http://www.whitehouse.gov/the-press-office/2014/11/11/us-china-joint-announcement-climate-change. Accessed 14 Jan 2015.

The White House. (2015). *Executive order – Establishing a federal flood risk management standard and a process for further soliciting and considering stakeholder input.* http://www.whitehouse.gov/briefing-room/presidential-actions/executive-orders. Accessed 4 Feb 2015.

United Nations Environment Programme (UNEP). (2009). *Global environment outlook: Environment for development, UNEP, Geneva.* www.unep.org/geo/. Accessed 23 Nov 2009.

Vurro, C., Russo, A., & Perrini, F. (2009). Shaping sustainable value chains: Network determinants of supply chain governance models. *Journal of Business Ethics, 90*, 607–621.

Walker, K. L., & Stohl, C. (2012). Communicating in a collaborating group: A longitudinal network analysis. *Communication Monographs, 79*, 448–474.

Wals, A. E. J., & Schwarzin, L. (2012). Fostering organizational sustainability through dialogic interaction. *The Learning Organization, 19*, 11–27.

Wenger, E. (1999). *Communities of practice: Learning, meaning, and identity.* Cambridge: Cambridge University Press.

Wexler, M. N. (2009). Strategic ambiguity in emergent coalitions: The triple bottom line. *Corporate Communications: An International Journal, 14*, 62–77.

Wolf, J. (2014). The relationship between sustainable supply chain management, stakeholder pressure and corporate sustainability performance. *Journal of Business Ethics, 119*, 317–328.

Chapter 8
Our Shared Journey Toward Sustainability

Abstract This final chapter discusses how the constitutive and pragmatic power of communication allows employees to exert agency and organizations and interorganizational collaborations to change societal-level *Discourses*. SMART messaging is discussed. Suggestions are provided to researchers and practitioners on next steps each might take. Remaining wisdom drawn from the interview transcripts is shared. Scholarly literature discussing wisdom and spirituality is reviewed. Both must be encouraged within and between our organizations if we are to move forward in creating and expanding truly meaningful sustainability initiatives.

Lewis and Clark's Story and Our Forward Momentum Although the Corps of Discovery members were not the first white men to cross the continent and generations of native peoples had already mapped the land, what happened to them "is a great story, brimming with energy and full of forward motion. In extraordinary settings, a remarkable cast of characters encountered adversity of epic proportions and struggled through one adventure after another" (Ronda 2003). One universal human story involves a journey. All people groups tell stories of their ancestors' journeys. Every individual tells stories of his or her own journey. "We are a people in motion . . .. And it was Lewis and Clark who gave us [the U.S.] our first great national road story." It's the story of a group of people representing various racial, ethnic, cultural, and social backgrounds who met and communicated with men and women representing multiple native cultures. Their journey provides us with a story of undaunted courage and can serve as a guiding metaphor for our journey in response to global climate change.

8.1 The Importance of Communication

Cox (2013) talks about how communication plays both constitutive and pragmatic roles during discussions involving the human–environment interface. In terms of its constitutive role, sensemaking theory (Weick 1995) describes how when people face environments filled with equivocal and uncertain information they focus on parts of their environment, engage in behaviors to reduce the uncertainty, and, if the

© Springer International Publishing Switzerland 2016
273

M. Allen, *Strategic Communication for Sustainable Organizations*, CSR, Sustainability, Ethics & Governance, DOI 10.1007/978-3-319-18005-2_8

behaviors sufficiently address the need, they retain them. Structuration theory (McPhee and Poole 2009) helps us realize that the rules and resources that guided our past behaviors both constrain and enable our present actions. But people have agency and can create new ways of behaving moving away, sometimes dramatically, from old ways of thinking and acting. Communication allows us to do this. Through communication, social constructions (e.g., what sustainability means in an organization) are created, maintained, and changed (Berger and Luckmann 1966). Over time, the *Discourses* about the human–environment interface have shifted from industrialism to include many others including sustainability. Norms within and across institutional fields are shifting as illustrated by the Portland Trail Blazers and the Green Sports Alliance's work with the Natural Resources Defense Council to make environmental issues a mainstream concern among US sports fans.

In terms of communication's pragmatic role, for decades scholars have investigated persuasion, social influence, behavioral change, citizen mobilization, the diffusion of innovation, and effective change-related communication. We know the characteristics of working environments where employees can import information, create, learn, and feel empowered. Rhetorical scholars show us how communication can mobilize large groups, and risk communication scholars inform us how to frame public messages that help citizens respond to environmental crises and challenges. Lindenfeld et al. (2012) quote Cantrill as writing, "While grave threats may in fact exist in the environment, the perception of such danger, rather than the reality thereof, is what moves people to action (1996, p. 76). In decision making, what knowledge gets counted as valid is almost as important as the knowledge itself. And change is occurring. Organizations are engaging in pro-environmental actions (e.g., eco-efficiencies, redesigned products and processes) which influence their institutional fields and their supply chains. Creative actions are breaking frames and helping us to develop new alternatives (constitutive) which shape others' behaviors (pragmatic). For example, The Sustainability Consortium's work, envisioned with the assistance of Walmart, is beginning to reverberate throughout entire supply chains. This trend will grow exponentially as sustainability indexes are developed for more product categories.

Frame breaking organizations must communicate about what they are doing, why they are doing it, and why others should do something similar. Communication at all levels is key to creating organizations and interorganizational groups capable of innovating within our communities, states, and nations. Sustainability-focused communication is important at the intrapersonal, interpersonal, group, organizational, interorganizational, and macro-levels. Intrapersonal communication focuses us on a communicator's internal use of language or thought. It is most directly influenced by his or her schemata, values and beliefs, worldview, information acquisition, sensemaking, goal congruence, and engagement. Interpersonal communication involves messages exchanged between several people and leads us to persuasion, social influence, framing, strategic ambiguity, information sharing, sensemaking, influential communicators, message strategies, and learning. Group communication at the organizational level directs us to decision making, problem solving, informal communication, innovation and creativity, and learning.

Organizational communication investigates communication within organizational contexts and points us toward changed structures which channel interactions, formal communication, diffusion of innovations, absorptive capacity, the communication of change, climate/culture, and learning. We also are directed toward corporate sustainability communication, certifications, external messaging (e.g., websites), collaboration, credibility, and an organization's position in and ability to influence its institutional fields. Interorganizational communication involves the communication occurring within a network of organizations and directs us to look at power, interorganizational collaboration, supply chain initiatives, certifications, network structures, and communities of practice. At the macro-level, we investigate communication related to social mobilization, *Discourses*, paradigms, legitimacy, diffusion of innovation, the global human–environment interface, and social justice debates. Every level influences every other level, especially its most proximal level. However, the most powerful levels rest at the organizational and interorganizational levels. Figure 8.1 illustrates the traditional view where employees, organizations, and interorganizational collaborations have limited ability to influence macro-environment discussions and actions regarding sustainability. Figure 8.2 highlights the role of empowered employees and the ability of organizational and interorganizational communication to continually redefine the major societal *Discourses* regarding sustainability.

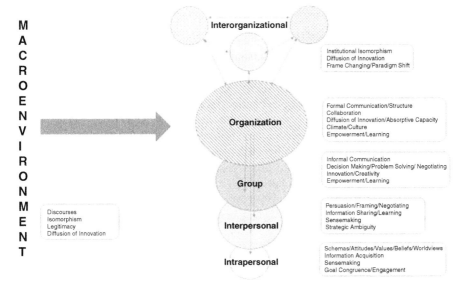

Fig. 8.1 External forces shape and constrain how organizations and interorganizational collaborations enact sustainability. Employees exert limited agency

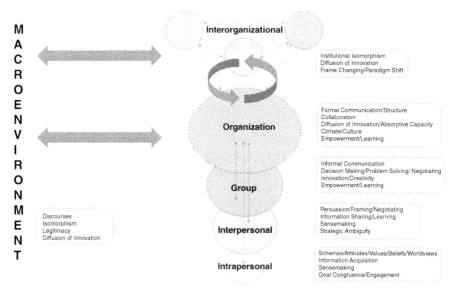

Fig. 8.2 The constitutive and pragmatic power of communication redefines sustainability. Employees exert agency

8.2 The Importance of Theory

In this book you have read about multiple communication, social psychology, management, sociology, and behavioral change theories. *Action Plan*: Look at the list of theories in the index. Identify 10 you believe are most useful to you at this point in your or your organization's sustainability journey. Certainly not all relevant theories and research are included in this book. Several recent edited books are worth noting although neither focuses us on the organizational context. Godemann and Michelsen (2011) also integrate theory and research from various disciplines including communication, psychology, and sociology. Servaes (2013) offers an overview of sustainability and communication from a development and social change perspective to address issues including community mobilization, mass media, and integrated communication approaches. If I've missed a theory, write about it within the context of sustainability and organizations. If you can develop a new one, please do so.

8.3 Messages to Researchers and Practitioners

The theories, research, and examples you have read about in this book have prepared you to engage in SMART messaging efforts. SMART messaging involves communication that is strategic, memorable, accurate, relevant, and trustworthy. Thus prepared, this section blends the scholarly literature with a list of actions researchers and practitioners may consider taking as they move forward on our joint sustainability-related journey.

8.3.1   Message to Researchers

> Sustainability challenges are beyond the capacities of our current institutions to address and require substantial change from systems [characterized by] crippling inertia. These problems are born of large-scale industrial economic policy, the rise of materialism, and the supremacy of profit over sustainability.... These problems have their roots in human behavior and institutional structures. We must strive to understand (as best we can) and grapple with system drivers and constraints, in short, the outcomes and the underlying causes. The problem constellations at hand are not the results of our systems; they are those systems.... Responsibility for those systems lies with individuals and collectives who collectively design, participate in, and accept the failures of the systems that govern our lives (van der Leeuw et al. 2012, p. 116).

We have multiple theories at our disposal, and now, as researchers and scholars, we need to act. Global warming presents us with an all-hands-on-deck situation because it is something that we collectively face. We need to become actively engaged in addressing serious problems since our:

> Education and creative and intellectual capital would ideally provide transition and solution options.... In global change, diagnosis is but the beginning of a long process in which strategic interventions are necessary to beget sustainability transitions (van der Leeuw et al. 2012, p. 116).

In the past, many researchers and scholars have not chosen to focus on "higher-level, socially relevant, life enhancing issues...[our research typically does not attempt] to frame communication or language within higher order constructs such as wisdom, enact public policy...or solve pragmatic communication-related malpractice" (Nussbaum 2012, p. 244). Today, communication researchers and scholars need to become:

> more involved in the goal of improving life on this planet as a member of a multidisciplinary team and begin to share our unique understanding of the impact of communication and language...with those noncommunication scholars who insist that communication does matter but lack the knowledge, sophistication, or will to move beyond their specific area of concentration (p. 244).

Already many action researchers *are* using their scholarship to improve the human condition and/or facilitate our communication about the human–environment interface. What about you? Are the questions you are addressing socially relevant and life enhancing? How might you contribute to addressing the issues we see looming on our horizon, if not sitting on our doorstep?

Moving forward we need increased public participation, improved integration of invested parties, more joint learning and knowledge sharing, and guidance focusing on negotiation and power sharing. Lindenfeld et al. (2012) describe a 5-year NSF funded collaboration that began in 2009. They worked with colleagues representing over 20 disciplines through the Sustainability Solutions Initiative (SSI) at the University of Maine. The team included social scientists representing environmental communication, psychology, economics, anthropology, human ecology, law, and policy. They developed models, tested theories, and researched ways collaboration and other forms of engagement can help stakeholders and communities overcome obstacles to developing a sustainable future. The SSI's goal was to help build Maine's capacity for addressing sustainability-related challenges by helping stakeholders identify and implement meaningful solutions that aligned with their values, needs, and desires. The researchers developed a portfolio approach to identify and characterize the uncertainty of alternative scenarios; help stakeholders understand the interplay among and complexity of diverse systems; and identify the limits, thresholds, and unintended consequences of various alternatives. Lindenfeld et al. write that they were inspired by the work of scholars who transcended their particular disciplinary framework and sought to develop local models designed to provide local solutions.

8.3.1.1   Actions for Researchers to Consider

- Work on truly meaningful research projects that have the potential to help influence our national *Discourses* and actions and/or prepare organizations and communities to adapt. In preparing this book I read article after article looking for useful clues on how we might proceed. Much of the research I reviewed, regardless of discipline, was interesting, but did not point us on a way forward in our journey.
- Join a multidisciplinary research team to work in your state. Many of us work at land grant schools charged with enhancing vitality within our state. Reach out to your natural science, sustainability science, management, law, public health, communication, and public administration peers to design joint research projects with important practical applications. Seek federal and/or state funding.
- Become an action researcher. If you analyze texts, analyze organizational texts focusing on sustainability. Use what you learn to help organizations craft honest, memorable, and mobilizing sustainability-related texts directed toward internal and external stakeholders. If you conduct survey research, investigate how to more effectively design and communicate sustainability-related change initiatives to employees or citizens. Assess their reactions.
- Become a facilitator. Help community groups collaborate to create resilient communities and ecosystems. Practitioners rely on their practical wisdom based on personal experience, industry norms, professional norms, and/or the predominant social *Discourse*. How might our theories help them solve the practical problems they face? Publish what you learn in outlets accessible to practitioners (e.g., trade journals and the popular press).
- Link up with important groups. Work with local, state, or national LEED groups, the Arbor Day Foundation, groups like Protect Our Winters and the Green Sports

Alliance. Work with NOAA to help them communicate climate-risk information to citizens; local and state governments to communicate around energy efficiency, recycling, and disaster preparedness; or your cooperative extension agents to help them better communicate with farmers about difficult regional environmental and climate-related issues.

- Plan cross-disciplinary mini-conferences and invite interested scholars representing communication, management, political science, public health, law, public policy, psychology, sociology, the natural sciences, and sustainability science. Apply for National Science Foundation funding to support such mini-conferences. Create partnerships between the Academy of Management (i.e., Organizations and the Natural Environment group), the International Communication Association and the National Communication Association (i.e., Environmental Communication and Organizational Communication groups), and any environmental groups in public policy and/or administration (e.g., the American Political Science Association, The American Society of Public Administration, the Policy Studies Organization, or the Association for Public Policy Analysis and Management).
- Create more graduate classes on your campus that link the natural sciences and communication with public policy to focus on interdisciplinary sustainability issues. Ask the ACUPCC to sponsor and/or publicize academic research that crosses disciplinary boundaries.
- Encourage capable social science M.A. students to earn environmental or sustainability science-related Ph.D. degrees. Encourage environmental or sustainability science B.A. students to earn an M.A. in communication or public policy.
- Help your students talk about sustainability-focused issues in their communities. For example, I ask my students to read the essay about George Marshall (see Sect. 2.2.1.2) and complete the Environmental Communication Scale (see Sect. 4.4.5). I provide them with some interesting information to discuss like which coastal states are preparing for sea level rise (see Sect. 7.4) and ask them to talk with a friend about that or a related issue. Finally, they write an essay describing what occurred. Their essays are interesting reading. Those conversations, even if congenial, do not flow very well. The only thing that will end this negotiated silence in our communities and nations is if we can learn to hold such conversations.
- Extend a theory or research stream you read about in this book. Make sure to address how whatever you learn can inform action and influence policy.

8.3.2   Message to Practitioners

Do we have all the tools that we need in the face of global climate change? Most definitely not. But we do have many technical, scientific, policy, network, and communication tools at our disposal. And people are continually developing more. We must link the practical to the theoretical in order to inform practice and use the practical to drive theory building. Partnerships are our primary model: interorganizational collaborations at the institutional field, disciplinary, industry, governmental, and transnational levels are key.

8.3.2.1   Actions for Practitioners to Consider

- If you are new on your sustainability journey, read Blackburn's (2007) very practical book. It crosses organizational sectors and is filled with explanations, practical strategies, checklists, forms, tips, and reference information. It is designed to be a *how to* guide (e.g., how to craft the business case for sustainability argument, how to design and implement internal processes to facilitate the achievement of sustainability initiatives). Wilhelm (2013) also provides a blueprint for successful implementation filled with suggestions offered by 40 Sustainability Directors.
- Partner with other organizations in your community to create a business (or organizationally focused) group that shares best practices and certifies sustainable organizations.
- If you are seeking inspiration regarding new sustainability initiatives, Smerecnik and Andersen (2011) offer some suggestions. *Enact sustainability management* by creating an environmental committee, a written environmental policy, an environmental impact assessment report, and a detailed program to reduce environmental impacts; hiring external consultants to help develop environmental policies or programs; sending officials to sustainability-focused conferences; measuring your greenhouse gas emissions or carbon footprint; and adopting nationally or internationally recognized sustainability certification programs. *Engage in more environmental communication* by providing environmental training of staff and environmental education of clients; including environmental statements in public messages; holding routine meetings to discuss environmentally related issues; supporting community environmental efforts, involvement, or advocacy; and talking with other organizations in your industry about environmental sustainability. *Manage pollution* by identifying the environmental pollution occurring on your property, seeking to prevent it, and work to repair or maintain local habitat and biodiversity. *Conserve resources* by ensuring separate collection of any hazardous waste, recovering food waste, composting organic and food waste, working with local recycling firms, purchasing recycled products [if possible] or products meant to be reusable; encouraging recycling among clients and staff; purchasing from local firms and

companies; and purchasing energy-saving materials and less hazardous materials. *Adopt water recycling* initiatives such as providing on-site wastewater treatment, reusing treated wastewater in the surrounding environment (e.g., landscaping irrigation), and adopting rainwater/snow runoff capture and reuse. *Conserve energy* by producing your organization's energy through solar, wind, or other renewable sources of energy; purchasing renewable energy from a local utility provider and renewable energy credits/green tags; using alternative fuel or hybrid vehicles in your transportation fleet, designing effective transportation to reduce environmental impact (e.g., design a plan for reducing car idling times), encouraging client and staff use of public transportation, providing employee carpool or alternative transportation incentives, enhancing building efficiency, utilizing sustainable materials and methods, and seeking LEED or Energy Star certification. *Improve office sustainability* by using energy saver control systems in rooms and energy-saving light bulbs and using sensor-activated lighting in locations that only require intermittent lighting.

- If you are a nonprofit, assess your own environmental performance. Nonprofits usually focus on strategies, their role in multistakeholder environmental planning, and their leadership dynamics (Dart and Hill 2010). Yet they also create environmental impacts.
- If you are a small- or medium-sized organization recognize there are many steps you can take. Cordano et al. (2010) found small- and medium-sized wineries developed simple environmental management programs which included environmental goals and policies, designated environmental responsibilities, provided employee training, set aside funds to support environmental innovation, and established environmental requirements for suppliers.
- Complete a reputable certification process to improve your sustainability. Communicate your certification in internally and externally directed messages. If a certification does not exist, work with your industry association to create one. Make it self-managing and large scale.
- Create or join local collaborative bodies made up of government, nonprofit, business, university, and citizen groups to identify how to create more resilient communities. Innovate and market your innovations.
- Investigate places where relevant pilot test projects are being done (e.g., the Transition Towns project). Find organizations that have a strong sustainability program and meet with them to find out more. Share what you learn with other organizations.
- Seek to change your organization's culture/climate to promote sustainability at the same time as you engage in eco-efficiency efforts.
- Help your employees become more sustainable in their own homes and personal lives using existing programs (e.g., the Home Energy Affordability Loan program, My Sustainability Plans).
- Become politically active. Create an industry association like Tom Kelly [Neal Kelly Company, OR] did or seek to influence your existing industry association. Form an activist arm like the Aspen Skiing Company did with Protect Our

Winters. Sign documents like the Climate Declaration BICEP launched (see Sect. 1.2).

- Use your position within your industry field to change standard operating norms. Work with your supply chain.
- Partner with your local university to increase and communicate your sustainability; stimulate your university to develop research programs that can enhance sustainability. Serve on university steering committees, fund Endowed Chairs, and financially support the creation of research centers.
- Reach out to natural science, communication, engineering, and public policy faculty. They can help you solve sustainability-related problems. In turn, you can help them identify and/or provide a setting where they can research pressing sustainability-related issues.

8.3.2.2 Suggestions Provided By Interviewees

This section provides some additional suggestions emerging in the comments my interviewees shared with me. I share them with you in hope that you will find some that you can enact in your own organization.

Neal Kelly Company (Small Business)

- Stay the sustainability course even during a recession or in the face of financial challenges.
- Work with consultants or take advantage of training provided by outside groups that fit your organization's goals and values (e.g., Natural Step). They can help you critically evaluate your existing policies and procedures (Portland Trail Blazers agrees) and identify your organization's biggest impacts on the environment. Rank order your priorities (WasteCap Nebraska and Portland Trail Blazers agree).
- Inform employees of your organization's goals and priorities so sustainability-related values become core to your organization, and so that employees can participate in decision making. Train employees on how to think through problems, generate options, and enact sustainability initiatives. Begin this training at their initial orientation.
- Enlist consensus among the managerial group to pursue the new course; then embed it throughout the organization through training and when developing strategic plans.
- Become a member of an advisory committee at your local university and provide input into student training and internship opportunities. They may become your new employees.
- Become an active member, leader, and/or innovator in sustainability-related associations in your regional trade association, local community, and state.

- Use your knowledge and expertise regarding sustainability to benefit your community.
- Actively lobby at the state level for sustainability-related initiatives. Be politically active (Portland Trail Blazers and Aspen Skiing Company agree).
- Redesign your organizational structure to be more in line with company values and goals.
- Attract skilled employees who share the same value set as your organization. (Arbor Day Foundation and the City of Portland agree).
- Involve employees in annual planning and be accountable at the annual meeting.
- Create opportunities for your suppliers to meet each other at events. Make sure your products are environmentally sustainable. Offer these products to end users making sure the price differential is as low as realistic.

ClearSky Climate Solutions (Small Business)

- Recognize that capital markets are necessary to any wise use of natural resources and that private funds investing in the protection of natural resources are necessary. (Ecotrust agrees).
- Find a communicator who can make science understandable and relevant to the lay public and who can lobby and convince others that change is in their best short-term interest.
- Associations need to conduct research to identify why the clients of their members would be interested in supporting sustainability-related initiatives and projects. Utilize their findings in your messaging to clients.
- Use the language of eco-efficiencies. It makes sense to businesses.

Bayern Brewing (Small Business)

- Get multiple uses out of your materials. Purchase equipment to support the multiple use of materials. Plan for the long run so that things last and are productive for as long as possible. Think of things others recycle as raw materials you can use.
- Promote your local and state economy by minimizing the money you send out of state to purchase supplies.
- Don't wait for government to mandate sustainable actions.
- Innovative even if it means dealing with risk and uncertainty (WasteCap Nebraska agrees). Be willing to jettison new ideas that don't work.
- Join a local sustainability organization so you can see what others around you, as well as in your own industry, are doing in terms of sustainability-related initiatives.
- Once your business is financially stable, fine-tune it to make it more sustainable and efficient. (Neal Kelly Company agrees).

Assurity Life Insurance (Mid-Sized Business)

- Build on your history of sustainability to construct or retrofit to LEED specifications. (Portland Trail Blazers, University of Arkansas—Fayetteville, Heifer International®, Arbor Day Foundation, and Ecotrust agree).
- Use one platform (e.g., a history of recycling) to build to another level (e.g., empower employees in their personal lives to be more sustainable).
- Create and support a Green Team. Task them with promoting awareness of sustainability among employees. Empower them.
- Recognize the transition to being a more sustainable organization doesn't happen overnight.

WasteCap Nebraska (Community Organization)

- Adapt and expand your focus as your business environment changes.
- Create a place so that peer networking and learning around sustainability can occur within your business community. (City of Portland agrees).
- Promote experiential learning around sustainability initiatives.
- Develop a catalyst to get your foot in the door so you can help people try one thing. Once they try one thing they are more likely to want to learn more about sustainability.
- Provide people with ways they can implement sustainability-related programs and see real short-term progress.
- Develop or implement a certification program that is self-reporting, includes education, and can be used to market to internal and external stakeholders. (Missoula Sustainability Council agrees).
- Maintain your reputation as a trusted resource for useful knowledge. Know other trusted resources you can partner with to provide services to clients you cannot perform alone.
- Don't worry about developing information—good training materials have already been developed and are available which you can tailor to your purposes.
- Realize that if you position yourself as a leader, people will follow.
- Encourage your organization's leader to share his/her values around sustainability repeatedly and display passion (Assurity Life Insurance agrees).
- Remember change takes a lot of communication, education, and training.

Missoula Sustainability Council (Community Organization)

- Identify if there is community interest in forming a group to promote sustainable and/or local business initiatives. (Portland Trail Blazers agrees).
- Bring in excellent speakers to build interest for sustainability. (WasteCap Nebraska and Sams Club agree). Develop forums where local people can ask questions of the speaker.

- Focus on the positive benefits that business can provide to sustainability and that sustainability can provide for business—vibrant local communities can offer a high quality of life based on environmentally friendly and socially aware practices.
- Identify resources. Ask local utilities to pay for and conduct Water Wise and Energy Wise trainings, assessments, and certifications for local organizations. Seek funding for small businesses to pay for upfront costs associated with sustainability.
- Prepare your communities for a resource constrained future. Emphasize creating a vibrant local economy where residents buy local and sustainable products.
- Provide a website where community members can learn about what your local businesses offer.
- Develop interest in your organization through your volunteers, board members, and existing clients using relationship-based marketing.
- Utilize interns from local schools and volunteers from the Senior Corps of Retired Executives.

Portland Trail Blazers (Sports Business)

- Encourage employees to attend summits and conferences related to sustainability topics.
- Don't just assume you are recycling—look around and see if you can do it better or do other things.
- Respond to top down, bottom up, and community pressures to become more sustainable.
- Create a sustainability charter or set of values, goals, and targets.
- Measure your carbon footprint (Sams Club agrees), conduct a Scope 3 analysis.
- Inspire your industry, community, state, and nation to be more sustainable.
- Conduct measurements, benchmark your metric, and continue to seek continuous improvements
- Investigate, instigate, and engage in district-scale systems and eco-efficiencies. Educate people about the idea of eco-districts and eco-cities.
- Co-found an industry group that promotes sustainability. Function as a role model or industry leader for other organizations to follow.
- Partner with groups like the Natural Resources Defense Council, the Department of Energy, and the Environmental Protection Agency.
- Support federal initiatives that address national energy policy and aim to reduce climate change impacts.
- Join the conversation that climate change matters. It is not a political issue but rather a human issue—sports require an environment with clean water and clean air.
- Work with your industry to create livable communities which embrace the fundamentals of sustainability. Inspire them to do better in terms of energy consumption, recycling, and food purchasing.

- Start with your biggest environmental impacts and make changes.
- Enlist your supply chain to help. (Bayern Brewing agrees).
- Ensure your employees know what your organization is doing in terms of sustainability. Discuss this in staff meetings. Train them to understand what you are doing, why you are doing it, what their role can be, and how to tell the story of what your organization is doing to your clients/customers. Provide them with talking points. (ClearSky Climate Solutions, The University of Arkansas, Fayetteville, and the University of Colorado, Boulder, agree).
- Position sustainability so that it is seen as something which enhances your organization's ability to achieve its main purpose or goals.
- Seek members for your internal sustainability committees who want to participate.
- Recognize that eco-efficiencies are subject to the law of diminishing returns. The medium and low-hanging fruit will quickly be picked.
- Acknowledge that making changes toward sustainability is a learning process. It takes time (Assurity Life agrees); there is a learning curve. It requires a large commitment if everyone is going to understand what is going on so they can be on the same page moving forward.
- Realize you are going to have to navigate different departments with different goals that may appear to conflict.
- Make sure that sustainability doesn't become an add-on, have-to layer in your organization (City of Boulder agrees).

Aspen Skiing Company (Sports Business)

- Continually educate your top management about sustainability and what you are trying to do in terms of how you communicate about sustainability.
- Send your management to meetings where they can meet respected others who embrace sustainability, give them reading materials, provide presentations, etc. (Portland Trail Blazers agrees).
- Be radically honest in terms of how you communicate about sustainability and don't just do it for marketing.
- Use your company's visibility to boycott larger companies to go in a more sustainable direction.
- Use multiple outlets (e.g., speeches, the popular press, journal articles, books) to speak about sustainability (Ecotrust agrees) and to drive change.
- Realize talking about catastrophe and gloom and doom isn't helpful. (City of Boulder agrees). Speak about disaster but don't make it the key point.
- People respond to stories of collaboration and bipartisanship and taking advantage of wasted resources.
- Mobilize your constituency to become part of a social movement.
- Talk about what your organization is doing in terms of sustainability policy and/or practices. We are not going to solve climate change if people are quiet. Be

honest and say what you are working on, that you aren't done yet, but you think it is important work.

- Remember employees have a mission to do something other than sustainability and they are only willing to do so much for something they see as ancillary. They need to see it as key to the business (Sams Club agrees).
- Operational greening isn't enough. Push people and industry groups to go beyond that.

State Farm Operations Center Green Team (Branch of Large Business)

- If you expect employee pushback on an idea think about how to get employee buy-in, educate them on the environmental problem, give them a reason for the change, offer an attractive alternative if possible, and make it easy to change.
- Be prepared to do a cost/benefit analysis when proposing new ideas.
- Post information about new practices on a website and on bulletin boards throughout your building(s).
- Keep employees engaged with new ideas and start them thinking about sustainability in their personal behaviors away from work. Provide employees with lots of information. Help them recycle things from home (e.g., old batteries).

Sams Club (Large Business)

- Place stories about sustainability in all major internal publications. Ensure it is discussed at all major meetings.
- Keep your leadership teams well informed and engaged; serve as an internal consultant to support their conversations; ensure they are connecting with key players globally and have input into the development of key goals.
- Provide buyers and suppliers with different communication forums (e.g., summits) for discussing sustainability, its importance, how it can improve business performance, and how to integrate it into decision making.
- Encourage buyer/supplier partnerships to improve sustainability performance and to address key performance indicators.
- Focus on intraorganizational eco-efficiencies but also on supply chain innovations and community involvement.
- Remember in order to create behavioral change, a new behavior must be easy to understand and execute, it must be understood, it's got to be desirable, it's got to be rewarding, and people have to be reminded when to engage in it.
- Integrate sustainability throughout the planning process, make it a part of routine reporting, and ensure some of your key performance indicators revolve around sustainability.
- Support the development of tools like the Sustainability Consortium's sustainability indexes which can be used to systematically and factually to vet the

sustainability-related claims being made by other organizations. Work with multiple stakeholders to develop sustainability indexes. (Tyson Foods agrees).
- When evaluating the right channels for getting your message out recognize you face limited resources and be strategic where you place your efforts.
- Recognize you are going into uncharted territories and will make mistakes along the way. Use mistakes to set better goals and strategies and execute better next time. Don't be afraid to admit when you make mistakes. Correct them. Be transparent. Correct misinformation quickly and forcefully by providing the truth (Tyson Foods agrees).

Tyson Foods (Large Business)

- Focus on what your organization does and then integrate sustainability throughout. Don't lose your focus—be deliberate about what you are seeking to achieve with your sustainability initiatives. Compare your core values against the principles of sustainability.
- Develop an honest story to share about your organization's place in global sustainability and how a sustainability focus fits your organization's history.
- Make sure the content is accurate, truthful, and honestly reflects what your company does and values. Work with lawyers and develop a fact book when writing your sustainability reports. Make sure your sustainability report, and related information, is user friendly, online, downloadable, and searchable.
- Measure and benchmark your activities using the Global Reporting Initiative against industry leaders and against top performing organizations in other industries.
- Seek a seat on the sustainability councils of your industry and your supply chain partner organizations. Serve on boards of all kinds (Ecotrust agrees).
- Become engaged in helping to create the sustainability standards which will guide your industry.
- Create an executive steering committee and a working council of other executives to focus on sustainability.
- Align your goals to your long-term strategic plan. Create a small number of focused, measurable achievable sustainability-related goals. (Sams Club agrees). Replace stale goals. Set goals that ensure something meaningful can be achieved. Set tight plans around a few key goals. Having plans in place to meet sustainability goals doesn't mean the plans are effective. Routinely assess their utility and effectiveness.
- Work with university teams to measure resource issues important to your organization (e.g., water scarcity challenges). Create internal councils to address these scarcities.
- Be prepared to quickly and honestly respond to critics in the media and elsewhere.

City of Fayetteville, AR

- Work with all departments to come up with well-defined metrics that meet the city's goals which can be used to evaluate internal projects. (City and County of Denver agrees).
- Work collaboratively with department heads. Find a benefit you can offer each which he/she values.
- Talk about life-cycle costs. Discuss bottom-line issues (e.g., cost containment).
- Don't use language (e.g., sustainability) that turns off people. Use language like livable and resilient. (City of Portland and Ecotrust agree).
- Create a sustainability working group that includes all the departments. This group identifies the issues the organization should work on and can get buy-in from the different departments. (Portland Trail Blazers and the City and County of Denver agree).
- Attend and contribute at weekly staff meetings with department heads and the mayor.
- Join a group like the Urban Sustainability Director's network (City of Portland agrees) that shares best practices and new ideas, hosts webinars and conference calls, and provides access to other city directors.
- Work to get your message across to people who are ambivalent, conservative, not interested in listening, and/or distracted by their daily lives.

City of Boulder, CO

- Move sustainability off the sideline and integrate it in larger departments like planning and transportation. (City of Portland agrees.) Team sustainability personnel up with planners and policy development personnel.
- Research the impact of sustainability initiatives on community members, consumers, and businesses.
- Be a process facilitator. Listen to your different constituencies; make sure the process is inclusive. Use the three I's—invite, include, and inspire.
- Seek to achieve sustainability in a way that sustains current or creates new livelihoods for people.
- Find an example you can use to illustrate how there is environmental quality, economic vitality, and social equity, but the sweet spot in the middle is sustainability. Think comprehensively and strategically about how you can advance all areas at the same time.
- Create a sustainability framework all departments can use as part of their master plans. Start with where they are at, have them articulate what sustainability means to them, and how what they do on their job contributes to sustainability.
- Ensure sustainability is routinely talked about by city decision makers.
- Create a skilled team to think strategically about providing clear and motivating communication. They can help you figure out what is important, what people

need to know, and how to present it in a way people will hear and act on the message.

- Stress that your organization has the opportunity to be a leader in creating a low-carbon economy and developing innovations to export elsewhere. (City of Portland and Ecotrust agree).
- Think critically about where your audience is in terms of understanding sustainability and how you might present information to them in an accessible way. Think about how you can encourage them to get involved. Engage in audience analysis (City of Portland and State of South Dakota agree).
- Enter into performance systems contracts (University of Arkansas, Fayetteville, agrees), prioritize projects, engage in deep retrofits, and then message around the results.
- Recognize you may need to engage in difficult conversations with your supply chain members.
- Join zero waste and resource use initiatives being promoted by regional networks and sustainability-focused professional organizations and city networks. (City of Portland agrees).

City and County of Denver, CO

- Identify one person in each agency who is your point of contact in terms of sustainability initiatives. Enlist each agency in supporting one or more of your city goals. These agencies act as facilitators to keep the agencies communicating and help them avoid overlaps and inefficiencies.
- Before deciding how to address a problem, have a clearly defined problem and clearly defined metrics for success. Have measurable goals, set baseline goals to reach the larger goals, evaluate current initiatives, develop new strategies, collect metric data, and share your results through an understandable dashboard. (State of South Dakota and the City of Boulder agree)
- Remember we need good practices. We may be incapable of developing best practices. Don't let the perfect become the enemy of the good.
- Research the best practices other cities are enacting so that your agencies have some options to consider. But compare yourself to the uncertainties of the future rather than to other cities.
- Create sustainability initiatives that are large scale enough to address the real challenges you anticipate your city is facing. (Missoula Sustainability Council agrees).

City of Portland, OR

- Message that sustainability isn't about the environment. It's about the environment and its connection to jobs, social equity and justice, personal health, etc. Message around what citizens care about. (State of South Dakota agrees)

- Form partnerships and seek input. The City acts with the school district, the county, all the corporations, and the universities. (Heifer International® agrees about the importance of partnerships).
- Provide certifications, awards, and celebration to recognize businesses that are seriously pursuing sustainability. (Missoula Sustainability Council and WasteCap Nebraska agree).
- Have policies requiring green procurement. (State of South Dakota and Assurity Life Insurance agree).
- Place sustainability initiatives within the individual work plans of city employees and evaluate their performance accordingly.
- Encourage the development of sustainability initiatives within your city. Serve as a living laboratory. Export these new businesses and processes globally. (Portland Trail Blazers and Bayern Brewing agree). Portland has the "We Build Green Cities" program.

State of South Dakota

- Seek operational cost savings, engage in eco-efficiencies, and retrofit buildings to reduce energy and water use.
- Use building design to bring about individual employee behavioral changes.
- Use signage at the physical locations where you want employees to engage in a pro-environmental behavior. (University of Boulder, CO, agrees).
- Utilize building representatives to communicate information to and from staff regarding things such as building comfort levels using the Center for the Built Environment survey from the University of California, Berkley. (University of Colorado, Boulder, agrees)
- Use strong data-specific messages that show why sustainable efforts just make sense.

Home Energy Affordability Loan Program (Nonprofit Organization)

- Work closely with the departments in an organization which know how to reach their own employees (e.g., human resources, communication). (Ecotrust agrees).
- Educate and seek the support of top decision makers. (The Natural Resources Defense Council agrees).
- Collect and use video testimonials from C-level people, as well as rank-and-file employees.
- Share with your audience how they can benefit from changing their behaviors (e.g., save money, be healthier).
- Create organizations which promote large-scale sustainability efforts. (Arbor Day Foundation agrees).
- Commit your organization or yourself to helping people rebuilt when environmental disasters strike.

- Develop innovative ways to help low income people practice sustainability initiatives that can improve their quality of life and/or save them money. (Heifer International® agrees).

Heifer International® (Nonprofit Organization)

- Utilize storytelling to inspire and motivate people (Sams Club and Ecotrust agree).
- Seek to create community assets and increase community resilience.
- Develop programs which require people to pass on the knowledge and resources they have gained to others in their community.
- Form partnerships in local communities to develop and enact sustainability initiatives. (Portland Trail Blazers agrees).
- Develop values-based partnerships with business partners being mindful of their product-related needs and goals. (The Natural Resources Defense Council and Arbor Day Foundation agree).
- Work to develop large-scale interventions impacting groups of at least 1,000 families.
- Remember decentralized and small-scale agriculture can feed the world, cool the planet, nurture the soil, and be flexible when local environmental conditions change.

Natural Resources Defense Council (Environmental Organization)

- Use research to guide legislators and help businesses develop cost competitive and effective ways to enact ecologically intelligent practices.
- When dealing with private sector groups remember to seek agreement on the facts, acknowledge their agenda, and address issues they are concerned about (e.g., supply chain reliability, product quality, vendor issues, branding and visibility issues). Work to help them identify the risks they face in a collegial nonthreatening way. Help them address their brand liability and reduce the environmental liabilities which threaten their brands.
- Promote a cultural shift in how people think about their relationship to the planet. Work with sports and the entertainment industries to make a different way of thinking about the environment mainstream.

Arbor Day Foundation (Conservation Organization)

- Build programs that are high impact—life changing, large scale, partner engaging, and sustainable. Build programs where a positive environmental impact happens by itself and does not need an annual budget and can manage itself. (Heifer International® agrees).

- Have a weekly standup where employees share positive occurrences and accomplishments. Be positive and inspiring (Ecotrust agrees). The important thing is to share how the problem was solved.
- Engage in clear upfront communication about your sustainability initiatives when engaging in vendor selection.
- Continually innovate around your teams, responsibilities, roles, processes, and vision.
- Do what you (as an organization) say you are going to do. Don't promise something you can't deliver. If we can't verify it, we don't promise it.
- Create programs where farmers can learn to limit erosion, cities and campuses can cool themselves and improve resident's quality of life, homeowners can cool their residences and increase their property values, utilities can manage peak energy production needs, our national forests can be replanted, and children can learn to appreciate their natural environment.

Ecotrust (Economic Development/ Conservation Organization)

- Bring likeminded businesses and organizations together so they can cross-pollinate each other's ideas.
- If you see a gap in the market for sustainable products, consider using a business model to deliver that product yourself while maintaining relevant ecosystems.
- Use research and partnerships to deliver services to those who need them. (Similar to the work of Heifer International®).
- Manage our environmental resources (e.g., water, forests) with an eye toward long-term viability.
- Let your employees identify the best way to communication with their external constituents.
- Use regular discussions within your organization to break up internal silos. Encourage employees to think of problems/solutions outside their silos and collaborate (City and County of Denver agrees).
- Put place back into the discussion of where people live their lives and do business. Share stories of how people are responding to the pressures in the places where they live.

University of Arkansas, Fayetteville

- Remember money spent on energy and resource inefficiencies influences your organization's ability to meet its goals and deliver its products. (State of South Dakota agrees).
- Broaden from promoting building efficiencies to integrating sustainability into other areas of your operations (e.g., curriculum, research agenda).
- Identify how lack of sustainability can influence your organization's reputation.
- Address social justice issues in your organization and pay people a livable wage.

- If you learn of new sustainability-focused initiatives in your industry, check them out. If acceptable, be one of the early adopters.
- Use savings from energy saving retrofits to reinvest in your organization. (HEAL program agrees).
- Coordinate and/or discuss sustainability initiatives occurring in your University and your City. Provide leadership, resources, and communication that support your commitment toward sustainability.
- Create a communication clearinghouse when people can see the depth and breadth of things occurring within their organization which involve sustainability initiatives.
- Stress that you are being wise stewards of state funds. (South Dakota agrees).
- Have a sustainability playbook. Create and retransmit what your organization is doing in terms of sustainability multiple times and in multiple ways.

University of Idaho Sustainability Center

- Empower your employees and students to be creative and take ownership of the design, implementation, and assessment of sustainability-related projects. (University of Colorado-Boulder agrees).
- Argue that the actions you are taking are commonsense and practical.
- Create highly visible demonstration projects which provide evidence that is proof-of-concept.

University of Colorado-Boulder

- Set multiple year goals. (Tyson Foods agrees).
- Provide cost justification and argue how the changes add value (State Farm-Lincoln and the University of Arkansas, Fayetteville agree).
- Make it easy for building users to report energy or water waste.
- Reward employees who offer suggestions that result in significant and verifiable energy savings.
- Talk about how sustainability initiatives align with your broader organizational goals and directives.
- Provide meaningful examples of the benefits of sustainability-initiatives. (South Dakota agrees).
- Provide opportunities for experiential learning when appropriate.
- Identify a champion. Require that the champion be accountable.
- Mention environmental, financial, social, and mission-specific benefits of sustainability in all your messages.

[handwritten marginalia: wisdom theo]

8.4 Wisdom and Spirituality

> Information is everywhere but wisdom appears short in supply when trying to address some of the key inter-related challenges of our time, such as runaway climate change, the loss of biodiversity, [and] the depletion of natural resources (Wals and Schwarzin 2012, p. 12).

I've been thinking a great deal about wisdom lately and if we have the wisdom we need to find our path ahead and walk confidently down it. People tend to rely on the *Discourses* of the past and call that wisdom. Deciding how to act was clear when we still believed ample resources existed to support the Dominant Social Paradigm and we were largely unaware of global climate change. But as circumstances change dramatically the old wisdom is no longer sufficient. As we learned from anthropology and genetics, in the past when faced with climate extremes only a few groups survived (see Sect. 5.5). What allowed some people who left their home territories to survive on the long journey? My guess is that it was a combination of skill with the hunting technology of their day, intelligence, and wisdom. Therefore, the scholarship surrounding the concept of wisdom and spirituality (as a source of wisdom) is reviewed. Both must be encouraged within and between our organizations if we are to move forward in creating and expanding truly meaningful sustainability initiatives.

8.4.1 What Is Wisdom?

Humans have been fascinated with the concept and its link to the positive side of our behavior for more than 5,000 years. Popular real and fictitious archetypes are abundant in contemporary culture (e.g., Gandhi and Gandalf) (Izak 2013). Plato talked about three senses in which wisdom could be approached: as a special quality of those who contemplate life, as the practical application of good judgment in matters of human conduct, and as scientific knowledge concerning the nature of things. Although a clear definition of wisdom eludes us and regular research on the concept is relatively young, these three senses continue to appear in contemporary definitions. According to balance theory, wisdom is the use of your intelligence and experience mediated by your values and directed toward the achievement of a common good. It requires a balance of intrapersonal, interpersonal, and extrapersonal interests over both the short and long term. The goal of wisdom is to achieve a balance among adapting to and shaping existing environments, and selecting new environments (Sternberg 2004).

[handwritten marginalia: theory + goal of wisdom]

Two main research streams have evolved over time (Schmit et al. 2012). One investigates laypeople's implicit theories and/or definitions and views wisdom as a personality characteristic with cognitive, affective, and reflective characteristics. Wise people are knowledgeable, mature, emphatic, intuitive, tolerant, generally older, and possess the ability to take creative action. The second research stream focuses on expert theorists' and researchers' conceptions and measures of people's degrees of wisdom or their wisdom-related performance. Of interest here is if a wise person knows something other people do not. In the 1980s, Baltes created the Berlin Wisdom Paradigm that defines wisdom as an expert knowledge system about the fundamental pragmatics of life (Izak 2013). It includes qualities including factual and procedural knowledge, value relativism, and the ability to cope with uncertainty. Scholars new to the wisdom literature will find the background reviews mentioned here (e.g., Izak 2013; Nussbaum 2012; Schmit et al. 2012) useful.

Building on both research streams and on prior conceptualizations in psychology and management, Schmit et al. (2012) developed and tested a multidimensional scale. They argue that wisdom is made up of seven dimensions: practical, reflective, openness, interactional aptitude, ethical sensibility, paradoxical tolerance, and experience. The practical dimension refers to an individual's ability to reason carefully, to screen incoming information, and to focus on what is most critical. A wise person must be able to form and understand logical arguments, look at problems from multiple perspectives, attempt to remove subjectivity from decision making, and question other's assertions or commonly accepted views. The reflective dimension of wisdom refers to a person's ability to review his/her past and present life. This allows us to learn how to react to unpleasant circumstances and acknowledge the reality of our present situation. The openness dimension refers to a person's creativeness, imagination, intellectual curiosity, and openness to and tolerance of alternative views and possible solutions. The interactional aptitude dimension involves a person's ability to regulate his/her emotions and behaviors and to understand other's emotions and behaviors. A wise person can understand other's expressions and build on that to understand their beliefs, attitudes, values, abilities, and inabilities. Wise people have strong interpersonal skills. The paradoxical dimension refers to a person's ability to tolerate uncertainty and ambiguity. A wise person can picture the long-term effects of alternative courses of action. The ethnical sensibility dimension involves ethics and virtue and the ability to make ethical judgments. Finally, the experience dimension refers to a person's experience with challenging life situations. The authors argue that the dimensions of experience and paradoxical tolerance are formative indicators of wisdom, while the other four dimensions are reflective indicators. We see within this list many of the desired characteristics of change agents (see Sect. 5.2.2.4) or collaborators (see Chap. 7).

Yet, modern conceptualizations of wisdom have been too narrowly focused on a limited number of philosophical and psychological frameworks at the individual level to really help us understand the role of wisdom in terms of organizational-level sustainability. Izak (2013) reviews recent organizational research into how to operationalize and transfer wisdom beyond the individual level. Corporate wisdom training seeks to improve organizational communication and team dynamics. The designers of the Social Practical Wisdom model, which builds on the Aristotelian conception of practical wisdom and the development of virtue and character, argue that wisdom can be enacted throughout an organization. In the organizational spirituality literature, managerial wisdom involves enabling others to act and encouraging the heart. Izak turns to foolishness, a commonly conceptualized antithesis of wisdom, for inspiration. He reviews Sternberg's description of the five attributes of foolishness: (1) unrealistic optimism—fools believe they are so smart they can do what they want; (2) egocentrism—fools focus on what benefits them and ignore their responsibilities to others; (3) omniscience—fools believe they know everything, rather than knowing what they do not know; (4) omnipotence—fools believe they can do what they want because they are all powerful; and (5) invulnerability—fools believe they can get away with whatever they do. Read back over the older *Discourses* and paradigms discussed in Chap. 2. Sounds like how we have treated the human–environment interface in the past, doesn't it?

But some elements of foolishness are necessary if we are to be wise (Izak 2013). Rationalist and objective frameworks for explaining social phenomena don't always work when we face unpredictable circumstances. March (1976, as cited in Izak) introduced the concept of technology of foolishness to help supplement the rationalist model of thinking. In circumstances where it is difficult to predict and plan for future circumstances, sometimes organizations and their managers must do things for no good reason—they must act before they think. The business case for sustainability arguments provides multiple rational reasons for organizations to modify their actions but sometimes organizations need "a strategy for suspending rational imperatives toward consistency" (p. 111). March provides a set of facilitating instructions including to treat goals as mere hypotheses, trust your intuition, see inconsistency between expressed values and behaviors as a transitory state, don't always trust your memory, and treat experience as a theory so you can experiment with alternative interpretations. He encourages playfulness and the development of the skills and attitudes of inconsistency within our organizations. Wisdom involves flexibility, improvising, and the ability to make exceptions because no action is wise in itself—it depends on the context within which it occurs. Izak writes,

> "Wisdom has to be allowed to be foolish. However, not in a paradoxical sense—instead, as a comprehensive attempt to expand the wisdom concept to include voices which are rarely heard. Such opening up of wisdom's framework can enable suspending the rationalistic drive to suppress heterogeneity of organizational realities, and, once again, render wisdom meaningful. It may, at least, provide a conceptual space in which rationalism and consistency may be challenged by creative and inclusive foolishness" (p. 113). Sounds a bit like dialogue, doesn't it?

The social dimension of wisdom has received limited scientific investigation. For example, the John Templeton Foundation funded a 5-year initiative to better understand the nature of wisdom, its benefits, and applications, and to advance policies directed at cultivating wisdom (Nussbaum 2012). The multidisciplinary project expanded our understanding of wisdom focusing on the brain, cognition, and public policy, but failed to include any communication scholars on the team. Nussbaum proposes communication and language should be placed at the core of the scientific study of wisdom and urges that we investigate how wisdom is enacted in our communication behaviors. He writes, "An understanding and exploration of the communication of wisdom can significantly advance the positive role wisdom plays at both the individual and societal levels" (p. 243). He suggests two communication theories might enhance our understanding of how communication and are used in wise ways: communication accommodation theory (CAT) and ~~ication privacy management theory~~ (CPMT). CAT investigates the different ways, motivations, and consequences of how individuals adapt their communication behaviors to others. The CAT concept of communicative respect appears appropriate when thinking about wisdom in social interactions. Those who exhibit communicative respect use the skills of empathetic listening, seek to understand their conversational partner's point of view, and communicate in a way that is seen as appropriate by their partner. In terms of the CPMT (see Sect. 5.4.2.4) an important social component of wisdom involves an individual's ability to appropriately share private information so as to achieve his/her interactive goals, while simultaneously assuring his or her conversational partner that his or her needs or society's needs are not being ignored. These theories focus on the skills and abilities communicators use rather than on the philosophical and psychological dimensions of wisdom. Through communication, acts emerging from joint wisdom are developed and shared.

As we proceed, certainly there will be challenges and our path will be shrouded in fog. You've read about challenges in terms of individual behavior change, organizational behavior change, industry/institutional change, and national and global change. You've read about climate deniers, conflicting ideologies, and systematically distorted communication. Certainly those things exist, but that does not mean we can't or shouldn't take this journey. Wisdom requires that more and more of our organizations act swiftly, collaboratively, and effectively.

8.4.2 Spirituality and the Right Thing to Do

Several years ago my graduate students interviewed 30 people who worked for one of two multinational organizations involved in a supply chain relationship. We asked them to define sustainability and discuss it within the context of their personal and professional lives. Repeatedly, people said sustainability is the "right thing to do." The rational business case for sustainability didn't appear to be what influenced this particular statement. It was something more. Was it faith, virtue, and/or a sense of fairness?

All the world's major religions urge their followers to care for the vulnerable poor and to be good stewards of our shared natural environment. At the societal level, the green politics *Discourse* focuses on global poverty and environmental justice, while the green consciousness *Discourse* includes eco-theology (Dryzek 2005). Eco-theology investigates the interrelationships between our religious/spiritual worldviews, the environmental degradation of nature, and our ecosystem management. Supporters of and contributors to eco-theology believe science and education are insufficient to inspire the change necessary in our current environmental crisis. A PBS Newshour (Facing Environmental Crisis 2014) broadcast indicated that as the Chinese government, now the world's #1 polluter, faces increasing environmental crises, they are beginning to encourage their citizens to return to Buddhist, Taoist, and Confucian temples in the hope of bringing back a pro-environmental lifestyle and compassion. A Tibetan monk was quoted as saying, "No matter if you are a newborn or an 80-year-old, you are all protectors. You are all responsible. And you have the responsibility. All life should be protected." And a Chinese politician was quoted as saying:

> Traditional Chinese culture promotes harmony between man and nature and encourages limited consumption and a simple way of life. We support this. We don't oppose taking from nature. We do oppose overexploitation. We want gold mountain, but we also want clear water and green mountain.

Meanwhile, some Christians read a *Green Bible*, an English version of the *New Revised Standard Version Bible* with a focus on environmental issues and teachings. In the USA more than 45 theological schools from the evangelical, mainline, and Roman Catholic traditions have joined the Green Seminary Initiative and/or the Seminary Stewardship Alliance to share best practices for teaching and practicing a pro-environmental ethic within the seminary context. The hope is that, ultimately, their students will nurture such values among their congregations (The Association of Theological Schools n.d.).

In terms of scholarship, studies and essays linking business and spirituality are widely represented in the business ethics, leadership, and management literatures. Divisions within our professional associations exist. Emerging fields focus on workplace spirituality, spiritual leadership, and conscious capitalism (Fry and Slocum 2008). Spirituality differs from religion because it is an individually interpreted, private experience whereas religion is an institutionalized and public process (Crossman 2011). Efforts to increase workplace spirituality have been relatively free of denominational politics and appear to positively influence an organization's triple bottom line. Zaidman and Goldstein-Gidoni (2011) suggest that scholars might view workplace spirituality as one form of organizational wisdom, although it is a form organizational members may reject as threatening because it clashes with existing assumptions of social order and rationality. Communication scholars (e.g., Molloy and Heath 2014) have investigated the relationship between spirituality and workplace communication, although the relationship is under-theorized.

Crossman (2011) provides her readers with ways to demonstrate and embed spiritual and environmental leadership into their organizations. Given the amount of time we spend at work, it is unrealistic to expect people to leave their spiritual selves at home. She also identifies commonalities in the underpinning values and associated discourse appearing in environmental and spiritual leadership literatures. Shared values include notions of the common and social good, stewardship, sustainability, servanthood, meaning, and connectedness. I am not arguing that spirituality should be a dominant argument in *all* organizations seeking to enact sustainability initiatives. But in organizations where the argument would resonate, it is a way to show employees how their spiritual beliefs overlap with their organization's goals and thereby influence goal congruence (see Sect. 6.3.1). Certainly it will be more motivational for some employees. My own spirituality is one reason I wrote this book. Like many of you, I believe we are responsible for helping the poor who face the worst climate-related stressors and that we are only briefly the trustees of our beautiful and complex planet. Addressing such issues are, for me, the "right thing to do" in my roles as scholar, researcher, and citizen.

8.5   Concluding Thoughts

In *The Hero with a Thousand Faces* Campbell (1949) argues that myths appearing across times and cultures share similar characteristics. He identified one reoccurring monomyth as involving a hero's journey. The monomyth has 17 stages which fit into three groupings (i.e., departure, initiation, and return). *Departure*: A normal person receives the call to head off on a quest into the unknown, but initially refuses it, possibly out of a sense of duty, fear, or inadequacy. But eventually our hero responds and receives guidance/assistance from a supernatural mentor. Our hero crosses over into unknown and dangerous territory where the rules and limits are unknown and is willing to undergo a metamorphosis. *Initiation*: In order for this metamorphosis to occur our hero must encounter and sometimes fail a series of tests and ordeals. Along the journey he or she experiences an all-powerful, all encompassing, unconditional love and faces temptations that may lead him or her to abandon or stray from the quest. Midway on the journey our hero confronts and is initiated by whatever or whomever holds the ultimate power in his or her life (i.e., meets someone of incredible power). He or she faces the difficult task of detaching from ego itself and must learn to experience faith and rely on mercy. The hero rests briefly in a state of knowledge, love, and compassion before achieving the goal of the quest. *Return*: The hero may not want to return to the normal world but does. During the return our hero encounters more adventures and challenges and, now weakened, must rely on powerful guides and rescuers. Once home, the hero must decide how to share the wisdom gained on the journey with the rest of the world. He or she experiences a balance between the material and spiritual worlds, learns to live in the moment, and feels free from the fear of death while experiencing the freedom to live.

We have many everyday heroes among us seeking to promote sustainability. You have read comments made by some of them in this book and others are very well known (e.g., the late Ray Anderson). Indeed, Rosteck and Frentz (2009) applied the monomyth to analyzing the film, *An Inconvenient Truth*. Do not be disheartened, we *are* accomplishing some of our national and global sustainability-related goals. However, we need even more heroes as the complexities we face increase. As the monomyth tells us, heroes must face their fears, reach out to wise guides, and share what they learn with those listening and hoping for solutions. Let's be heroes together and use our organizations and our interorganizational collaborations to protect the world's children, create more resilient communities, preserve and strengthen our existing plants and animals, and halt the wanton destruction of our biosphere. We have business case arguments supporting our actions, it is the wise course to take, and it is the morally right thing to do.

References

Berger, P. L., & Luckmann, T. (1966). *The social construction of reality: A treatise in the sociology of knowledge*. Garden City: Anchor Books, NY.

Blackburn, W. R. (2007). *The sustainability handbook: The complete management guide to achieving social, economic and environmental responsibility*. London: Earthscan.

Campbell, J. (1949). *The hero with a thousand faces*. Princeton: Princeton University Press.

Cordano, M., Marshall, R. S., & Silverman, M. (2010). How do small and medium enterprises go "green"? A study of environmental management programs in the U.S. wine industry. *Journal of Business Ethics, 92*, 463–478.

Cox, R. (2013). *Environmental communication and the public sphere* (3rd ed.). Washington, DC: Sage.

Crossman, J. (2011). Environmental and spiritual leadership: Tracing the synergies from an organizational perspective. *Journal of Business Ethics, 103*, 553–565.

Dart, R., & Hill, S. D. (2010). Green matters? An exploration of environmental performance in the nonprofit sector. *Nonprofit Management & Leadership, 20*, 295–314.

Dryzek, J. S. (2005). *The politics of the earth: Environmental discourses* (2nd ed.). Oxford: Oxford University Press.

Facing Environmental Crisis, Can Buddhist Values Offer Non-religious China a Green Path? (2014). http://www.pbs.org/newshour/bb/can-buddhist-values-offer-non-religious-china-a-greener-path/. Accessed 14 Feb 2015.

Fry, L. W., & Slocum, J. W., Jr. (2008). Maximizing the triple bottom line through spiritual leadership. *Organizational Dynamics, 37*, 86–96.

Godemann, J., & Michelsen, G. (Eds.). (2011). *Sustainability communication: Interdisciplinary perspectives and theoretical foundation*. New York: Springer.

Izak, M. (2013). The foolishness of wisdom: Towards an inclusive approach to wisdom in organization. *Scandinavian Journal of Management, 19*, 108–115.

Lindenfeld, L. A., Hall, D. M., McGreavy, B., Silka, L., & Hart, D. (2012). Creating a place for environmental communication research in sustainability science. *Environmental Communication: A Journal of Nature and Culture, 6*, 23–43.

McPhee, R. B., & Poole, M. W. (2009). Structuration theory. In S. W. Littlejohn & K. A. Foss (Eds.), *Encyclopedia of communication theory* (Vol. 2, pp. 936–940). Los Angeles, CA: Sage.

Molloy, K. A., & Heath, R. G. (2014). Bridge discourses and organizational ideologies: Managing spiritual and secular communication in a faith-based, nonprofit organization. *Journal of Business Communication, 51*, 386–408.

Nussbaum, J. F. (2012). The communication of wisdom: The nature and impact of communication and language change across the life span. *Journal of Language and Social Psychology, 32*, 243–260.

Ronda, J. P. (2003). *Why Lewis and Clark matter*. *Smithsonian.com*. http://www.smithsonianmag.com/history/why-lewis-and-clark-matter-87847931/?no-ist. Accessed 15 Jan 2015

Rosteck, T., & Frentz, T. S. (2009). Myth and multiple readings in environmental rhetoric: The case of An Inconvenient Truth. *Quarterly Journal of Speech, 95*, 1–10.

Schmit, D. E., Muldoon, J., & Ponders, K. (2012). What is wisdom? The development and validation of a multidimensional measure. *Journal of Leadership, Accountability and Ethics, 9*, 39–54.

Servaes, J. (Ed.). (2013). *Sustainability, participation and culture in communication: Theory and praxis*. Chicago, IL: University of Chicago.

Smerecnik, K. R., & Andersen, P. A. (2011). The diffusion of environmental sustainability innovations in North American hotels and ski resorts. *Journal of Sustainable Tourism, 19*, 171–196.

Sternberg, R. J. (2004). What is wisdom and how can we develop it? *The Annals of the American Academy of Political and Social Science, 591*, 164–174.

The Association of Theological Schools (n.d.). *Going green: Member schools share commitment and best practices.* http://www.ats.edu/uploads/resources/publications-presentations/docu ments/going-green.pdf. Accessed 14 Feb 2015

van der Leeuw, S., Wiek, A., Harlow, J., & Buizer, J. (2012). How much time do we have to fail? The urgency of sustainability challenges vis-à-vis roadblocks and opportunities in sustainability science. *Sustainability Science, 7*(Suppl 1), 115–120.

Wals, A. E. J., & Schwarzin, L. (2012). Fostering organizational sustainability through dialogic interaction. *The Learning Organization, 19*, 11–27.

Weick, K. E. (1995). *Sensemaking in organizations: Foundations for organizational science.* Thousand Oaks, CA: Sage.

Wilhelm, K. (2013). *Making sustainability stick: The blueprint for successful implementation.* Pearson: FT Press.

Zaidman, N., & Goldstein-Gidoni, O. (2011). Spirituality as a discarded form of organizational wisdom: Field-based analysis. *Group & Organization Management, 36*(5), 630–654.

Index

A

Absorptive capacity, 146, 148, 200, 275
Abstract *vs.* concrete action, 131
Action plan, 13, 79, 80, 83, 88, 95, 100, 122, 144, 150, 223, 276
Additional persuasive arguments, 132–133
Administrative rationalism, 31, 33
Affective organizational commitment (AOC), 202, 211
Agenda 21, 42, 53
Alliances, 252
Altruism, 107, 117–118, 129
Annual meetings, 71, 74–76, 166, 283
Appreciative inquiry, 179, 239–240
Architecture, 88–95
Association for the Advancement of Sustainability in Higher Education (AASHE), 45
Attitude–behavior link, 113
Attitude–behavior model, 116
Attitudes, 55, 56, 107, 113, 115–118, 134, 165, 213, 235, 265, 296

B

Barriers, 113, 118, 122, 123, 126, 148, 151, 160, 190, 194, 265, 266
  related to practicality, 114
  to action, 113
Beliefs, 65, 107, 117, 129, 177, 191, 201, 211, 249

Best practices, 7–11, 29, 42, 56, 73, 79, 93, 112–114, 118, 121–122, 126, 134, 147, 148, 150, 157, 168, 172, 181, 198, 209, 220, 234–237, 239, 241, 245, 247–252, 254–255, 255, 263, 290
The Board of directors, 66, 151–152
Business case arguments, 220
Business case for sustainability, 41, 81, 132, 147, 220, 263, 280, 297

C

Campaign interventions, 110, 125–126, 129
Certifications, 71, 84–88, 261, 265, 268, 285, 291
Change, 4, 65–68, 139, 161–164
Channels, 12, 49, 99, 143–146, 152, 165, 173, 218, 288
Citizenship behaviors, 15, 81, 87, 109, 190, 213, 243
Climate action plans, 6, 17, 45, 167
Climate change, 3, 4, 6, 11, 13, 18, 26, 30, 38, 56, 62, 78, 81, 94, 140, 158, 160, 175, 183, 241, 285, 295
  denial, 6, 26, 53–54, 94
Climate declaration, 5–7, 22, 109, 282
Collaborations, 49, 88, 151, 160, 231–269
Collective identity, 249
Collective information sampling, 240
Communication plans, 224

© Springer International Publishing Switzerland 2016
M. Allen, *Strategic Communication for Sustainable Organizations*, CSR, Sustainability, Ethics & Governance, DOI 10.1007/978-3-319-18005-2

CPSIA information can be obtained
at www.ICGtesting.com
Printed in the USA
LVOW04*1741270816

502122LV00004B/5/P